SAFECOMP 96

Springer
London
Berlin
Heidelberg
New York
Barcelona
Budapest
Hong Kong
Milan
Paris
Santa Clara
Singapore
Tokyo

International Programme Committee:

E. Schoitsch *(IPC Chairman)* (A)
R.E. Bloomfield *(Chair EWICS)* (UK)
S. Bologna (I)
G. Cleland (UK)
F. Dafelmair (D)
G. Dahll (N)
P. Daniel (UK)
J. McDermid (UK)
W. Ehrenberger (D)
R. Genser (A)
J. Gorski (PL)
W. Halang (D)
D. Inverso (USA)
J. Järvi (SF)
K. Kanoun (F)
F. Koornneef (NL)
H. Kopetz (A)
V. Maggioli (USA)
Ch. Mazuet (F)
M. v. der Meulen (NL)
A. Pasquini (I)
G. Rabe (D)
J. Rainer (A)
F. Redmill (UK)
B. Runge (DK)
G. Schildt (A)
I. Smith (UK)
L. Strigini (I)
A. Weinert (D)
C.W. Wyatt-Millington (UK)
J. Zalewski (USA)

Local Organising Committee:

G. Sonneck *(Chairman)*, G. Herrmann, T. Kadi,
G. List, J. Rainer, E. Schoitsch, E. Vöhringer

SAFECOMP 96

The 15th International Conference on Computer Safety, Reliability and Security

Vienna, Austria
October 23 - 25, 1996

Edited by
ERWIN SCHOITSCH

EWICS TC7 Reliability Safety Security

Organized by

Austrian Research Centre Seibersdorf

Co-Organized by

Austrian Federal Ministry of Science, Transport and the Arts

Sponsors

European Workshop on Industrial Computer Systems
Technical Committee 7
(EWICS TC 7)
Federal Research and Testing Centre Arsenal

Co-Sponsored by

OCG (Austrian Computer Society)
IFIP Technical Committee 5 WG 5.4
IFIP Technical Committee 10 WG 10.4
European Joint Research Centre Ispra
ENCRESS

Springer

Erwin Schoitsch
Österreichisches Forschungszentrum Seibersdorf Ges.m.b.H.
A-2444 Seibersdorf, Austria

ISBN-13 978-3-540-76070-2 e-ISBN-13: 978-1-4471-0937-2
DOI 10.1007/978-1-4471-0937-2

British Library Cataloguing in Publication Data
A catalogue record for this book is available from the British Library

Softcover reprint of the hardcover 1st edition 1997

Typesetting: Camera ready by contributors

34/3830-543210 Printed on acid-free paper

Preface

Computer systems are indispensible in modern technical systems such as railways, airtraffic, aerospace and transport telematic systems, security systems, energy suppliers, telecommunication or process industry. They perform tasks that few decades ago were very difficult if not impossible. As they perform these tasks with increasing efficiency, more and more tasks are shifted from hardware to software which means that the dependability of computer systems becomes crucial for the safety, security and reliability of technical systems.

The issue of dependable computer systems has as number of aspects including reliability, safety, security, development, testing, validation, verification, assessments, safety case, regulation, research and industrial practice, and, not to forget because of increasing importance from a (holistic) systems view, human factors, sociological and legal aspects.

All these are addressed in this book which reproduces the 36 papers from 12 countries presented at the 15th International Conference on Computer Safety, Reliability and Security, SAFECOMP '96, held in Vienna, Austria, October 23 - 25, 1996. SAFECOMP '96 continues the successful series of SAFECOMP conferences which was originated by the European Workshop on Industrial Computer Systems, Technical Committee 7 on Safety, Security and Reliability (EWICS TC7).

Therefore, this international conference reflects not only the state of the art, experience and new trends in technology of dependable systems, but also reflects the increasing strength of Europe in this area, in competition with the USA and Japan, which is due to the efforts of the European Commission, its member states (Austria has become member of the EU just recently, with an overwhelming vote of 66.6% demonstrating the high expectations and hopes of the people), academia and industry. SAFECOMP '96 takes into account the work done in EWICS TC7, a group of experts from European (and some overseas) experts from all major sectors of industry, research and from assessment organisations and authorities, from two recent European projects: ENCRESS (European Network of Clubs for REliability and Safety of software-intensive Systems) and OLOS (A holistic approach to the dependability analysis and evaluation of control systems involving hardware, software and human resources), and some other relevant European projects in specific fields such as REAIMS, CASCADE, CONVERGE-SA, PASSPORT, DRIVE Safely, EMCATT etc.

As Chairman of the International Programme Committee (IPC) I would like to thank all authors who submitted their work, the presenters, the members of the IPC and the Local Organizing Committee, the session chairmen and the sponsors and co-sponsors for their efforts and support. I want to thank also the chairman Martyn Dowell (JRC Ispra) and IPC chairman Gerd Rabe (TÜV Nord) of the previous SAFECOMP for their advice as well as the Federal Ministry of Science, Transport and the Arts and the Austrian Research Center Seibersdorf for their contribution to the organisation of this conference.

Erwin Schoitsch

Seibersdorf, Austria
July 1996

Contents

ROUND TABLE

INVITED PAPER

SESSION 3: Reliability and Safety Assessment

SESSION 4: Industrial Applications and Experience

SESSION 5: Railway Applications and Experience

SESSION 6: Management and Development

SESSION 7: Human Factors

INVITED PAPER

SESSION 8: The Safety Case Legal Aspects

SESSION 9: Security

List of Contributors

T. Andersen
Det Norske Veritas AS
Veritasveien 1
N-1322 Hovik
Norway

G. Aprea
c/o P. Firpo
Sciro Electra s.r.l.
Via N. Bixio 4/9
I-16128 Genova
Italy

J. Arlat
Centre National de la Recherche Scientifique
Laboratoire d'Analyse et d'Architecture des Systems
7, Avenue du Colonel Roche
F-31077 Toulouse
France

W. Artner
Frequentis Nachrichtentechnik GesmbH
Spittelbreitengasse 34
A-1120 Wien
Austria

A. Avižienis
UCLA Computer Science Department
University of California
Los Angeles
CA 90095
USA
and
Vytautas Magnus University
Kaunas, Lithuania

C. Bernardeschi
Dipartimento di Ingegneria della Informazione
Univ. di Pisa
I-56126 Pisa
Italy

A. Bertolino
Istituto di Elaborazione della Informazione
CNR
Via S. Maria, 46
I-56126 Pisa
Italy

R. Bloomfield
ADELARD
Coborn House
3 Coborn Road
London, E3 2DA
UK

M. Borcherding
Institute of Computer Design and Fault Tolerance
University of Karlsruhe
D-76128 Karlsruhe
Germany

J. Bowers
Department of Psychology
Manchester University
Manchester
UK

A.C. Brombacher
Faculty of Mechanical Engineering
Eindhoven University of Technology
P.O. Box 513
NL-5600 MB Eindhoven
The Netherlands

R. Budde
GMD - SET-EES
Forschungszentrum
Informationstechnik GmbH
Schloß Birlinghoven
D-53754 Sankt Augustin
Germany

J.V. Bukowski
c/o W. M Goble
Moore Products Co.
Sumneytown Pike
Spring House
PA 19477
USA

F. Charron
RST - Reliable Software Technologies
Corporation
Suite 250
21515 Ridgetop Circle
Sterling
VA 20166
USA

Y. Chen
Dept. of Computer Science
University of the Witwatersrand,
Johannesburg
2050 Wits
South Africa

B. Ciciani
University of Rome La Sapienza
Via Salaria 113
I-00198 Roma
Italy

P. Colantuoni
c/o P. Firpo
Sciro Electra s.r.l.
Via N. Bixio 4/9
I-16128 Genova
Italy

Y. Crouzet
LAAS-CNRS
7, avenue du Colonel Roche
F-31077 Toulouse Cedex
France

F.J. Dafelmair
TÜV Bayern Sachsen
Dep. G2-ETL 30
Westendstraße 199
D-80686 München
Germany

L. Emmet
ADELARD
Coborn House
3 Coborn Road
London E3 2DA
UK

R. Eriksen
Det Norske Veritas AS
Veritasveien 1
N-1322 Hovik
Norway

L.-H. Eriksson
Logikkonsult NP AB
Swedenborgsgatan 2
S-118 48 Stockholm
Sweden

A. Fantechi
c/o S. Gnesi
Istituto di Elaborazione della
Informazione - C.N.R.
Via S. Maria, 46
I-56126 Pisa
Italy

C. Feyling
NSB Gardermobanen
N-0107 Oslo
Norway

P. Firpo
Sciro Electra s.r.l.
Via N. Bixio 4/9
I-16128 Genova
Italy

N. Fota
LAAS-CNRS
7, Av. du Colonel Roche
31077 Toulouse
France
and
SOFREA VIA
3 Carrefour de Weiden
F-92441 Issy les Moulineaux
France

S. Gnesi
Istituto di Elaborazione della Informazione - C.N.R.
Via S. Maria, 46
I-56126 Pisa
Italy

W.M Goble
Moore Products Co.
Sumneytown Pike
Spring House
PA 19477
USA
and
TU-Eindhoven
Reliability of Mechanical Equipment Group WOC
Building W-hoog 3-121
P.O. Box 513
NL-5600 MB Eindhoven
The Netherlands

W. Goerke
University of Karlsruhe
Institut für Rechnerentwurf und Fehlertoleranz
Kaiserstrasse 12
D-76128 Karlsruhe
Germany

J. Górski
Franco-Polish School of New Information and Communication Technologies
Mansfelda 4
P.O. Box 31
PL-60 854 Poznan 6
Poland

J. Griffyth
Mathematics/Computing Department
Open University
Walton Hall
Milton Keynes MK7 6AA
UK

W.J. Gutjahr
Department of Statistics, Operations Research and Computer Science
University of Vienna
Universitätsstr. 5
A-1010 Wien
Austria

W.A. Halang
Fern Universität
Faculty of Electrical Engineering
D-58084 Hagen
Germany

M. Heisel
FB Informatik - FG Softwaretechnik
Technische Universität Berlin
Franklinstraße 28-29
Sekr. FR 5-6
D-10587 Berlin
Germany

K.M. Hobley
School of Computer Studies,
Safety Critical Computing Group,
University of Leeds
Leeds LS2 9JT
UK

G.R.J. Hockey
Department of Psychology,
University of Hull
Hull HU6 7RX
UK

M.J.M. Houtermans
TU-Eindhoven
Reliability of Mechanical Equipment
Group WOC
Building W-hoog 3-121
P.O. Box 513
NL-5600 MB Eindhoven
The Netherlands

P.H. Jesty
School of Computer Studies
Safety Critical Computing Group
University of Leeds
Leeds LS2 9JT
UK

M. Kaâniche
LAAS-CNRS
7, avenue du Colonel Roche
F-31077 Toulouse
France

K. Kanoun
LAAS-CNRS
7, avenue du Colonel Roche
F-31077 Toulouse
France

J.L. King
Department of Information and
Computer Science
University of California
Irvine, CA. 92717
USA

F. Koob
Bundesamt für Sicherheit in der
Informationstechnik
Postfach 200363
D-53133 Bonn-Dottendorf
Germany

W. Kuhn
FZ Seibersdorf
A-2444 Seibersdorf
Austria

J.-C. Laprie
LAAS-CNRS
7, avenue du Colonel Roche
F-31077 Toulouse Cedex
France

R. Lido
c/o P. Firpo
Sciro Electra s.r.l.
Via N. Bixio 4/9
I-16128 Genova
Italy

C. Loftus
Department of Computer Science
University of Wales, Aberystwyth
Dyfed SY23 3DB
Wales
UK

F. Long
Department of Computer Science
University of Wales, Aberystwyth
Dyfed SY23 3DB
Wales
UK

P.G. Marsden
EDS
1 & 3 Bartley Way
Bartley Wood, Hook
Hampshire RG27 9XA
UK

C. Mazuet
Schneider Electric
SES/ST/ETS
F-38050 Grenoble Cedex 9
France

A. Merceron
GMD - SET-EES
Forschungszentrum
Informationstechnik GmbH
Schloß Birlinghoven
D-53754 Sankt Augustin
Germany

H.-P. Meske
Fern Universität
Faculty of Electrical Engineering
D-58084 Hagen, Germany

K. Miller
Department of Computer Science
Sangamon State University
Springfield, IL
USA

S. Mitra
Lloyd's Register, London
Lloyd's Register House
29 Wellesley Road
Croydon CR0 2AJ
UK

B. Nowicki
Franco-Polish School of New
Information and Communication
Technologies
Mansfelda 4
P.O. Box 31
PL-60 854 Poznan 6
Poland

A. Pasquini
ENEA (sp 088)
Via Anguillarese 301
I-00060 ROMA
Italy

D. Pellegrino
c/o P. Firpo
Sciro Electra s.r.l.
Via N. Bixio 4/9
I-16128 Genova
Italy

H.-J. Petersen
Siemens Transportation Systems
D-38023 Braunschweig
Germany

A. Peytavin
CENA
7 Av.Edouard Belin, BP. 4005
F-31055 Toulouse
France

D. Pugh
Department of Computer Science
University of Wales, Aberystwyth
Dyfed SY23 3DB
Wales
UK

I. Pyle
Department of Computer Science
University of Wales, Aberystwyth
Dyfed SY23 3DB
Wales
UK

M. Rapone
c/o P. Firpo
Sciro Electra s.r.l.
Via N. Bixio 4/9
I-16128 Genova
Italy

D.J. Richardson
Department of Information and
Computer Science
University of California
Irvine
CA 92717
USA

A. Rizzo
University of Siena
Laboratorio Mutimediale
Via del Giglio, 14
I-53100 Siena
Italy

N. Rowden
Siemens Integra Verkehrstechnik AG
Industriestraße 42
CH-8305 Wallisellen
Switzerland

H. Selami
FZ Seibersdorf
A-2444 Seibersdorf
Austria

F. Senesi
c/o P. Firpo
Sciro Electra s.r.l.
Via N. Bixio 4/9
I-16128 Genova
Italy

I.D.R. Shannon
Opal Engineering
Macmillan House NP22 E500
Paddington Station
London W2 1FT
UK

D.E. Sniezek
Lockheed Martin Federal Services
1801 State Route 17C
M/D 0604
Owego
NY 13827-3998
USA

H. Stöckl
Siemens AG Österreich
PSE EZE PNZ
Friedrich-Hillegeiststraße 1
A-1020 Wien
Austria

L. Strigini
Centre for Software Reliability
City University
Northampton Square
London EC1V 0HB
UK

C. Sühl
FB Informatik - FG Softwaretechnik
Technische Universität Berlin
Franklinstraße 28-29
Sekr. FR 5-6
D-10587 Berlin
Germany

K.-H. Sylla
GMD - SET-EES
Forschungszentrum
Informationstechnik GmbH
Schloß Birlinghoven
D-53754 Sankt Augustin
Germany

P. Thévenod-Fosse
LAAS-CNRS
7, avenue du Colonel Roche
F-31077 Toulouse Cedex
France

C.S. Turner
Department of Information and
Computer Science
University of California
Irvine
CA. 92697-3425
USA

M. Ullmann
Bundesamt für Sicherheit in der
Informationstechnik
Postfach 200363
D-53133 Bonn-Dottendorf
Germany

S. Viller
Computing Department
Lancaster University
Lancaster
UK

S. Visram
NATS
London Area Programme
CAA House
45-59 Kingsway
London WC2B 6TE
UK

J.M. Voas
RST - Reliable Software1 Technologies Corporation
Suite 250
21515 Ridgetop Circle
Sterling
VA 20166
USA

S.J. Westerman
Department of Psychology
University of Hull
Hull HU6 7RX
UK

S. Wittmann
Bundesamt für Sicherheit in der Informationstechnik
Postfach 200363
D-53133 Bonn-Dottendorf
Germany

Z. Żurakowski
Institute of Power Systems Automation
ul. Wystawowa 1
P-51-616 Wroclaw 12
Poland

Invited Paper

Systematic Design of Fault-Tolerant Computers

Algirdas Avižienis

UCLA Computer Science Department
University of California, Los Angeles, CA 90095, U.S.A.
and
Vytautas Magnus University, Kaunas, Lithuania

Abstract

The origin of the concept of fault tolerance and the evolution of guidelines for the systematic design of fault-tolerant systems is reviewed. The current formulation of the guidelines, called a design paradigm, is presented. The problem of using off-the-shelf subsystems in a fault-tolerant system is discussed. In conclusion, an analogy of complex fault-tolerant systems and living organisms is suggested as a means to advance the understanding of fault tolerance.

1 Origin of the Fault Tolerance Concept

The concept of fault tolerance originally appeared in technical literature in 1967 as follows [1]:

"We say that a system is fault-tolerant if its programs can be properly executed despite the occurrence of logic faults."

The creation of the fault tolerance concept was the consequence of the convergence of three developments.

First, there were the ongoing efforts by computer designers to increase the reliability of computing by the use of various practical techniques. As soon as the first computers were built and turned on, it became apparent that physical defects and design mistakes could not be avoided by careful design and by the choice of good components alone. For this reason, designers of early computers employed redundant structures to mask failed components, error control codes and duplication or triplication with voting to detect or correct information errors, diagnostic techniques to locate failed components, and automatic switchovers to replace failed subsystems [2, 3, 4, 5].

Second, in parallel with the evolution of the engineering techniques, the general problem of building reliable systems from unreliable components was addressed

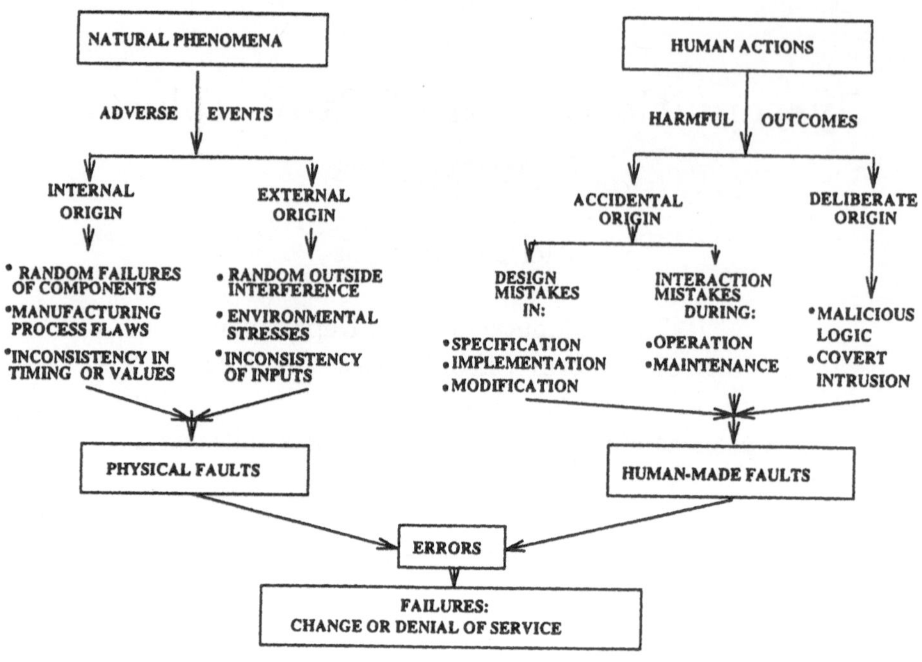

Figure 1: Why Systems Fail

by some of the founders of computer science, most eminently J. von Neumann in 1952 [6], E.F. Moore with C.E. Shannon in 1956 [7], and others.

Third was the new challenge of building unmanned spacecraft for interplanetary exploration that was assigned by NASA to Caltech's Jet Propulsion Laboratory (JPL) in late 1958. Mission lengths of up to ten years and more were being considered, and on- board computing was a prerequisite for their success. Design of computers that would survive a journey of several years and then deliver their peak performance at a distant planet was an entirely unexplored discipline.

Existing theoretical studies of the long-life problem indicated that large numbers of spare subsystems offered a promise of longevity, given that all spares could be successfully employed in sequence. The JPL problem was to translate the idealized "spare replacement" system model into a flightworthy implementation of a spacecraft guidance and control computer.

A proposal to design such a computer, called "A Self-Testing-And-Repairing System for Spacecraft Guidance and Control," and designated by the acronym "STAR" was presented in October, 1961 [8]. It was supported by JPL and accepted by NASA management, and the research effort continued for more than ten years, culminating with the construction and demonstration of the laboratory model of the JPL- STAR computer [9]. The originality of the concept was recognized by U.S. Patent No. 3,517,671 "Self Testing and Repairing Computer," granted on June 23, 1970 to A. Avižienis and assigned to NASA. A flight model of the JPL-STAR was designed for a 10-15 year space mission, but its building was halted when NASA discontinued the Grand Tour mission for which it was intended [10].

The longevity requirement led to the study of all accessible engineering solutions and theoretical investigations of reliability enhancement. The variety of existing theories and techniques motivated the definition of the unifying concept of fault tolerance that merged diverse approaches into a cohesive view of all system survival attributes, and greatly facilitated the design of the JPL-STAR computer. It also was a logical step for JPL, the birthplace of the fault tolerance concept, to be the co-sponsor and to take the initiative to organize the first International Symposium on Fault-Tolerant Computing that was held in Pasadena, CA on March 1-3, 1971.

During the following twenty five years we have seen a continuing growth of the set of faults that are to be tolerated by fault- tolerant systems. The original concept dealt with transient and permanent logic faults of physical origin. Faults due to human mistakes in design were added when the growing complexity of software and of logic on VLSI chips made the removal of all *design faults* prior to operational use not certain. Experience also led to the addition of *interaction faults*, inadvertently introduced by humans during the operation or maintenance of a computer.

Finally, consequences of malicious actions intended to alter or to stop the service being delivered by a system were recognized as being *deliberate design faults*. This concept establishes a common ground for the unified treatment of security and fault tolerance concerns in system design. The assurance of full compatibility and integration of security and fault tolerance techniques is a major challenge for contemporary designers. An overview of the current set of faults to be tolerated is presented in Figure 1.

In retrospect, it may be said that the concept of fault tolerance has served well during the past quarter of a century in facilitating the appearance of successively more dependable systems for the control and support of various essential functions of contemporary society: computing and communications, transportation, nuclear power, financial transactions, health care delivery, etc. [11].

2 The Evolution of Design Principles

Thirty years of experience have shown that the building of dependable systems requires the balanced use of both fault avoidance and fault tolerance techniques. An imbalance in either direction leads to an ineffective use of resources and severely limits the attainable dependability. The definition of the concept of fault tolerance initiated the evolution of principles for the systematic design of fault-tolerant systems. The specification and design of the STAR computer at JPL involved much improvisation and experimentation with design alternatives. It became apparent that the lessons learned during this process could serve as the foundation for a more orderly approach that would utilize a set of guidelines for the choice of fault masking, error detection, fault diagnosis, and system recovery techniques.

The first effort to devise such guidelines was presented at the 1967 Fall Joint Computer Conference in the paper "Design of Fault-Tolerant Computers" [1]. That paper first introduced the term "fault-tolerant computer" and the concept of "fault tolerance" into technical literature. It also presented a classification of faults and outlined the alternate forms of masking, diagnosis, and recovery techniques along with some criteria for choices between "massive" (i.e., masking) and "selective" application of redundancy. The design of the JPL- STAR computer was used to illustrate the application of these criteria in choosing the fault tolerance techniques for a spacecraft computer that had long life and autonomy requirements with strict weight and power constraints. The 47 references covered the most relevant published work to mid-1967.

The earlier book "Failure-Tolerant Computer Design" by W.H. Pierce [12] served as an important reference; however, Pierce's definition of "failure tolerance" corresponded exactly to fault masking in logic circuits, including voting, adaptive, and interwoven logic, redundant relay contact networks, and application of error correcting codes as a masking technique. It is a definitive work on the masking forms of redundancy that were known at that time. However, neither error detection, nor fault diagnosis, nor recovery techniques were included as elements of Pierce's "failure-tolerant" computers.

The 1967 paper was the first of a sequence of publications that formulated an evolving view of how to attain dependable computing by the judicious introduction of fault tolerance during system design. Two different classes of faults - those due to physical causes and those due to human mistakes, oversights, and deliberate actions are considered. This evolving view was presented in a series of papers on guidelines for fault-tolerant system design and implementation, supported by specific discussions of the techniques, scope, and aims of fault tolerance and fault avoidance in hardware, software, communication, and man/machine interfaces. Milestones of this series have been the papers: [13, 14, 15, 16, 17]. Strong motivation for the effort came from the increasing number of successful fault-tolerant systems that offered new design insights and more operational experience.

3 The Design Paradigm for Fault-Tolerant Systems

The unifying theme of the above referenced work over the past thirty years has been the evolution of a design paradigm for fault-tolerant systems that guides the designer to consider fault tolerance as a fundamental issue throughout the design process. The word "paradigm" is used here in the dictionary sense of "pattern, example, model" in place of the word "methodology" that implies a study of methods, rather than a set of guidelines with illustrations that is discussed here.

Taken in order of appearance, the papers show a progressive refinement of concepts and an expansion of the scope to include the tolerance of "human made" design and interaction faults. Other recently introduced themes are the balancing of performance and fault tolerance objectives during system partitioning, and the integration of subsystem recovery procedures into a multi-level recovery hierarchy. Strong emphasis has been directed to the application of ***design diversity*** in a multichannel system in order to attain tolerance of design faults, [14, 17], including the tolerance of deliberate design faults [18].

Fault tolerance has now been recognized as the key prerequisite of dependability for very large systems, such as the Advanced Automation System for air traffic control in the USA [19,23]. Because of their great functional complexity, such systems pose the most severe challenge yet in the design of fault-tolerant systems. The introduction of fault tolerance into very complex, distributed systems is most likely to succeed if a methodical approach is employed. This approach begins with the initial design concepts and requires the collaboration of performance and fault tolerance architects during the critical tasks of system partitioning, function allocation, and definition of inter- subsystem communication and control. Such a design approach is presented here as the ***design paradigm*** for fault-tolerant systems.

The design paradigm is an abstraction and refinement of observed design processes, in which the various steps often overlap. Its objective is to minimize the probability of oversights, mistakes, and inconsistencies in the process of meeting the specified goals of dependable service with respect to defined classes of faults by means of the chosen implementation of fault tolerance. The paradigm is stated for the implementation of a new design. If the goal is the improvement of an existing design, then each step is a reexamination, possibly leading to changes of previously made decisions.

The paradigm partitions the system building process into three activities: *specification, design, and evaluation.* Design consists of *system partitioning, subsystem design, and system integration* steps. *Evaluation* takes place during and after each design step. The principal steps of the paradigm are summarized in Figure 2.

3.1 Specification

The specification activity begins with the detailed definition of *system requirements* that describe the services to be delivered in terms of functionality and performance. Dependability-related aspects of service that need to be identified are:

1. The existence of different *mission phases*, characterized by different environments and conditions of operation during the lifetime ("mission") of the system.

2. The varying *criticality* of service delivery and system *safety* and *security* goals during different mission phases.

3. The acceptability of possible different *modes of service* (full, reduced, degraded, emergency, safe shutdown, etc.) during each phase.

4. System *availability* goals for each mission phase, expressed in terms of the maximum allowed frequency and duration of service mode reductions due to fault occurrence, including time bounds on subsequent recovery actions.

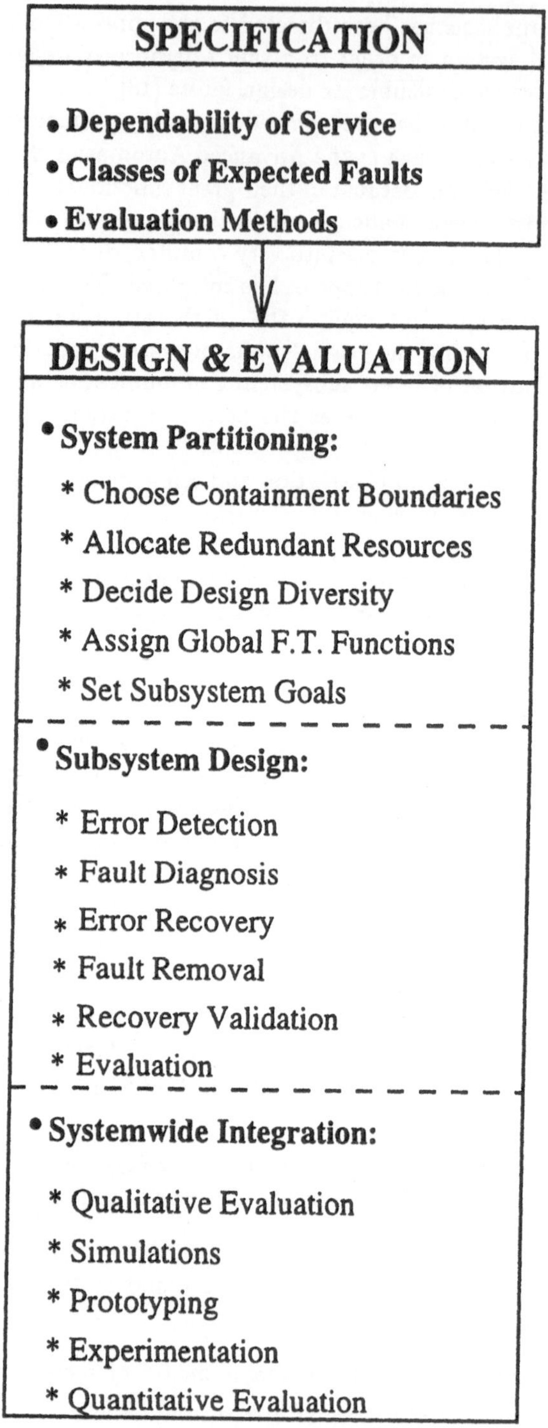

SPECIFICATION

- **Dependability of Service**
- **Classes of Expected Faults**
- **Evaluation Methods**

DESIGN & EVALUATION

- **System Partitioning:**
  * Choose Containment Boundaries
  * Allocate Redundant Resources
  * Decide Design Diversity
  * Assign Global F.T. Functions
  * Set Subsystem Goals
- **Subsystem Design:**
  * Error Detection
  * Fault Diagnosis
  * Error Recovery
  * Fault Removal
  * Recovery Validation
  * Evaluation
- **Systemwide Integration:**
  * Qualitative Evaluation
  * Simulations
  * Prototyping
  * Experimentation
  * Quantitative Evaluation

Figure 2: A Design Paradigm

5. Conditions for the *renewal of resources* (external repair by replacement and remote fault tolerance support) for different mission phases: "on demand" renewal, periodic renewal, no renewal at all.

6. System *longevity goals*, expressed in terms of reliability and maintainability requirements.

The *second* aspect of the specification activity is the identification of the *classes of faults* expected to occur during each mission phase. Fault classes may differ because of differences in environmental conditions and operating interfaces. There are two parts of this problem:

1. Internal faults that can arise within the system during each mission phase.

2. External faults that can introduce errors into the system via the interfaces (I/O links, external fault tolerance support, human/machine interaction) during each mission phase.

The *third* aspect of specification is to choose the *evaluation methods* that will be applied to assess the likelihood that the design will meet the specified dependability goals: availability, reliability, maintainability, safety, security. The independent means of validating the evaluation also need to be agreed upon and specified at this time. Fault and error occurrence scenarios need to be defined for the purpose of evaluation, including two (or more) independent near- coincident faults with overlapping effects, and the existence of latent errors.

3.2 Partitioning

The first stage of design is the definition of system architecture, expressed by the structure of its building blocks and communications links. This process is here called partitioning. Functionality, performance and fault tolerance requirements, available technologies, and past experience all affect the choice of the hardware, software, communication and interface subsystems that comprise the system being designed.

This step of the paradigm requires a close collaboration between the two groups of system architects: those responsible for performance and those responsible for dependability and fault tolerance. The partitioning that is accomplished should be a balanced solution that accommodates both sets of requirements (performance and dependable operation) at the same time.

Several fundamental decisions regarding the implementation of fault tolerance need to be made here. First, partitioning defines a set of fault and error containment boundaries. The purpose of the boundaries is to assure the independence of failure of individual subsystems.

Second, the choice is made of the methods to introduce redundant hardware and software resources. Major options are: (1) multiple channel computation: duplex, triplex, etc., (2) spare subsystems, (3) degradation by exclusion of failed subsystems. Combinations of these techniques may be selected as well. The choice of time redundancy (repetition of computations) is also a possibility.

The third major decision is whether design diversity of redundant resources is to be implemented in some or all software and/or hardware subsystems. Design diversity is an especially effective means to assure safety in critical applications and to protect the most vital subsystems of complex systems and networks.

The fourth decision is whether detection, diagnosis, and recovery functions (discussed in the next section) are to be (1) localized within individual subsystems; (2) shared by a set of functionally identical subsystems; (3) extended over a group of functionally diverse subsystems; or (4) made a global function of all subsystems. A hierarchical arrangement of two, three, or all four of these approaches offers other solutions of the detection and/or recovery problem.

After the preceding decisions have been made, every subsystem is characterized by its own error and fault containment requirements, as well as the participation requirements in set, group, and/or global detection and recovery functions discussed in the preceding paragraph. At this time the system-level goals for availability, reliability, maintainability, safety, and security (discussed in 3.1 above) can be apportioned among the independent sets of functionally equivalent subsystems that compose the entire system.

3.3 Subsystem Design

At the completion of partitioning each subsystem is characterized by a set of requirements for functionality, performance, and dependability. The dependability requirements include classes of faults to be tolerated, boundaries for error and fault containment, diversity objectives, modes of service, availability in terms of limits on the frequency and duration of service mode reduction and/or complete outage, and the bounds on recovery time. Furthermore there are the goals for subsystem reliability, maintainability, safety and security that are derived from the system-wide goals. Finally, the subsystem may be required to take part in *external* (i.e., group, set, or global) error detection, fault diagnosis, and/or recovery. The dependability goals are supported by diverse or "cloned" replication at subsystem level and by built-in (*internal*) error detection, fault diagnosis (location), and recovery algorithms.

Errors are undesired states that are caused by active faults and may lead to subsystem failure. Error detection methods fall into two categories: (a) *concurrent* with delivery of service; and (b) *preemptive*, requiring that service delivery should be suspended during the application of detection procedures. Error detecting codes are an example of the former, and memory "scrubbing" an example of the latter method.

In addition to detection, it is necessary to record the detected erroneous state for analysis that will facilitate fault diagnosis. The main methods of fault diagnosis are: (a) analysis of a detected erroneous state; (b) repetition of previous operation(s); (c) application of test patterns; (d) use of special circuitry: microdiagnostics, scan-in/scan-out, etc.; (e) use of independent (monitoring) subsystems. A special error detection and fault diagnosis requirement is that it is necessary to detect errors produced by faults in the fault tolerance mechanisms themselves. Furthermore, provisions are needed to check spare subsystems for dormant faults and to detect latent errors in stored information that is not subject to other forms of error detection.

The choice of specific error detection and fault diagnosis methods strongly depends on the function of the subsystem. For example, error control codes are very often used in memory, while duplication and comparison is common for processors. The choice and location of error detection mechanisms must minimize the probability of undetected error propagation across subsystem boundaries. Additional detection and diagnosis techniques may be incorporated to support externally implemented (group, set, global) fault tolerance functions.

Every error signal that originates in a subsystem must invoke an *error recovery* procedure that attempts to restore the system to a valid state without known errors. Error recovery can be: (a) *backward*, returning the system to a previous, error-free state; or (b) *forward*, constructing a valid, error-free new state from existing (usually redundant) information.

A persistent fault may prevent successful error recovery. In this case, fault diagnosis identifies a faulty subsystem, and fault removal is performed by: (a) substituting a good spare subsystem; or (b) reconfiguring the system to continue functioning in spite of or without the faulty subsystem. Finally, an *independent validation* that the recovery was successful is a desirable attribute for every subsystem.

Recovery algorithms are usually hierarchical: a fast, local recovery is attempted first. If it does not succeed, a more extensive and time-consuming recovery procedure is invoked, and so on, until either recovery is accomplished, or the system is safely shut down by the last procedure of the recovery hierarchy. The time that is available to complete a recovery strongly depends on the application of the system. For example, a complete repetition that is acceptable in financial transaction handling would be too slow in real-time flight control. In addition to its internal recovery procedures a subsystem may also incorporate features that support external (group, set, global) cooperative recoveries.

The adequacy of the chosen internal error detection, fault diagnosis and recovery methods is assessed by an evaluation that takes place during subsystem design and may lead to modifications of initial choices.

3.4 Systemwide Integration of Fault Tolerance

Large and complex fault-tolerant systems have experienced unanticipated, failu inducing improper interactions of otherwise well-designed fault-tolerant subsystems that have not been perfectly integrated into a distributed, fault-tolerant system. Especially difficult to integrate are the various hierarchical fault diagnosis and recovery sequences that support the localized fault tolerance of subsystems and those that support the fault tolerance of functional services ("threads") that are provided by two or more subsystems.

A second major integration problem is presented by two or more *nearly concurrent fault manifestations*. The large size and distributed nature of new systems lead to the possibility of two or more independent fault manifestations occurring close in time, most likely because of the previously undetected existence of latent errors and dormant faults. This, in turn, will require two or more recovery algorithms to be concurrently active, with the resulting risks of mutual interference, deadlocks, and behavior that is very difficult to anticipate.

A successful integration of several fault-tolerant subsystems requires in-depth analysis as well as extensive experimentation, including *incremental demonstrations* of the capabilities, protection, completeness, and consistency of all fault tolerance functions and their specific implementation not only under single, but also under *multiple overlapping* fault conditions. The growing complexity of fault-tolerant systems in critical applications requires extensive use of the most advanced analytic and experimental evaluation techniques in order to eliminate inadequacies of global (systemwide) detection and recovery algorithms.

4 Evaluation

The evaluation of the adequacy of the chosen fault tolerance techniques is a continuous process during the steps of system partitioning, subsystem design, and system integration. At each step analytic and experimental evaluation is an important design tool that facilitates the choices between alternative fault tolerance techniques and assesses the likelihood of meeting the dependability goals.

Successful completion of the design of a fault- tolerant system requires a convincing verification of two properties: the *completeness* of the design and its *potential to meet the stated goals* of availability, reliability, maintainability, and safety. The verification therefore must consist of the application of two distinct evaluations: first *qualitative*, then *quantitative*, or *numerical* [20].

The *qualitative evaluation* of fault tolerance attributes verifies that all the necessary defenses against the expected classes of faults have been properly incorporated into the design. A study of the preceding steps of the design paradigm leads to the guidelines for a structured qualitative evaluation. It is a "pass/fail" type of examination that employs an "inverse" of the design paradigm, called the *qualitative evaluation paradigm* [20], to apply a series of searching questions about the completeness and appropriateness of fault tolerance techniques.

It is essential that the qualitative examination and evaluation should be satisfied prior to generating numerical predictions of system reliability and availability. Otherwise, unreasonably optimistic predictions can be obtained because of unjustified simplifications that remain unnoticed. Examples are: (1) insufficient fault assumptions, such as omission of design faults and man/machine interaction faults, disregard of nearly concurrent manifestations of two or more faults, disregard of latent errors and dormant faults, etc.; (2) overestimation of the effectiveness and speed of error detection, fault diagnosis, system state recovery, and reconfiguration methods; (3) implicit assumptions that all fault tolerance functions do not contain design faults themselves, that they are fully protected against physical faults, and that they always interact perfectly on a systemwide basis.

The *numerical evaluation* verifies that the quantitative requirements of availability, maintainability, reliability, and safety can be met by the design that has passed the qualitative evaluation. A design first must be described in terms of a system dependability model. This model is characterized by sets of physical, structural, repair, fault tolerance, and performance parameters for every subsystem, including subsystem communication links.

The most critical and difficult task is the determination of the most likely ranges of coverage and execution time parameters for all fault tolerance functions of every subsystem. These functions are: error detection, fault location and removal, state restoration, reconfiguration, and recovery validation. Their execution must be coordinated on a systemwide basis. For this reason, the coverage and execution time parameters have to be predicted for subsystem interactions, assuming that one or more than one overlapping fault manifestation can occur, or that latent errors may be encountered in the course of an attempted recovery.

In order to be able to predict the impact of faults on system performance, the availability of the complete system must be estimated in terms of two distinct measures:

1. The *Mean Time between Mode Reductions* that indicates the expected frequency of transitions into modes of operation below full service;

2. The Duration of *Mode Reduction* that indicates the expected length of stay in the less desirable modes of operation, expressed in various measures (mean value, 99th percentile, worst case, etc.).

These measures are derived using the system reliability model that has been validated by means of the qualitative evaluation paradigm. They are very sensitive to the values of the *coverage* and *execution time* parameters of fault tolerance functions. These parameters need to be obtained by means of a sequence of progressively more specific and precise methods: estimates, analyses, simulations, and incremental experimentation with prototypes of critical elements under fault conditions. Proofs of certain design properties also may prove to be useful here. Finally, it is also necessary to assess the validity of evaluation tools and models that were used in an evaluation.

5 The "Off-the-Shelf" Problem

The preceding sections discuss the design of fault- tolerant systems that are composed of an integrated set of fault-tolerant subsystems. However, practical consideration of cost and development time often lead to the use of pre-existing or "off-the-shelf" subsystems, such as microprocessors, displays, sensors, workstations, operating systems, application software packages, etc., as building blocks of systems that are expected to be highly dependable. The off-the-shelf items usually either have limited fault tolerance or none at all.

An example of this kind of system is a Picture Archiving and Communications System (PACS). This system provides radiology data to a hospital and research community at UCLA, archiving images from X-rays, MRI and other advanced scanning devices. PACS consists of a networked collection of medical imaging devices, high performance servers for image processing and storage, optical disk storage devices, and numerous workstations for display and analysis of radiological data. In addition it is integrated with other networks that make up the Hospital Information System, and Radiology Information System [21].

The PACS system requires a high degree of availability, since downtime makes crucial records such as images unavailable. Downtime of more than a few seconds is very costly. It also must protect the integrity of its database of patient records, many of which are legally required to be kept for many years. Loss of or unauthorized access to any of these records can have serious medical (and potentially legal) consequences. Currently PACS is not fault-tolerant and costly outages often occur.

The problem of "retrofitting" PACS and similar systems for fault tolerance is very difficult. The common approach is to design a "monitor" software subsystem that attempts to check all subsystems for indications of failure, records abnormal symptoms, and initiates reconfiguration when needed. The weakness of this approach is the lack of protection for the monitor software itself, since it must reside and execute in an off-the-shelf processor.

A more fundamental solution for a large heterogenous distributed system such as PACS is to make it more dependable by implementing a smaller, highly fault-tolerant hardware system that monitors its operation, assures the protection of data integrity, and manages recovery when part of the system fails by switching in spare resources or reconfiguring the system [22]. The Test-and-Repair Processor of the JPL-STAR computer [9] is the first prototype of such a hardware monitor.

6 A Conceptual Model for the Future

It is evident that the complexity of today's computing and communication systems is already, in a general sense, comparable to the complexity of living organisms. However, a significant difference is noticeable in the specifications. While living organisms carry out the requirement "survive and reproduce," the requirements of the AAS air traffic control system[19,23], for example, fill hundreds of pages with text, diagrams, and equations.

The evolution of living organisms and computing systems began at opposite ends. For living organisms, survival of individuals and species came first, and higher cognitive functions evolved gradually over billions of years, culminating with the emergence of homo sapiens. On the other hand, modern man built the computer to emulate his intellectual functions, and only grudgingly paid attention to assurance of the survival of the computer's programs in the presence of faults.

The concept of fault tolerance gathered diverse techniques that were used to cope with the consequences of faults into a cohesive set. Since then, fault tolerance has evolved as the survival attribute of computing and communication systems, especially of those in critically important applications.

As the years go by, one fundamental principle is becoming more and more evident: the more good our sophisticated computing and communication systems can contribute to the wellbeing and the quality of life of the human race, the more harm they can cause when they fail to perform their functions, or perform them incorrectly. Let us just consider the control of air, rail, and subway traffic, the emergency response systems of our cities, the flight controls of airliners, the safety systems of nuclear power plants, and most of all, the rapidly growing dependence of health care delivery on high-performance computing and communications. And the list goes on and on...

At the same time, the challenges to dependable operation are progressively growing in scope and severity. Complex systems suffer stability problems due to unforeseen interactions of overlapping fault events and mismatches of defense mechanisms. "Hackers" and individuals with criminal intent invade systems and cause disruptions, misuse, and damage. Accidents lead to the severing of communications links that serve entire regions. Finally, it may be foreseen that "info-terrorists" will attempt to cause similar damage with malicious intent.

Fault tolerance is the only guarantee that those vitally important systems will not, figuratively speaking, turn against their builders and users by failing to serve as expected because of physical, design, or human-machine interaction faults, or even because of malicious attempts to disrupt their essential services. Past experience has shown that fault tolerance is most effective when it is an integral function of every subsystem as well as a hierarchically organized function of the entire system.

Yet, it is alarming to observe that the explosive growth of complexity, speed, and performance of single-chip processors has not been paralleled by the inclusion of more on-chip error detection and recovery features. It is likely that the incorporation of these most advanced, high performance processors into high-availability life-critical systems would pose problems similar to those encountered by the AAS: can the outputs be trusted to be error-free?

One solution is the design of multiple channel systems, possibly with diverse processors. While it is suitable for modest-sized dedicated applications, such as flight control, the cost and complexity would be unacceptable for AAS or PACS-like systems and their even more complex successors in the 21st century.

The urgent problem remains: we need to convince the builders of chips as well as the buyers of AAS-like systems that the elegant ideas of building reliable systems from unreliable components that were conceived for components such as a NAND gate [6] or a relay contact [7] do not scale up to distributed systems in which each one of hundreds or thousands of component nodes is a high-performance computer itself. Redundancy alone is not the answer.

The solution that I propose here is that the model of a living organism, in which distributed, specialized survival functions protect the capacity for cognitive action, offers the best analogy for the very complex and highly dependable distributed systems that we are expected to build for the benefit (or entertainment) of our fellow humans in the near future.

Why is this model for fault-tolerant systems attractive? First, we are setting up an analogy with the most dependable information processing systems in existence - the various species of highly evolved living creatures that have survived over millions of years on our continually changing planet. There is much to be learned there, and it is not fault avoidance.

Second, the analogy reaches out to appeal to a wide spectrum of intelligent people, since it does not require knowledge of computer science or technology, or an a priori belief in the utility of fault tolerance.

Third, the information age is only beginning, and we should set our goals high if we are to succeed in our role as the protectors of the defining resource of the coming millennium. There is no doubt that the threats of disastrous failures are real and serious.

Four specific analogies that elaborate the model are suggested below. The first observation (focusing on humans) is that the immediate defense mechanisms of a living body (immune system, lymphatic system, pain sensors, healing processes, etc.) are autonomous (i.e. in "hardware" or "firmware") and do not require cognitive function ("software") support. However, higher-level protection (medication, imaging, physical therapy, surgery, etc.) can be invoked by the brain's decision. While this appears self-evident, most computing subsystems built today (microprocessors, disk drives, memory chips, etc.) lack a sufficient set of local error detection and recovery attributes, and software must be involved in managing their fault conditions. That is analogous to expecting cognition to compensate for the absence of immunity or of the sense of pain - not an effective substitution.

The second observation is that the immediate defense mechanisms discussed above are distributed and their generic services are shared by the specialized subsystems of the body. The analogous generic defense mechanisms are error detecting or correcting codes, totally self-checking logic, comparisons, local rollback or voting, etc.

The third observation is that diversity is the key attribute of a species that protects it against extinction due to genetic defects in the individual members, while the diversity among species has assured the continuity of life on earth. The self-evident analogy here is the use of hardware and software design diversity in fault- tolerant systems.

Finally, it may be useful to view the dependability aspects of networks of computing systems as similar to those of social structures, in which maintaining a consistent view of common reality and avoiding the distribution of (accidentally or deliberately) contaminated information are major objectives. The analogous techniques are the protocols for fault-tolerant consensus and robust data structures for data integrity.

The most immediate objective of the analogies identified above is to communicate the nature and the advantages of fault tolerance attributes to the community of users with the need for dependable communications and computing.

Customer demand is the force that will motivate the manufacturers of microprocessors and other subsystems to take a pause in their desperate race for more function, speed, and storage capacity in order to introduce at chip and other subsystem level the fault tolerance features that will allow the building of dependable systems that do not depend almost exclusively on software services to assure dependable operation.

I conclude with the confident prediction that while the speed of computing will be ultimately limited by the laws of physics, the demand for more dependability will not cease as long as humans will use computers to enhance the quality of their lives.

Acknowledgement

The year 1995 marked 40 years since the author's first professional work on system dependability at JPL. It has been my good fortune to work with and learn from many colleagues at the University of Illinois, Caltech's Jet Propulsion Laboratory and the UCLA Computer Science Department. University of Illinois professors James E. Robertson and David E. Muller were my mentors and role models for my entire career.

The work with my friends in the IEEE CS Technical Committee on Fault-Tolerant Computing and IFIP Working Group 10.4 on Dependable Computing and Fault Tolerance has been a rewarding experience and a source of inspiration for this paper.

Many thanks to my son Rimas Avižienis, EECS student at UC Berkeley, for preparing this text with care and dedication.

References

[1] Avižienis, A., "Design of Fault-Tolerant Computers," *AFIPS Conference Proceedings, 1967 Fall Joint Computer Conference*, Vol. 31, Washington D.C.: Thompson, 1967, pp. 733-743.

[2] " *Proceedings of the Second Symposium on Large-Scale Digital Calculating Machinery,* " Sept. 13- 16, 1949, Annals Computation Lab., Harvard University, Vol. XVI, Cambridge, MA: Harvard University Press, 1951.

[3] *Proceedings of the Joint AIEE-IRE Computer Conference*, Dec. 10-12, 1951.

[4] "Information Processing Systems - Reliability and Requirements", *Proceedings of Eastern Joint Computer Conference* , December, 1953.

[5] Session 14: "Symposium: Diagnostic Programs and Marginal Checking for Large Scale Digital Computers," in *Convention Record of the IRE 1953 National Convention*, part 7, New York, N.Y., March 1953, pp. 48-71.

[6] von Neumann, J., "Probabilistic Logics and the Synthesis of Reliable Organisms from Unreliable Components," Automata Studies, C.E. Shannon and J. McCarthy editors, *Annals of Math Studies* No 34, Princeton, NJ: Princeton University Press, 1956, pp. 43-98.

[7] Moore, E.F., Shannon, C.E., "Reliable Circuits Using Less Reliable Relays," *Journal of the Franklin Institute* 262, No. 9 and 10, Sept. Oct. 1956, pp. 191-208 and 281-297.

[8] Avižienis, A., Rennels, D.A., "The Evolution of Fault Tolerant Computing at the Jet Propulsion Laboratory and at UCLA: 1960-1986," in *The Evolution of Fault-Tolerant Computing*, Springer- Verlag: Vienna and New York, 1987.

[9] Avižienis, A., Gilley, G.C., Mathur, F.P., Rennels, D.A., Rohr, J.A., Rubin, D.K., "The STAR (Self-Testing-and-Repairing) Computer: An Investigation of the Theory and Practice of Fault- Tolerant Computer Design," *IEEE Trans. on Computers,* Vol. C-20, No. 11, November 1971, pp. 1312-1321; also in *Digest of the 1971 International Symposium on Fault Tolerant Computing*, Pasadena, CA, March 1971, pp. 92-96.

[10] "TOPS Outer Planet Spacecraft," *Astronautics and Aeronautics*, Vol. 8, September, 1970.

[11] Special Issue, *Digest of the 25th International Symposium on Fault-Tolerant Computing*, Pasadena, CA., June 27-30, 1995.

[12] Pierce, W.H., Failure-Tolerant Computer Design, Academic Press: New York and London, 1965.

[13] Avižienis, A., "Architecture of Fault-Tolerant Computing Systems," *Digest of FTCS-5, the 25th International Symposium on Fault-Tolerant Computing*, Paris, June 1975, pp.3-16..

[14] Avižienis, A., "Fault-Tolerance: The Survival Attribute of Digital Systems," *Proceedings of the IEEE*, October 1978, Vol. 66, No. 10, pp. 1109-1125.

[15] Avižienis, A., Laprie, J.C., "Dependable Computing: From Concepts to Design Diversity," *Proceedings of the IEEE*, Vol. 74, No. 5, May 1986, pp. 629-638.

[16] Avižienis, A. "A Design Paradigm for Fault Tolerant Systems," *Proceedings of the AIAA Computers in Aerospace VI Conference*, Wakefield, MA, October 1987, pp. 52-57.

[17] Avižienis, A., "Software Fault Tolerance," in *Information Processing 89, Proceedings of the IFIP 11th World Computer Congress*, San Francisco, CA, G.X. Ritter (Ed.), Elsevier Science Publishers, B.V. North Holland, 1989, pp. 491-498.

[18] Joseph, M.K., Avižienis, A., "A Fault Tolerance Approach to Computer Viruses," *Proceedings of the 1988 IEEE Symposium on Security and Privacy*, Oakland, CA, April 18-21, 1988, pp. 52-58.

[19] Avižienis, A., Ball, D.E., "On the Development of a Highly Dependable and Fault Tolerant Air Traffic Control System", *Computer*, Vol. 20, No. 2, February 1987, pp. 84-90.

[20] Avižienis, A., "The Dependability Problem: Introduction and Verification of Fault Tolerance for a Very Complex System," *Proceedings of the 1987 Fall Joint Computer Conference*, Dallas, Texas, October 1987, pp. 89-93.

[21] Avižienis, A., Huang, L.J., He, Y., Valentino, D.J., "Software and System Engineering for a Large- Scale PACS" Paper No. 2435-54 in the *Proceedings of SPIE Conference 2435*, "PACS Design and Evaluation: Engineering and Clinical Issues," San Diego, CA, Feb. 26 - Mar. 2, 1995.

[22] Avižienis, A., "Fault Tolerance by Means of External Monitoring of Computer Systems," *AFIPS Conference Proceedings*, Vol. 50, May 1981, pp. 27-40.

[23] Cristian, F., Dancey, R.D., Dehn, J.D., "Fault Tolerance in the Advanced Automation System," *Digest of FTCS-20, the 20th International Symposium on Fault Tolerant Computing*, June 1990, pp. 6-17.

Session 1
Formal Methods and Models

Formal Verification of Safety Requirements on Complex Systems *

Cinzia Bernardeschi
Dipartimento di Ingegneria della Informazione, Univ. di Pisa
Pisa, Italy

Alessandro Fantechi
Dipartimento di Sistemi e Informatica, Univ. di Firenze
Firenze, Italy

Stefania Gnesi
Istituto di Elaborazione della Informazione - C.N.R.
Pisa, Italy

Abstract

In this paper we present a logical characterization, by means of ACTL formulae, of safety requirements to be formally verified over safety critical complex systems. In this class of systems the formal verification of requirements is often hardened by state explosion problems. To deal with this problem, the characterization we propose allows the satisfiability of a safety requirement over a complex system to be derived by its satisfiability over those component subsystems that are directly involved in the given requirement. The proposed methodology has been successfully used for the formal verification of safety requirements of a particular system, that is a railway computer based signalling control system.

1 Introduction

Formal methods are mathematically-based techniques that can offer a rigorous and effective way to model, design and analyze computer systems. In the field of safety critical systems, formal methods are generally recognized as a fault avoidance technique which can increase dependability by removing errors at the requirements, specification and design stages of development [1]. Formal methods also offer a means of handling the design fault problem in safety critical systems because the often quoted 10^{-9} faults per hour reliability is well beyond the range of quantification and there is no choice but to develop life-critical systems in the most rigorous manner available. Besides achieving safety by avoiding design errors, formal methods can also be used to formally verify that a system satisfies safety requirements.

Moreover, the use of formal methods is increasingly required by the international standards and guidelines for the development of safety critical computer-

*Work partially supported by OLOS HCM Network-EC Contract No. CHRX-CT94-0577

controlled systems. Formal methods have been a topic of research for many years and the question now is whether these methods can be effectively used in industrial applications. Tool support is necessary for a full industrialization process and there is a clear need for improved integration of formal method techniques with other software engineering practices. As far as the verification of the correctness of a system is concerned, several automatic tools have been recently developed to this purpose. Such tools are generally based on a "theorem prover" or a "model checking" approach.

In theorem prover approaches, the system state is modeled in terms of set-theoretical structures on which invariants are defined, while operations on the state are modeled by specifying their pre- and post-conditions in terms of the system state. Properties are described by invariants, which must be proved to hold through the system execution, by means of theorem proving.

Model checking approaches, on the other hand, work on a finite state representation of the behavior of the system. Verification is usually carried out by checking the satisfiability of some desired properties over the model of the system by model checking algorithms [2, 3] or equivalence relations [4]. In particular, safety requirements may be expressed as temporal logic formulae and may be checked on the model of the system. A case study in this direction is presented in [5]. There, the specification of a railway interlocking signalling control system is given using both process algebras and automata formalisms. The generation of the model of the behavior of the system and the verification of the safety requirements is obtained through the use of the tools in the JACK environment [6].

Model checking approaches give an automatic verification method of system properties. Unfortunately, this approach has the drawback that systems composed of several subsystems can be associated a finite state model with a number of states which is exponential in the number of the component subsystems. Moreover, systems which are highly dependent on data values, share the same problem producing a number of states exponential in the number of data variables. The so called "state space explosion" problem, together with the relatively low maturity of the verification tools, has favoured the theorem prover approaches in the first industrial experiences of application of formal methods (see, for an example of successful application of a theorem prover approach, [7]).

Basing on the case study in [5], in this paper we propose a model checking verification methodology which allows the state explosion problem to be partially solved by "zooming" those parts of the system on which the safety requirements have to be verified. The verification is then performed on the subsystems focused by this "zooming" technique, and the results of the verification are then extended to the global system.

The paper is organized as follows: Section 2 presents an overview of the formalisms used in our approach to describe systems and properties and to proof the satisfaction of the safety requirements. Section 3 gives a description of the methodology developed to make the formal verification of safety requirements on complex systems tractable. In Section 4, we give an example on the application of the methodology to the railway interlocking control system. Finally, in Section 5 some conclusive remarks on the work are reported.

2 An introduction to our formalisms

2.1 The CCS/MEIJE process algebra

Process algebras [4, 8] are generally recognized as a convenient means for describing sequential or concurrent interacting systems (processes) at different levels of abstraction. They rely on a small set of basic operators, which correspond to primitive notions of concurrent systems, and on one or more notions of behavioral equivalence or preorder. The operators are used to build complex systems from more elementary ones. The behavioral equivalences are used to study the relationships between descriptions of the same system at different levels of abstractions (e.g., specification and implementation). The semantic models of process algebras are Labelled Transition Systems (L.T.S.) which describe the behavior of a process in terms of states and labelled transitions, relating states.

Among process algebras we used CCS/MEIJE [9]. In CCS/MEIJE a system consists of a set of communicating processes; each process executes input and output actions and synchronizes with other processes to execute its activities. Some of the CCS/MEIJE operators are reported in Figure 1.

Syntax	Operator	Meaning
$a?:P$ $a!:P$ $\tau:P$	Action prefix	Action a is performed, and then execute P Actions are names followed by either the suffix "!" or the suffix "?". There exists a special action, called **tau**, that represents an internal state transition of a process.
$P+Q$	Nondeterministic choice	Alternative choice between the behavior of P and that of process Q
$P \parallel Q$	Parallel composition	The interleaved executions of the P and Q synchronize on complementary input and output actions (i.e. actions with the same name but a different suffix)

Figure 1: Some of the CCS/MEIJE operators

2.2 The ACTL temporal logic

ACTL [10] is the action based version of CTL [11] whose semantic models are labelled transition systems. ACTL is a branching time temporal logic that is suitable, embedding the idea of "evolution in time by actions", for describing the behavior of systems that perform actions during their working time. The syntax and the informal semantics of some of the ACTL operators are shown in Figure 2.

In the figure, α is an action of the set of input and output actions that a given system can perform. An execution (*path*) is a (finite or not) sequence of

Action formulas

χ	::=	*true*	"any observable action"
		false	"no observable action"
		α	"the observable action α"
		$\sim\chi$	"any observable action different from χ"
		$\chi \mid \chi'$	"either χ or χ'"
χ'	::=	χ	

State formulas

ϕ	::=	*true*	"any behavior is possible."
		false	"no behavior is possible."
		$\sim\phi$	" ϕ is impossible"
		ϕ & ϕ'	"ϕ and ϕ'"
		$E\gamma$	"there exists a possible execution in which γ"
		$A\gamma$	"for each of the possible executions γ"
		$<action>\phi$	"there exists a next state reachable with *action*, in which ϕ"
		$[action]\phi$	" for all next states reachable with *action*, ϕ is true"
ϕ'	::=	ϕ	

Path formulas

γ	::=	$G\phi$	"at any time ϕ"
		$F\phi$	"there is a time in which ϕ"
		$[\phi\{\chi\}U\{\chi'\}\phi']$	"at any time χ is performed and *also* ϕ, *until* χ' is performed and then ϕ'"

Figure 2: Some of the ACTL operators

actions. A *state* represents a time in which a single action has been completed and a new next action may be performed. It is possible that there is more than one action that the system can perform, when its execution reaches a state. Each of these actions represents the beginning of an alternative continuation of the execution. A *state formula* gives a characterization about the possible ways an execution could continue after a state has been reached, while a *path formula* states some properties of a *single* execution.

ACTL can be used to define *safety* and *liveness* properties of concurrent systems in terms of the actions a system can perform. Safety properties claim that anything bad does not happen; i.e. we can prove that there are no path in the LTS such that a given action sequence occurs. Liveness properties claim that something good eventually happens; i.e. we can prove that there exists a path in the LTS in which a given action is allowed to occur.

As an example of ACTL safety and liveness properties, we can consider the following formulae. The formula

```
AG [a]  E [ true {b} U <c> true ]
```

means that if an action `a` is performed, then there is no execution of a system in which the action `b` can be performed until the action `c` has been executed. The formula well characterizes safety requirements and, for example, it is false

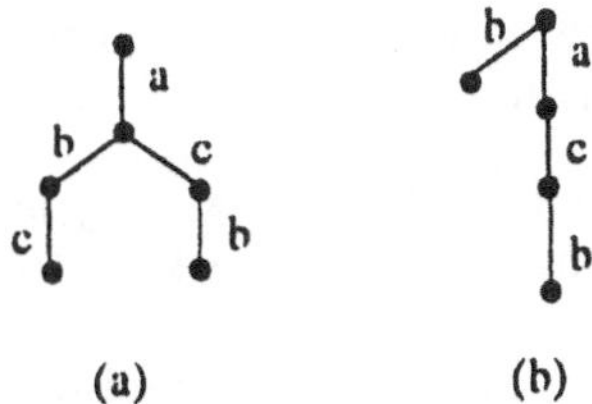

Figure 3: Two LTSs.

on the system modeled by the LTS in figure 3 (a), since in the left path of the tree the action `b` is executed after the action `a` but before the action `c`. The formula is, instead, true on the system modeled by the LTS in figure 3 (b).

A typical ACTL formula stating a *liveness* property is `EF <b> true`, which means that on every execution of a system modeled by the LTS, the action `b` can be executed. The formula is obviously true on the systems modeled by the LTSs in figure 3.

3 The "zooming" technique

Safety requirements expressed by temporal logic formulae can be automatically verified on systems specified in process algebras and modeled by LTSs, using specific model checking algorithms. Unfortunately, this verification approach has the drawback that complex systems composed of several subsystems can be associated a finite state model with a number of states which is exponential in the number of the component subsystems. This problem, called "state space explosion", makes the verification unfeasible by automatic tools.

In this section we show a methodology that can be followed to make the problem of safety requirements validation tractable by current tools. The methodology consists in the following steps:

1. The system is specified as a collection of subsystems, by process algebra terms and automata;
2. The safety requirements are expressed as temporal logic formulae;
3. The safety requirements are analyzed to enlight the subsystems directly involved in their satisfiability; typically, only few subsystems are engaged in the definition of a single requirement. Possibly, this activity can require a decomposition of a formula in subformulae that should be checked on the model of the system as well;
4. The subsystems are defined using suitable "abstraction techniques", which allow to "abstract away" all the details of the whole system with respect to the property to be checked. The part of the system ignored becomes a "free" external environment. This means that it can have any possible behavior. In some cases the behavior of this external environment can

be restricted to follow some general conditions directly derived from the expected behavior of the real system;

5. Any formula produced in steps 2) and 3) is then verified on the corresponding submodel, using model checking tools.

The proposed methodology can be applied only if we guarantee that any property ϕ we prove on a subsystem *Sub* can extend to the whole system S. Formally, we have to assure that:
$Sub \models \phi$ implies $S \models \phi$.
This is not true, in general, for any formula ϕ. Consider for example a liveness property, that says that eventually an action **act** is performed by the subsystem *Sub*. This is expressed by the ACTL formula ϕ_{live} = `EF <act> true` . If we prove, by model checking, that $Sub \models \phi_{live}$, this means that there exists a path in *Sub* in which the action **act** is eventually performed. If we immerse our subsystem in the global system S, the behavior of *Sub* is constrained by the behavior of other processes of S. This means that some parts of its behavior are lost or cut. In particular, it could be the case that the execution of *Sub* in which **act** is performed is lost or cut, and therefore $S \not\models \phi_{live}$.

Safety properties, on the contrary, tell that something undesired is not true in any state of the system. This should assure that if a safety property ϕ_{safe} is true on *Sub*, it is also true on S. Actually, proving this fact requires a careful characterization of the class of safety properties considered.

We can consider two typical forms of ACTL formulae that express safety properties:

1. ϕ_{safe1} = `AG [act1]` $\sim$ `E[true {` $\sim$ `act2} U {act3}` ϕ_0 `]`, that reads "For all states, when an action **act1** occurs, then there exists no execution in which **act2** is not performed before **act3** is performed".
   We recall that step 3) guarantees that ϕ uses a set of actions which are possibly performed by *Sub*, but not by any other subsystem of S.
   If $Sub \models \phi_{safe1}$, then in all the states reached by **act1**, there exists no execution in which the path formula between square brackets holds. In particular, there exists no execution in which the action **act3** is eventually performed. Since the behavior of *Sub* may have had some executions lost or cut when immersed into S, no execution in which the action **act3** is eventually performed either exists in S. Therefore, $S \models \phi_{safe1}$. (A similar reasoning applies for formulae of the type:
   `AG [act1]` $\sim$ `E[true {` $\sim$ `act2} U` ϕ_0 `]`).

2. ϕ_{safe2} = `AG [act1] A[true {` $\sim$ `act2} U {act3} true]`, that reads "For all states, when an action **act1** occurs, then for all executions **act2** is not performed before **act3** is performed".
   If $Sub \models \phi_{safe1}$, then in all the states reached by **act1**, all executions satisfy the path formula in the square brackets. Some of these executions may be lost in considering *Sub* immersed into S, and this does not affect the truth of the formula. Some of these executions could instead be cut, and this could affect the truth of the formula because it might be that

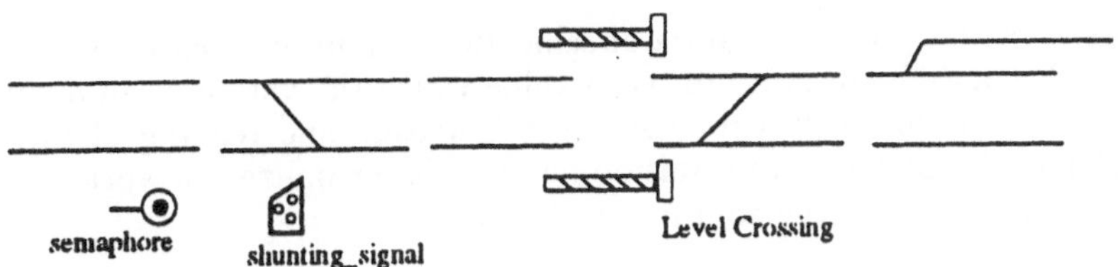

Figure 4: Interlocking a level crossing with a shunting route mechanism

act3 is never performed along this execution. If we restrict to those executions which are not cut, the formula is still true. We can say therefore that:
$Sub \models \phi_{safe1}$ implies $Sub \models_{FC} \phi_{safe2}$,
where $\models_{FC}$ reads "satisfies under fairness constraints" (in the sense of [2]), that is, it satisfies if we consider only those executions which respect some fairness constraints. Here the fairness constraint says that act3 should occur infinitely often. Again, this corresponds to accept a "loss of liveness", but still assuring the safety of those executions which are live. This is reasonable in the case we can prove liveness separately, or when we adopt a "fail safe" attitude, in which the possible loss of liveness is accepted as the price we pay for safety.

The considered forms of safety properties can be extended considering sequences of actions instead of single actions or more complex action formulae instead of formulae given by single actions.

4 An example

This methodology has been tested on the specification of the railway signalling control system presented in [12]. The subject of the experiment was a computer based interlocking system, produced by Ansaldo Trasporti. This system, known as CBI, can carry out, with very high reliability, all control, interlock and clearing functions performed by relay interlockings, both for mainline railways and underground transport applications.

The state of the system is represented by variables corresponding to the physical or to the logical entities. Each operation includes a list of verifications on some variables and a list of value assignments to other variables (operation's output). The control of each physical/logical entity (the level-crossing and the shunting route mechanism in our example) is realised by the set of operations related to the entity. In particular, each of such entities can be seen as a collection of *operations*, that describe the behavior of the entity, *variables* that are used to store global information about the entity, and *attributes*, that are statical configuration parameters for the entity itself.

The formalization of the CBI system, started from the industrial semiformal specification, through the use of the tools in the JACK environment [6]. Variables and operations in the original specification are mapped onto processes in the process algebra specification. A process corresponding to a variable has a different state for each value which the variable can assume; the actions possible for such a process are the writing and reading of each value. We refer to [5] for more information about this formalization work which corresponds to step 1) of the zooming technique.

At the step 2) the safety requirements have been expressed by using temporal logic formulae. As an example, we report a safety constraints which must be verified by the system.

Requirement : The shunting route mechanism does not send the proceed command to the shunting signal if the level crossing is not closed.

This safety requirement can be expressed more precisely as propositions on the model of the system, in the following way: *if in* any *state of the model, it is true that the position of the level crossing control is* open or undefined, *then* there is not *an execution of the processes* starting from such a state, *in which the position of the level crossing control never becomes* closed *and the shunting signal is sent the proceed command.*

This property can be translated into the following ACTL formula:

```
AG ( [undefined_pos? | opened_pos?]
(~E[true {~ closed_pos?}U (<raise_shunt_sign!> true)] ) )
```

We can notice that this safety property has the form of the safety properties described at point 1 in the previous section.

In step 3), the given requirements are analyzed. We can notice that the one presented above concerns the safety of the interlocking between a level crossing control, a physical control entity, and the shunting route mechanism, a logical control entity aimed at reserving track segments in front of a train before giving permission for it to move, see figure 4.

Therefore, as step 4), we generate the model of the subsystem composed by the level crossing control and the shunting route mechanism.

At step 5) of the methodology we checked the satisfaction of the safety requirements on the model generated at step 4) before. The requirement is proved to be true on the model, as shown by the following trace of the model checker AMC, available inside the JACK environment:

```
|=  AG ( [undefined_pos? | opened_pos?]
    (~E[true {~ closed_pos?}U (<raise_shunt_sign!> true)] ) )
The formula is TRUE in state 0 time: (user: 0.00 sec, sys: 0.00 sec)
```

According to what said at point 1 in the previous section, we can derive from this result the validity of the safety requirement on the global system.

5 Conclusions

The zooming technique presented in this paper allows the satisfiability of safety requirements over complex systems to be checked, deriving it from the satisfiability of the same requirement over some selected component of the whole system. The selection is performed looking at the requirements, and enlightening the parts of the system directly involved in the requirements. The correctness of this process is guaranteed by the structure of logical characterization of the safety properties.

Obviously, the possibility of applying successfully this methodology was due to the following characteristics of the class of systems we have considered:

- The system is composed of a number of subsystems, each devoted to the control of a physical or logical entity (such as level crossing, pointworks or shunting route control). The interactions among entities, being devoted to the interlocking function, develop along the spatial contiguity of the controlled physical entities. Although several entities compose a system that control a railway plant, the controller of a level crossing, for example, has to interact only with the controllers of the routes passing that crossing, and not (directly) with the controllers of the pointworks on the other side of the railway plant yard. Hence, the safety requirements concerned with the interlocking system are related only to two subsystems: the level crossing subsystem and the related shunting route subsystem.

- Each subsystem can be modeled with a reduced number of states. In fact, the space of data values is restricted. A subsystem is related to the different states that the controlled physical entity can assume, and such entities exhibit a discrete behavior, rather than a continuous behavior which requires a large amount of different values to be represented.

References

[1] Bowen, J.P., Hinchey, M.G, Seven More Myths of Formal Methods, *IEEE Software*,12, July 1995, pp. 34-41.

[2] E. M. Clarke, E. A. Emerson, A. P. Sistla. Automatic Verification of Finite-State Concurrent Systems Using Temporal Logic Specification. *ACM Transaction on Programming Languages and Systems*, 8(2), April 1986, pp. 244 - 263.

[3] J. R. Burch, E.M. Clarke, K. L. McMillan, D. L. Dill, L. J. Hwang. Symbolic model checking: 10^{20} states and beyond. *Information and Computation* 98(2), June 1992, pp. 142-270.

[4] R. Milner. Communication and Concurrency. Prentice Hall, 1989.

[5] A. Anselmi, C. Bernardeschi, A. Fantechi, S. Gnesi, S. Larosa, G. Mongardi, F. Torielli. An experience in formal verification of safety properties of a railway signalling control system. *Proceedings of the SAFECOMP'95 Conference*, Belgirate, Springer - Verlag, 1995, pp. 474-488.

[6] A. Bouali, S. Gnesi, S. Larosa. The integration Project for the JACK Environment. *Bulletin of the EATCS*, n.54, October 1994, pp.207-223.

[7] C. Da Silva, B. Dehbonei, F. Mejia. Formal Specification in the Development of Industrial Applications: Subway Speed Control System. *Formal Description Techniques*, V (C-10) M. Diaz and R. Groz (Editors) Elsevier Science Publishers B, V, (North-Holland), 1993.

[8] Hoare, C.A.R., *Communicating Sequential Processes*, Prentice Hall Int., London, 1985.

[9] Boudol, G Notes on Algebraic Calculi of Processes. *Notes on Algebraic Calculi of Processes*, NATO ASI Series F13, 1985.

[10] De Nicola, R., Vaandrager, F W., Action versus State based Logics for Transition Systems, in *Proceedings Ecole de Printemps on Semantics of Concurrency* Lecture Notes in Computer Science 469, Springer-Verlag, 1990, pp. 407-419.

[11] Emerson, E. A., Halpern, J. Y., "Sometimes" and "Not Never" Revisited: on Branching Time versus Linear Time Temporal Logic, *Journal of ACM*, 33, 1986, 151-178.

[12] G. Mongardi. Dependable Computing for Railway Control Systems, in *Dependable Computing for Critical Applications 3*, Dependable Computing and Fault-Tolerant Systems 8, Springer-Verlag, 1992, pp. 255-277.

Formal Specification of Safety-Critical Software with Z and Real-Time CSP

Maritta Heisel, Carsten Sühl
Technische Universität Berlin
FB Informatik – FG Softwaretechnik
Franklinstr. 28-29, Sekr. FR 5-6, D-10587 Berlin
email: {heisel, suehl}@cs.tu-berlin.de

Abstract

A method for the specification of software for safety-critical applications is presented. It is based on a combination of the formal specification languages Z and real-time CSP. Guidelines for the development and validation of specifications are provided. Specification development is supported by reference architectures that can serve as templates for the specification of concrete systems. Validation is performed by inspection of specifications according to specific criteria and by proof of properties. These proofs rely on a common semantic model for the two languages. An example serves to illustrate the method.

1 Introduction

Failures of safety-critical systems may cause severe damage. It is therefore worthwhile to apply formal methods to specify software components for safety-critical systems. Only they can guarantee preciseness and unambiguity of the specification. Moreover, the possibility to mathematically prove that the specified safety constraints are fulfilled makes formal methods a highly desirable complement to conventional ones in this context.

This paper presents a method for specifying software for safety-critical applications based on formal specification techniques. It takes into account the special conditions of this application domain, namely the necessity not only to model the software part of the system to be developed but also the environment in which the software operates.

Since safety-critical systems are reactive in general, their reactive behavior is of major importance. That aspect cannot be adequately described with algebraic or model-based specification languages. On the other hand, the software must maintain an internal representation of the system's state that can be changed by operations. This aspect must also be specified. These considerations lead us to choose a combination of the formal specification languages Z [10] and real-time CSP [1]. The functional aspects of the system are specified in Z, whereas real-time CSP serves to specify the behavioral aspects.

While the use of an adequate language is a *necessary* condition for developing satisfactory specifications of safety-critical software, it is far from being *sufficient.* To make formal specification techniques applicable in practice, they must be complemented with a methodology that guides their use. To this end, we define reference architectures that capture different kinds of safety-critical systems. For these architectures, detailed guidelines can be given how to set up a formal specification. They serve to draw the specifier's attention to aspects important for safety.

Finally, the formal nature of the developed specification makes it possible to rigorously demonstrate that the specification fulfills the necessary safety constraints. Such proofs usually refer to the Z part as well as the CSP part of the specification. It is therefore necessary to assign a common semantics to a combined specification.

In Section 2, we describe the kind of systems we want to specify. The method itself, consisting of a software model and two reference architectures, is presented in Section 3. There, an example is presented, too. Section 4 introduces the common semantic model for the combined language. Finally, we compare our approach to related work (Section 5), summarize our achievements, and point out directions of further research (Section 6).

2 Underlying System Model

We assume that there is a technical process whose control component is at least partially realized by software, see Figure 1 and [5]. Such a software component affects certain process variables (*manipulated variables*) by means of sending commands to actuators. By evaluating the current states of certain process variables which are measured by sensors (*controlled variables*), the control component is able to approximate the current state of the real process in order to verify the effect of the commands sent to the actuators within the process (feedback control).

The behavior of the technical process does not only depend on internal conditions within the process, e.g. the state of the manipulated variables, but it is also influenced by external disturbances. The basic objective of process control is to achieve the process control function in spite of disturbances from the environment. From these considerations, we are able to infer which subsystems of a technical process have to be modeled when following the proposed method designed to achieve system safety:

- *all* parts of the process-control component, i.e. software components, mechanical and electrical components, and interfaces to human operators,
- sensors, determining the projection of the real process state to the internal state of the control component, and
- actuators, which realize the execution of commands given by the control component within the real process.

At this point, the essential difference between *correctness* and *safety* becomes clear. The term correctness is defined as the property of a software

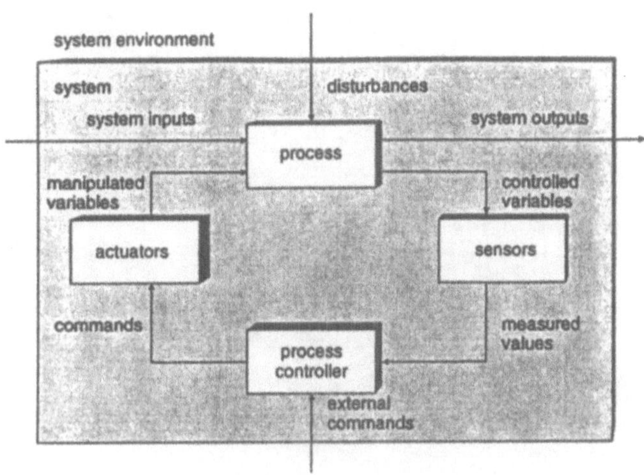

Figure 1: System Model

component to fulfill the relation between inputs and outputs prescribed in the component specification. Thus, incorrect measurements of process variables or measurements of process variables arriving from the sensors at the wrong time as well as the failed realization of given commands by the actuators are not relevant in the context of correctness. In contrast, the notion of safety is defined as the system property to be free from accidents or losses. In this case, the thorough examination of the above situations is a necessary condition.

The development of software-based components of a control system takes place in a number of stages that are applied with repetitions and in an interleaved manner. These are hazard analysis, formal specification, validation of the formal specification, design, and program synthesis. In this paper, we focus on the development of the formal specification. More details on the other stages can be found in [11, 3].

3 Formal Specification

In general, the control component of a technical process refers to a *reactive* system, which is characterized to be mainly event-triggered. It continuously reacts to events occurring within the environment by invoking internal operations and subsequently emitting resulting events into the environment. In accordance with Harel [2], we split the specification of a software component into two parts.

1. In the *structural* and *dynamic* part the reactive behavior of the software component is specified, i.e. its reaction to the occurrence of events within the real process (detected by sensors) which is realized by invoking internal operations and giving commands to the actuators. In this part, real-time requirements and the ordering of events are crucial.

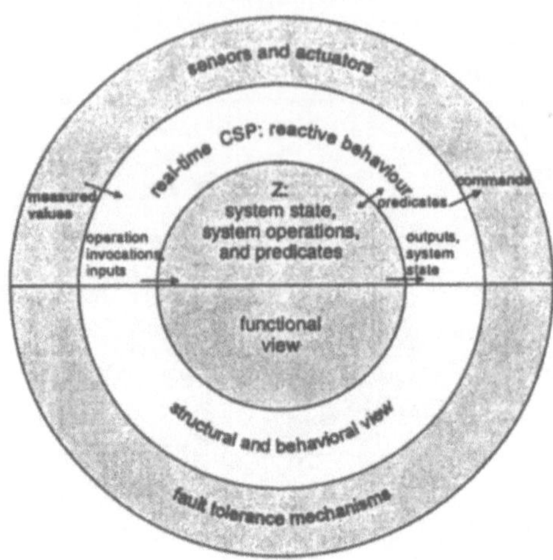

Figure 2: Software Model

2. In the *functional* part the properties and the structure of the possible system states, i.e. data structures as well as system operations applied to these states are specified. System operations are defined by relations between inputs, outputs, and the system states before and after the execution of the respective operation.

The specification languages Z and real-time CSP provide constructs to adequately express both aspects.

3.1 Software Model

To achieve a suitable combination of both parts of the formal specification of a software component formulated in Z and real-time CSP, we propose the software model shown in Figure 2.

1. The innermost component which is expressed in Z specifies the functional aspects, i.e. the structure and the properties of the valid system states as well as the requirements for system operations.
2. Around this innermost component, a CSP process specifies the reactive behavior, i.e. the absorption of values provided by the sensors, the invocation and termination of internal operations, and the transmission of the operation results to the actuators.
3. The outermost component models the required behavior of the sensors and actuators. It offers the possibility to specify fault tolerance mechanisms, e.g. the redundant arrangement of sensors and actuators.

Both the Z specification and the sensors and actuators form the environment of the CSP process.

Two different styles of specifying the reactive behavior in the CSP part can be distinguished. First, a term of the syntax of real-time CSP can be given to model the dynamic behavior in a constructive manner which is amenable to further refinement. Second, predicates can be used to constrain the set of possible behaviors. This is a more abstract way of specification. Both approaches are semantically equivalent and can thus be combined arbitrarily.

Informally, the relation between the elements of the Z part and the CSP part of a formal specification can be explained as follows. For each system operation *Op* specified in the Z part which is intended to be externally available, the CSP part is able to refer to the events *OpInvocation* and *OpTermination*, whose occurrences represent the invocation of the system operation *Op* by the software component and the operation termination, respectively. The two different events mark the execution interval of an operation. This makes it possible to specify requirements for the maximal duration in terms of assumptions about the environment which constrain the availability of these events. An example can be found in Section 3.2.1. Alternatively, if the duration of the execution is assessed to be negligible, only one event *OpExecution* can represent the whole execution of the operation *Op*.

For each input (*in*? : *Type*) of a system operation *Op*, there is a communication channel *in* within the CSP part onto which an input value possibly derived from sensor measurements is written before operation invocation. The alphabet of this channel is identical to the type of the operation input.

Analogously, for each output (*out*! : *Type*) of a system operation, there is a communication channel *out* in the CSP part from which the output value of the operation is read after termination and possibly used to derive commands to the respective actuator.

The dynamic behavior of a software component may depend on the current system state (e.g. operational mode). To achieve this, a process of the CSP part is able to refer to the current system state via predicates which are specified in the Z part by schemas. This link between both parts contributes to a clear separation of the system states from the dynamic aspects.

The connection between the CSP part and the specification of the intended behavior of the sensors and actuators is as follows. The CSP part is linked with every sensor via a communication channel from which the measured values of the respective sensor are read. Analogously, the CSP part is connected with every actuator via a communication channel onto which the commands to the respective actuator are written.

Furthermore, the specification of communication channels in terms of CSP processes makes it feasible to model aspects of a distributed communication, for example the delay of transmission or the redundant arrangement of unreliable communication channels.

3.2 Architectures

For conceptual reasons, it is useful to distinguish between (at least) two architectures defining the manner in which activities of the control component take place and which system components trigger these activities. These architectures serve as frameworks supporting the specifier with general structures to be instantiated in the context of the specific application. The specification of a concrete control component might necessitate the appropriate combination of both architectures. In the following, one architecture is delineated in terms of its generic structures. The other architecture is illustrated by means of an example.

3.2.1 Centralized Coordination of Passive Sensors

For this architecture it is assumed that all sensors are passive, i.e. they cannot cause activities of the control component, and their values are permanently available. There is only one control operation which is executed at time instants uniquely defined by the current system state (e.g. equidistant points of time). A further assumption is that all actuators are able to perform the possible commands at arbitrary time instants.

Functional aspects. The functional aspects of the control operation, i.e. the relation between the system states before and after execution, inputs, and outputs of this operation, are specified by an operation schema *ControlOperation* within the Z part. It is assumed that the controller is always in one of the operational modes *Mode*1, ..., *ModeK* that are defined with respect to the needs of the technical process. Within distinct modes, which can model different environmental or internal conditions, the behavior of the control component might be totally different. The behavior within an operational mode *ModeI* is specified by the internal operation *OpModeI*. This yields the following general schema:

```
── ControlOperation ──────────────────────────────
│ ΔSystemState
│ input1? : IType1; ... ; inputN? : ITypeN
│ output1! : OType1; ... ; outputM! : OTypeM
├──────────────────────────────────────────────────
│ mode = Mode1 ⇒ OpMode1
│ ∧ ... ∧
│ mode = ModeK ⇒ OpModeK
└──────────────────────────────────────────────────
```

The operation has all relevant inputs from the sensors at its disposition. These inputs and the current operational mode determine the successor mode which is specified by the internal operations *OpModeI*. The outputs serve to give commands to the actuators.

Dynamic aspects. The following CSP process *ControlComponent* serves as a general template to specify the reactive behavior of the control component as well as its structural connection to other system components. Its behavior is cyclic which is modeled by the recursive process definition. Before invoking the control operation, all associated input values are read from the respective sensor channels (*sensor*1, ..., *sensorN*) in an arbitrary order which is modeled by the use of the parallel operator ||. When the control operation has terminated, all output values are written to the respective actuator channels (*actuator*1, ..., *actuatorM*). The parallel process *Wait INTERVAL* delays the process execution so that the next invocation of the control operation will happen exactly *INTERVAL* time units after the current invocation.

$$\begin{array}{l} \mathit{ControlComponent} \mathrel{\widehat{=}} \mu\, X \bullet \\ \quad ((\mathit{sensor}1?\mathit{valueI}1 \to \mathit{input}1!\mathit{valueI}1 \to \mathit{Skip} \parallel \ldots \parallel \\ \quad \mathit{sensorN}?\mathit{valueIN} \to \mathit{inputN}!\mathit{valueIN} \to \mathit{Skip}); \\ \quad \mathit{ControlOperationInvocation} \to \mathit{ControlOperationTermination} \to \\ \quad (\mathit{output}1?\mathit{valueO}1 \to \mathit{actuator}1!\mathit{valueO}1 \to \mathit{Skip} \parallel \ldots \parallel \\ \quad \mathit{outputM}?\mathit{valueOM} \to \mathit{actuatorM}!\mathit{valueOM} \to \mathit{Skip}) \\ \parallel \\ \quad \mathit{Wait}\ \mathit{INTERVAL});\ X \end{array}$$

In addition to the process term, the predicate *EnvironmentalAssumption* specifies an assumption about the duration of the operation execution. The maximal time distance between the invocation and the termination are *INTERVAL* time units[1]. Moreover, the invocation of the control operation must be possible at any time.

$$\begin{array}{l} \mathit{EnvironmentalAssumption} \mathrel{\widehat{=}} (\forall\, t : [0, \infty) \bullet \\ \quad \mathit{ControlOperationInvocation}\ \mathsf{open}\ t\ \wedge \\ \quad \mathit{ControlOperationInvocation}\ \mathsf{at}\ t \Rightarrow (\exists\, t' : (t, t + \mathit{INTERVAL}] \bullet \\ \qquad \mathit{ControlOperationTermination}\ \mathsf{open}\ t')) \end{array}$$

The above shows that (i) it is possible to give fairly detailed guidelines concerning the shape of formal specifications in a given context, and (ii) that both styles of specifying behavior are useful and should be combined as appropriate.

3.2.2 Decentralized Coordination of Active Sensors

In this architecture, there may be several control operations, and the sensors may actively invoke actions of the control component.

Architecture. This architecture is illustrated by means of an example which is concerned with the software controller of a gas burner. It is a simplified version of the case study presented by Ravn et al. [8]. There are two actuators: The gas actuator controls the emission of gas and receives the commands to start or stop the emission of gas at arbitrary time instants from the controller. The ignition of escaping gas is controlled by the ignition actuator which can be instructed to start or stop the ignition at arbitrary time instants, too. There are two sensors. The thermometer sensor measures the temperature in the vicinity of the burner and actively reports to the controller decreases below as well as

[1] e open t means that the environment of the process is ready to participate in event e at time t. e at t means that event e happens at time t.

increases above a certain point, indicating a disappearance or an appearance of the flame, respectively. Requests by the user to activate or deactivate the gas burner are made available to the controller by the thermostat sensor. These sensor reports cause an immediate reaction of the controller. The gas burner system is safety critical, because a persistent escape of unburned gas or a failure to realize a request to deactivate the flame could lead to major accidents.

The reactive behavior of the controller depends on the current operational mode. These modes and the possible transitions between them are illustrated in Figure 3.

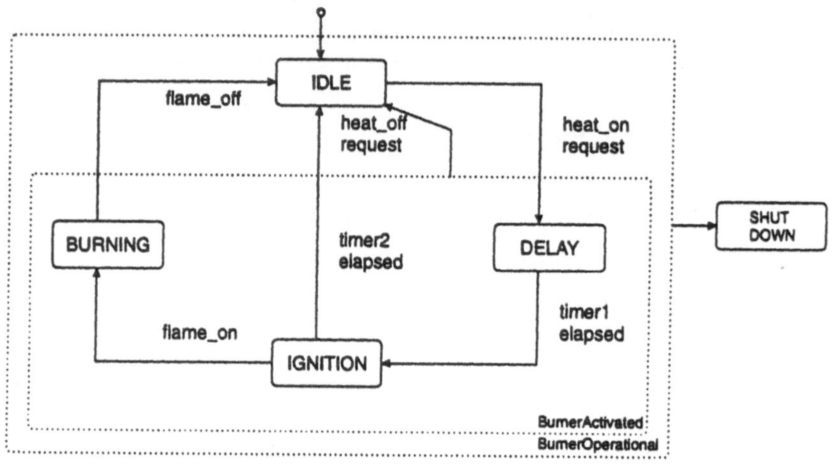

Figure 3: Possible mode transitions

In the *IDLE* mode the controller waits for an activation request without an emission of gas or an activation of the ignition source. After the occurrence of an activation request the controller changes to the *DELAY* mode waiting for 30 seconds to ensure that two different attempts to ignite the gas are sufficiently separated. This delay is realized by a process *Timer*1 which is one component of the controller. After the change to the next mode *IGNITION*, the controller tries to activate the burner by starting the emission of gas and activating the ignition source. In the beginning of this phase, a timer process *Timer*2 is set which produces an alarm event after 2 seconds if not being reset in case of the detection of a flame. This is necessary, because a persistent escape of unburned gas is dangerous to the environment. If a flame is detected, the controller changes to the *BURNING* mode in which gas further escapes but the source of ignition is switched off. Otherwise the controller returns to the *IDLE* mode. In the case of a disappearance of the flame in the *BURNING* mode a transition to the *IDLE* mode results. Furthermore, the occurrence of a request to deactivate the gas burner will cause a change to the *IDLE* mode from every other mode with priority. The *SHUT_DOWN* mode can be entered from each of the other modes if safety can no longer be guaranteed.

Functional aspects. The set of operational modes is defined by the type *Modes*. *YesNo* is the type consisting of the two constants *Yes* and *No*.

$$Modes ::= IDLE \mid DELAY \mid IGN \mid BURNING \mid SHUT\_DOWN$$

The abstract system state of the gas burner controller is specified by the schema *GasBurner*. There is one major system variable *mode* representing the current operational mode. The other system variables *gas*, *ignition*, and *flame* can be deduced from this system variable.

The commands to the actuators are modeled as outputs of system operations. They can directly be deduced from the current system state.

GasBurner
$$mode : Modes$$
$$gas, ignition, flame : YesNo$$

$$gas = Yes \Leftrightarrow mode \in \{IGN, BURNING\}$$
$$flame = Yes \Leftrightarrow mode = BURNING$$
$$ignition = Yes \Leftrightarrow mode = IGN$$

Actuators
$$GasBurner'$$
$$gas!, ign! : YesNo$$

$$gas! = gas' \land ign! = ignition'$$

After initialization the controller is in the *IDLE* mode.

$$GasBurnerInit \mathrel{\widehat=} [\, GasBurner' \mid mode' = IDLE \,]$$

The system operations are straightforward. They follow the state transition diagram of Fig. 3. We present three of them.

$$HeatOffRequest \mathrel{\widehat=} [\, \Delta GasBurner;\ Actuators \mid mode \in \{DELAY, IGN, BURNING\} \land mode' = IDLE \,]$$

$$IgnitionOK \mathrel{\widehat=} [\, \Delta GasBurner;\ Actuators \mid mode = IGN \land mode' = BURNING \,]$$

$$ShutDown \mathrel{\widehat=} [\, \Delta GasBurner;\ Actuators \mid mode' = SHUT\_DOWN \,]$$

The following schemas define predicates on the system state that are used to model the interdependencies between the dynamic behavior and the current system state.

$$BurnerIsDeactivated \mathrel{\widehat=} [\, GasBurner \mid mode = IDLE \,]$$
$$IgnitionIsActivated \mathrel{\widehat=} [\, GasBurner \mid ignition = Yes \,]$$
$$FlamePresent \mathrel{\widehat=} [\, GasBurner \mid flame = Yes \,]$$

Dynamic view. The dynamic behavior of the gas burner controller is defined by the real-time CSP process *GasBurnerControl*. After initialization represented by the event *GasBurnerInit* the behavior is defined by three parallel subprocesses. The subprocess $GasBurnerControl_{READY}$ coordinates the execution of system operations depending on the occurrence of external events. To accomplish this task, it cooperates with two parallel subprocesses *Timer*1 and *Timer*2, modeling two different timer components.

$GasBurnerControl \mathrel{\hat{=}} GasBurnerInit \rightarrow$
$(GasBurnerControl_{READY} \parallel Timer1 \parallel Timer2)$

The process *HighPriority* specifies the reaction of the controller to the occurrence of the event *heat_off_request* which has a higher priority than the other external events. The process *Wait* ϵ models the fact that technical components cannot react immediately. The if statement defines a consistency check. If there is a request to turn off the gas burner and at the same time the internal state is *IDLE* then there is an inconsistency between the external signals and the internal state of the burner. Safety can no longer be guaranteed, and the shutdown operation is invoked. Otherwise, the timers are reset, and the *IDLE* mode is entered.

HighPriority $\hat{=}$
heat_off_request → *Wait* ϵ; if ¬ *BurnerIsDeactivated*
then *reset_timers* → *HeatOffRequestExecution* → *ActuatorCtr*;
$GasBurnerControl_{READY}$
else *ShutDownExecution* → *ActuatorCtr*; *Stop* fi

The dynamic behavior of the controller after initialization is specified by the process $GasBurnerControl_{READY}$. As has been suggested, there are two classes of external events: high priority and low priority events. If a high priority and a low priority event occur simultaneously (i.e. with a time difference of at most ϵ) then only the high priority event is treated, and the other one is ignored. This is modeled with the timeout operator $\triangleright\{\epsilon\}$. The event *heat_on_request* is neither a high nor a low priority event as it is excluded that it occurs simultaneously with the high priority event. Again, consistency checks between the external signals and the internal system state are performed.

$GasBurnerControl_{READY} \mathrel{\hat{=}} \mu X \bullet$
HighPriority
□
heat_on_request → *Wait* ϵ; if *BurnerIsDeactivated*
then *set_timer1* → *HeatOnRequestExecution* → *ActuatorCtr*; *X*
else *ShutDownExecution* → *ActuatorCtr*; *Stop* fi
□
flame_on → (*HighPriority* $\triangleright\{\epsilon\}$if *IgnitionIsActivated*
then *reset_timers* → *IgnitionOKExecution* → *ActuatorCtr*; *X*
else *ShutDownExecution* → *ActuatorCtr*; *Stop* fi)
□
flame_off → (*HighPriority* $\triangleright\{\epsilon\}$if *FlamePresent*
then *FlameFailureExecution* → *ActuatorCtr*; *X*
else *ShutDownExecution* → *ActuatorCtr*; *Stop* fi)
□
timer1_elapsed → (*HighPriority* $\triangleright\{\epsilon\}$
set_timer2 → *IgnitionExecution* → *ActuatorCtr*; *X*)
□
timer2_elapsed → (*HighPriority* $\triangleright\{\epsilon\}$
IgnitionFailureExecution → *ActuatorCtr*; *X*)

The behavior of the controller is cyclic which is expressed by the recursive process definition. The process continuously reacts to incoming events from the sensors. Its reaction in terms of the execution of a system operation and the output of commands to both actuators can depend on the current system state. This dependence is modeled by the use of predicates on the system state, defined by Z schemas. After execution of a system operation the controller sends the specified outputs to the particular actuators defined by the process *ActuatorCtr*.

The two timer components are represented by the real-time CSP processes *Timer*1 and *Timer*2 that consist of a *set_timer* event, followed by a wait process and a *timer_elapsed* event. This process can be interrupted at any time by a *reset_timers* event.

Proof of safety constraints. One safety constraint for the gas burner is that in each interval of 30 seconds there may be an escape of unburned gas of at most $2 + 2\epsilon$ seconds, where ϵ is the response time for the technical components. According to the definition of the state schema *GasBurner*, an escape of unburned gas exists only in the *IGN* mode. Hence, within an interval of at most 30 seconds, the sum of the lengths of all subintervals which begin with the occurrence of the event *IgnitionExecution* (marking the begin of an ignition phase) and end with the next occurrence of an event from the set of operations (marking the end of the ignition phase) must not exceed $2 + 2\epsilon$ seconds.

The following argumentation of the enforcement of the safety constraint is separated into two parts. First, it is shown that every period within the *IGN* mode lasts at most $2 + 2\epsilon$ seconds, which is essentially the purpose of the *Timer*2 process. Let t be an arbitrary point in time.

$$\begin{array}{lll} & \mathit{IgnitionExecution} \text{ at } t & \\ \vdash & & [\mathit{GasBurnerControl}_{READY}] \\ & \mathit{set\_timer2} \text{ at } t & \\ \vdash & & [\mathit{Timer2}] \\ & (\exists\, t' : (t, t+2] \bullet \mathit{reset\_timers} \text{ at } t') \lor & \\ & \mathit{timer2\_elapsed} \text{ at } (t+2) & \end{array}$$

First case:

$$\begin{array}{lll} & \exists\, t' : (t, t+2] \bullet \mathit{reset\_timers} \text{ at } t' & \\ \vdash & & [\mathit{GasBurnerControl}_{READY}, \mathit{HighPriority}] \\ & \exists\, t' : (t, t+2] \bullet & \\ & \quad \mathit{IgnitionOKExecution} \text{ at } t' \lor \mathit{HeatOffRequestExecution} \text{ at } t' & \end{array}$$

Second case:

$$\begin{array}{lll} & \mathit{timer2\_elapsed} \text{ at } (t+2) & \\ \vdash & & [\mathit{GasBurnerControl}_{READY}, \mathit{HighPriority}] \\ & \mathit{IgnitionFailureExecution} \text{ at } (t+2+\epsilon) & \end{array}$$

$$\vee (\exists t' : [t+2, t+2+2\epsilon] \bullet$$
$$\quad \mathit{HeatOffRequestExecution} \text{ at } t' \vee \mathit{ShutDownExecution} \text{ at } t')$$

In all cases, the *IGN* mode is left after at most $2 + 2\epsilon$ seconds. Second, we show that two different periods in the *IGN* mode are separated by at least 30 seconds. Let $t1$ and $t2$ be two arbitrary points in time at which the event *IgnitionExecution* occurred.

$$t1 < t2 \wedge \mathit{IgnitionExecution} \text{ at } t1 \wedge \mathit{IgnitionExecution} \text{ at } t2 \wedge$$
$$(\forall t : (t1, t2) \bullet \neg\ \mathit{IgnitionExecution} \text{ at } t)$$
$$\vdash \qquad \text{[possible mode transitions]}$$
$$\exists t : (t1, t2) \bullet \mathit{set\_timer1} \text{ at } t \wedge \mathit{mode} = \mathit{DELAY} \qquad (\text{at } t)$$
$$(\forall t : (t1, t2) \bullet \neg\ \mathit{IgnitionExecution} \text{ at } t)$$
$$\vdash \qquad [\mathit{Timer1}]$$
$$\exists t : (t1, t2) \bullet (\neg\ (\exists t' : [t, t+30) \bullet \mathit{timer1\_elapsed} \text{ at } t')) \wedge$$
$$(\forall t : (t1, t2) \bullet \neg\ \mathit{IgnitionExecution} \text{ at } t)$$
$$\vdash \qquad [\mathit{GasBurnerControl}_{READY}]$$
$$\exists t : (t1, t2) \bullet (\neg\ (\exists t' : [t, t+30) \bullet \mathit{IgnitionExecution} \text{ at } t')) \wedge$$
$$(\forall t : (t1, t2) \bullet \neg\ \mathit{IgnitionExecution} \text{ at } t)$$
$$\vdash \qquad \text{[predicate logic]}$$
$$(\forall t' : (t1, t1+30) \bullet \neg\ \mathit{IgnitionExecution} \text{ at } t')$$
$$\vdash$$
$$t2 \geq t1 + 30$$

As a consequence, there can only be one subinterval of $2 + 2\epsilon$ seconds with the escape of unburned gas in any interval of at most 30 seconds. Thus the safety constraint is guaranteed by the software controller.

This proof has been carried out in a mathematical style. It appeals to the intuitive understanding of the formal specification. If formal proofs, i.e. derivations in a formal system, are to be performed, a common semantic model of the two languages is indispensable.

4 Semantic Model

In this section, we outline the formal definition of the semantics associated with a combined specification as explained informally in the previous sections.

The basis of this definition is the semantic function of the timed failures model of real-time CSP [1]. This model associates with each process term a set of *timed failures* which represents possible observations of the process. A timed failure consists of a *timed trace* which is a sequence of events, where each event has a time stamp associated with it. Moreover, a timed failure comprises a *timed refusal*, i.e. a set of timed events that can be refused by the system in case the corresponding timed trace was observable.

$$timed\,failures : Process \rightarrow \mathbb{P}(\mathrm{seq}\ TimedEvents \times \mathbb{P}\ TimedEvents)$$

Analogously, the set of all possible observations of a system specified by a combination of Z and real-time CSP has to be determined. In this context, a third component is of importance, namely the evolution of the system state within the observation interval. Hence an observation for a combined specification is a tuple consisting of a timed trace, a timed refusal, and a so-called *timed state.* A timed state is defined as a function which maps every time instant of the observation interval to the respective system state observed.

The Z part of a specification is characterized by a state schema *State*, an initial state schema *InitState*, a set of external operation schemas $Op1, \ldots, OpN$, and a set of predicates on the system state $Pred1, \ldots, PredM$. The set *RESTR_RTCSP_PROCESS* contains all process terms of real-time CSP that do not allow subprocesses to perform a state changing operation in parallel with other subprocesses accessing the system state. Thus the signature of our semantic function is as follows.

$$\begin{aligned} &timed\,failures\,states : SCHEMA \times SCHEMA \times \mathbb{P}\,SCHEMA \times \\ &\quad \mathbb{P}\,SCHEMA \times RESTR\_RTCSP\_PROCESS \rightarrow \\ &\quad \mathbb{P}((\mathrm{seq}\ TimedEvents \times \mathbb{P}\ TimedEvents) \times (TIME \nrightarrow STATES)) \end{aligned}$$

A possible observation $((s, X), tstate)$ of the behavior of the specified system can be interpreted in the following sense: the timed failure (s, X) consisting of the timed trace s and the timed refusal X is defined by the semantic function *timed failures* as a possible observation of the CSP process, and the timed state *tstate* maps each instant of the observation interval to such a system state that can be reached, starting from an initial state and proceeding in accordance with the operation events contained in the timed trace s. The formal definition of the function *timed failures states* can be found in [11].

5 Related Work

The use of Z to specify safety-critical software is not uncommon. Jacky [4] uses this language to define a framework for safety-critical systems that emphasizes safety interlocking. McDermid and Pierce [6] define a graphical notation based on a variant of statecharts [2] that is translated into Z for the purpose of mechanical validation. This notation is used to specify and develop software for Programmable Logic Controllers. Heisel [3] describes several phases in the development of safety-critical software where Z is used in the specification phase.

The work presented here is distinguished from these approaches in that it is intended to be used for systems where the exclusive use of Z does not lead to satisfactory results. The expressive power of Z does not suffice to specify the behavior of sophisticated real-time systems adequately. Other researchers share our goal to provide more powerful constructs to express behavioral and real-time requirements.

Ravn et al. [8] use the duration calculus to express functional requirements and safety constraints. The duration calculus is a specialized formalism designed to express requirements on the duration of states. These durations are expressed as integrals. In contrast, our approach uses less specialized formalisms that are more easily accessible and more widely used.

Weber [12] combines Z and statecharts for purposes similar to ours. Since statecharts are a semi-formal specification technique, the resulting specification is not completely formal. Using a formal language like CSP, however, yields completely formal combined specifications, as shown in Section 4.

Rottke [9] votes for a combination of the functional language ML and Colored Petri Nets. His basic objective is to obtain executable specifications, whereas our goal is to develop specifications as concise and abstract as possible.

6 Conclusion

With the work presented here, we have provided an elaborate methodology for the formal specification of software for safety-critical applications:

- The system model underlying most of these applications was taken into account by explicitly referring to it in the methodology.
- Two formal languages were chosen which, in combination, provide adequate constructs for the specification of safety-critical software components.
- A software model for the combined use of the two languages was presented.
- This model was further refined into two reference architectures that capture frequent cases of safety-critical systems. These architectures can be instantiated for concrete systems, thus providing detailed guidance for specifiers.
- The feasibility of the approach was illustrated by means of an example, and it was shown that (albeit informal) proofs can be performed to demonstrate that safety constraints are indeed fulfilled by the specification.
- In order to show that the combined language has a well-defined semantics and to lay the basis for formal proof, a common semantic model was defined.

To further enhance the applicability of this method, we intend to develop a calculus that allows one to perform formal proofs on and refinements of combined specifications and to implement this calculus in a generic prover like Isabelle [7]. Additionally, we will investigate how programs can be synthesized from combined specifications. A starting point is the work described in [3], where the synthesis of programs from Z specifications is described. With the described enhancements, all stages from specification to program synthesis could be carried out formally and with machine support.

Acknowledgment. Thanks to Thomas Santen for his comments on this work.

References

[1] Jim Davies. *Specification and Proof in Real-Time CSP*. Cambridge University Press, 1993.

[2] David Harel. Statecharts: a visual formalism for complex systems. *Science of Computer Programming*, 8:231–274, 1987.

[3] Maritta Heisel. Six steps towards provably safe software. In G. Rabe, editor, *Proceedings of the 14th International Conference on Computer Safety, Reliablity and Security (SAFECOMP), Belgirate, Italy*, pages 191–205, London, 1995. Springer.

[4] Jonathan Jacky. Specifying a safety-critical control system in z. *IEEE Transactions on Software Engineering*, 21(2):99–106, February 1995.

[5] Nancy Leveson. *Safeware: System Safety and Computers*. Addison-Wesley, 1995.

[6] J.A. McDermid and R.H. Pierce. Accessible formal method support for PLC software development. In G. Rabe, editor, *Proceedings of the 14th International Conference on Computer Safety, Reliablity and Security (SAFECOMP), Belgirate, Italy*, pages 113–127, London, 1995. Springer.

[7] L. C. Paulson. *Isabelle*. LNCS 828. Springer-Verlag, 1994.

[8] A.P. Ravn, H. Rischel, and K.M. Hansen. Specifying and verifying requirements of real-time systems. *IEEE Transactions on Software Engineering*, 19(1):41–55, January 1993.

[9] Thomas Rottke. Validierung und Verifikation während des Requirements Engineering. In F. Saglietti, editor, *Proceedings 5th German ENCRESS Workshop*, Hamburg, 1996. Institut für Sicherheitstechnologie.

[10] J. M. Spivey. *The Z Notation – A Reference Manual*. Prentice Hall, 2nd edition, 1992.

[11] Carsten Sühl. Eine Methode für die Entwicklung von Softwarekomponenten zur Steuerung und Kontrolle sicherheitsrelevanter Systeme. Master's thesis, Technical University of Berlin, 1996.

[12] Matthias Weber. Combining Statecharts and Z for the design of safety-critical systems. In M.-C. Gaudel and J. Woodcock, editors, *FME '96 — Industrial Benefits and Advances in Formal Methods*, LNCS 1051, pages 307–326. Springer Verlag, 1996.

Safety Analysis Based on Object-oriented Modelling of Critical Systems

Janusz Górski and Bartosz Nowicki
Franco-Polish School of New Information and Communication Technologies
Mansfelda 4, 60-854 Poznań 6, Poland
{gorski, nowicki}@efp.poznan.pl

1 Introduction

Many of the contemporary domains of human activity are associated with high risk, e.g. energy production and distribution, transport, medicine, chemical industry, etc. As computers are commonly applied in those domains, more and more responsibility for maintaining the risk within the acceptable limits is put on a computer system and its software. This raises the problem of computer system safety, understood as the level of guarantee which can be built into the system and assessed independently that the misoperation of the system will not result in a hazardous state.

In this article we present an approach where the computer system is modelled within the context of its application. The model is then used to formally specify the required properties which, if fulfilled, lead to the reduction or elimination of risks associated with the application. To follow this approach we had to solve the following problems:

- *the underpinning modelling framework should enable natural and adequate representation of the system together with its application context.* Our choice here was the object-oriented modelling as it is relatively matured and widely used to represent requirements and is applicable to most of real-life problems.
- *the model should enable expression of the properties which refer to the application rather than to the computer system itself.* This is because safety is the application domain concept and as such is best understood when expressed in the application terms. Again, as object-orientation provides for easy modelling of the reality it was the natural choice to meet this requirement.

In the subsequent sections we present our approach in more detail. During this presentation we refer to a simple example: the gas burner system. After developing the model of the system, two basic analyses are performed:

- the functional requirements are analysed to verify if the system fulfils its intended mission. An example questions asked during this type of analysis would be:
  * *is it guaranteed that (in the absence of draught) the system ignites the gas in the response to the operator's command?*
  * *is it guaranteed that there will not be any spontaneous ignition without the explicit operator's command?, etc.*

- the analysis of safety requirements verifies if the system excludes the occurrence of the hazardous states. Within the context of our application the hazard occurs if there is too high concentration of gas in the burning chamber.

The distinction of the above steps is in accordance with the international recommendations [1, 2] where the separation of the design and safety analysis processes is strongly recommended. In our approach we apply the *Object Modelling Technique (OMT)* [3] to develop the object oriented model of the gas burner. OMT employs *statecharts* notation [4] to represent dynamic properties of the system. The analysis of reachability of states has been performed with the help of the STATEMATE [4] tool.

2 The Model of the Gas Burner

Our example is a gas burner system which mission is to supply heat to a technical process. The main component of the system is a burning chamber, together with the associated gas supplying pipe and the ignition device. A temperature sensor is used to detect that the gas is burning. The valve on the gas supplying pipe and the ignition device are controlled by a computer based controller. The controller receives external signals (from the system operator) to start and to stop the heating process.

During further analysis it is assumed that the increase of gas concentration in the burning chamber is proportional to the time period during which the valve remains open (assuming that there is no ignition). Similarly, if the valve closes, the gas concentration decreases proportionally to time, due to ventilation. It is also assumed that the minimal gas concentration sufficient to gas inflammation is known. The ignition device generates sparks in response to the command received from the controller. A spark is generated with the delay necessary for charging the ignition device. The signals from the controller received during the charging process are ignored. The temperature inertia of the sensor is known.

Not every spark generated while the gas concentration is sufficient for ignition will result in gas inflammation, due to the possibility of a temporary draught. A draught can also blow the flame away. It is assumed that during normal exploitation of the system gas is always available.

Object oriented approach assumes three different perspectives of description of the reality being the subject of modelling: the *structural view*, the *behavioural view* and the *transformational view* [3]. Those perspectives are essentially different and each of the related models concentrates on just a single distinguished set of system properties. While taking the structural view we concentrate on the permanent properties of the problem. The key concept here is that of *object* and *object class* and the model shows which are the key objects in the problem domain and what are their static relationships. The behavioural view concentrates on how the objects interact. The interactions are based on *events* and the model shows how the events are generated by objects and how the object's behaviour is affected by the events sent by other objects. The transformational view considers the input-output relations in the considered problem. The key concept is *data transform (function)* and the model shows how the input data submitted to the system are transformed to produce the required outputs. As the three views provide different perspectives of the same real problem, some consistency among them must be preserved. The relationship between the object and dynamic models is straightforward: for each concrete class of the object model, there is a *state diagram* that specifies the set of possible states and sequences of states. Different state diagrams execute concurrently. The data processing performed in the system is specified by means of *data flow diagrams*. As in this paper we concentrate on the control-dominated applications, this view of the system is of the less importance.

The object model of the considered example system is given below.

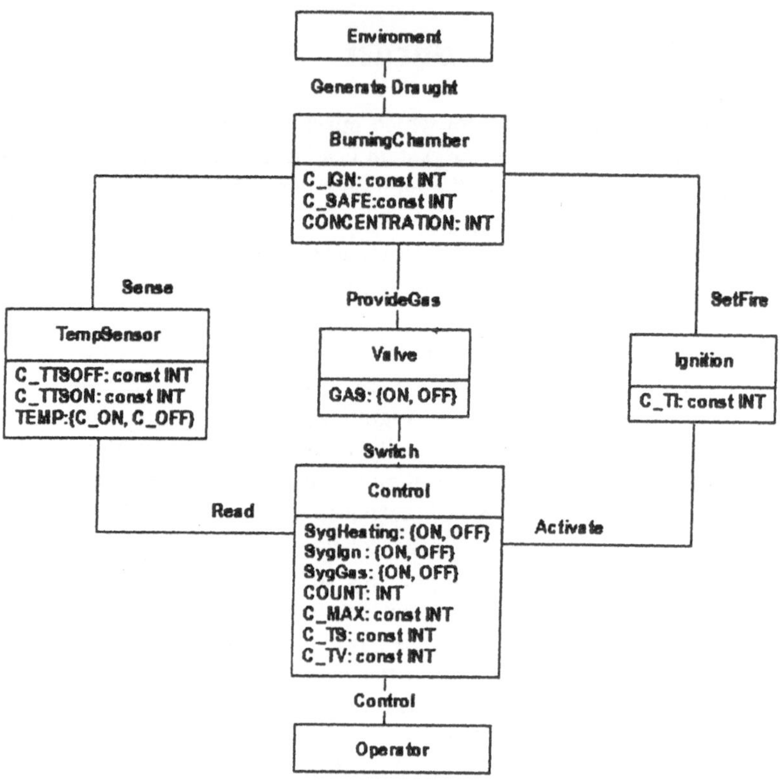

Fig.1. The object model of the gas burner.

The following table gives a description of the objects in the system.

OBJECT	ATTRIBUTE	DESCRIPTION
Burning Chamber		the gas burning device
	C_IGN	minimal concentration of gas sufficient for gas inflammation
	C_SAFE	maximal, safe concentration of gas
	CONCENTRATION	current concentration of gas

TempSensor		temperature sensor
	C TTSOFF	the time necessary to cool the sensor
	C TTSON	the time necessary to heat up the sensor
	TEMP	temperature measured by the sensor
Valve		the gas valve
	GAS	the current position of the valve
Ignition		the ignition device
	C TI	the charging time of the ignition device
Control		the (computerised) controller
	SygHeating	external command to start/stop heating
	SygIgn	the signal controlling the ignition
	SygGas	the signal controlling the gas valve
	COUNT	the counter of ignition attempts
	C MAX	the maximal count of ignition attempts
	C TS	the synchronisation delay
	C TV	the ventilation time of the chamber
Environment		the object generating draught
Operator		the system operator

The model presented in Fig.1. is general and includes objects representing the control system (Control), the controlled system (BurningChamber, TempSensor, Ignition, Valve) and the broader environment (Environment, Operator). The model is *closed* in a sense that it includes all the elements taken into account during analysis. The answer to the question if the scope of the model is sufficient depends on the kind of analysis aimed at. This problem (although extremely important) will not be discussed in this paper. However, we can observe that the scope is sufficient to represent the relationships occurring in a real system:

- Temperature sensor *Senses* heat produced in the burning chamber.
- The values measured by the sensor are *Read* by the control system.
- Controller *Switches* the valve which in turn *Provides Gas* to the chamber.
- Controller *Activates* the ignition device which in turn *Sets Fire* in the chamber.
- Operator *Controls* the activities of the controller .
- Environment *Generates Draught* which may prevent gas inflammation.

3 The Dynamic Model

The state diagrams representing the behaviour of the objects of Fig.1. are shown in the Figures 2,3,4 and 5.

Figure 2 presents behaviour of the ignition and the valve. Initially the ignition device is in the CHARGING state and after the time C_TI it moves to the READY state. It remains in this state until the SYGIGNON event occurs. Then it goes back to the CHARGING and the SPARK is generated.

Initially the valve is OFF. In response to the event SYGGASON the valve becomes ON and the gas is supplied to the burning chamber (GASON). Closing the valve (SYGASOFF) moves the object to the OFF state again and gas is no longer provided to the chamber (GASOFF).

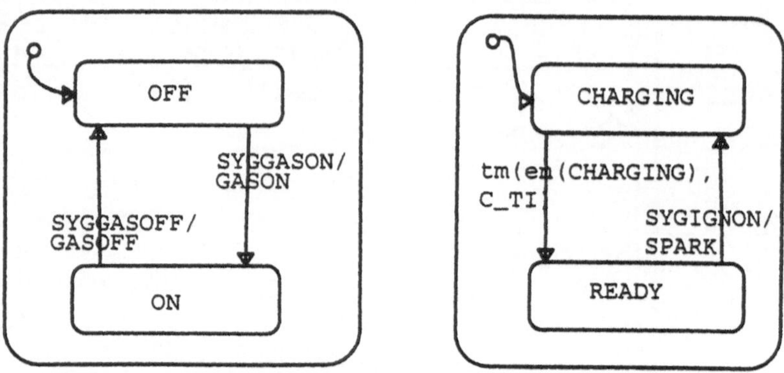

Fig.2. The dynamic models of the ignition and the gas valve.

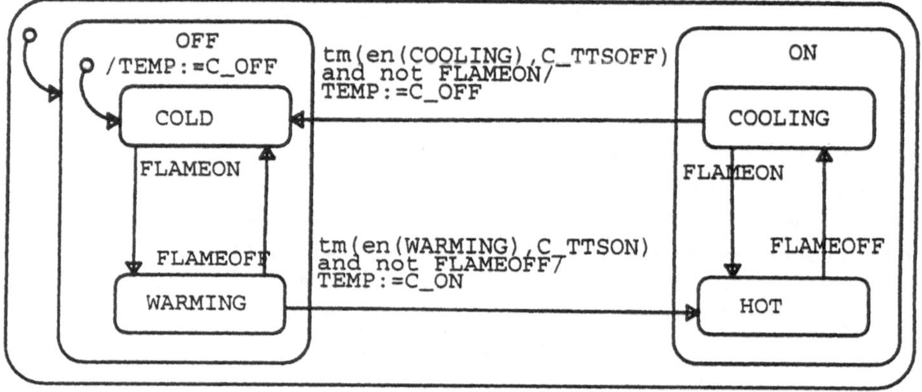

Fig.3. The dynamic model of the temperature sensor

The temperature sensor, presented in Fig. 3., detects that heat is produced. The sensed temperature is represented by the attribute TEMP. Initially the sensor is COLD. Gas inflammation (FLAMEON) moves the sensor to the WARMING state. After the time C_TTSON the sensor becomes HOT. After putting the flame out (FLAMEOFF) the sensor is COOLING and then after the time C_TTSOFF becomes COLD again.

Figure 4 presents the burning chamber behaviour. The value of the CONCENTRATION attribute represents the gas concentration in the burning chamber. This value changes as the time is passing, increasing in the COLLECTINGGAS state (CONCENTRATION++), and decreasing in the VENTILATION state (CONCENTRATION--). The burning chamber has two basic sates: BURNING and IDLE. The initial state NOGAS (substate of IDLE) represents the situation where there is no gas in the burning chamber. When the gas is let in to

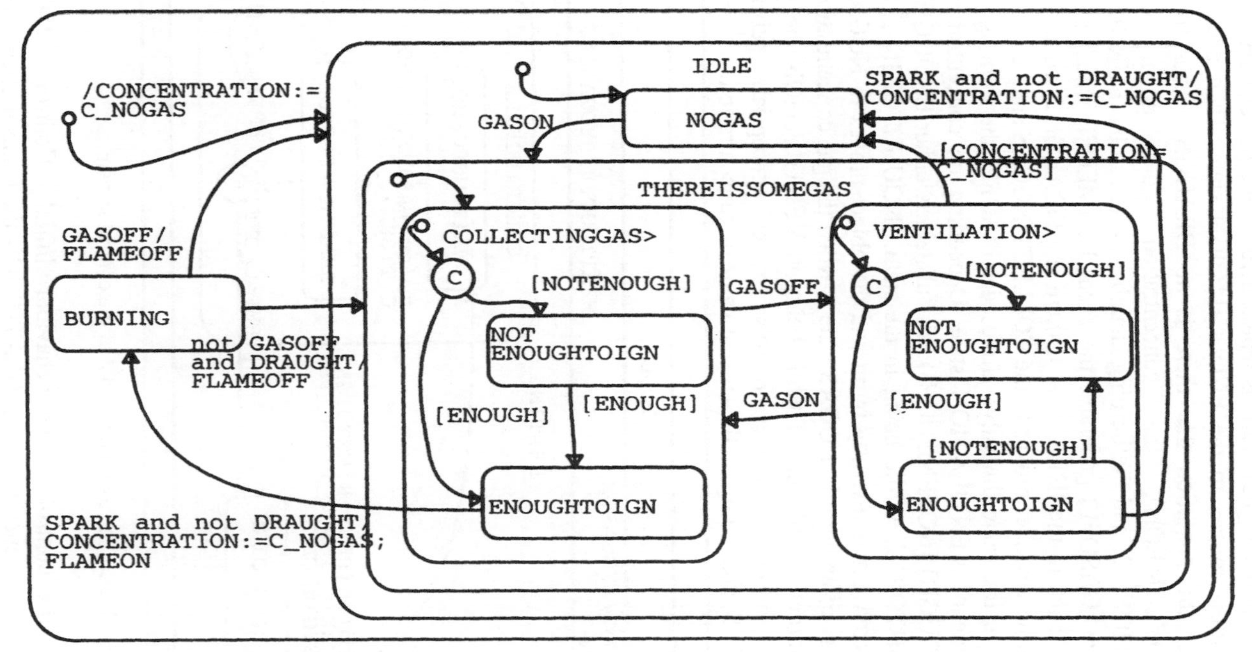

Fig.4. The dynamic model of the burning chamber.

the chamber (GASON) the burning chamber moves to the COLLECTINGGAS state (more precisely to its substate NOTENOUGHTOIGN which represents the situation where the concentration of gas is not enough to its inflammation). After the concentration goes above C_IGN (this is represented by the [ENOUGH] condition) the burning chamber moves to the ENOUGHTOIGN state. Then each SPARK occurrence under the condition that there is no DRAUGHT leads to the gas inflammation (FLAMEON), the concentration goes down to zero (CONCENTRATION:=0) and the burning chamber moves to the BURNING state. Closing the valve (GASOFF) generates the event FLAMEOFF which moves the burning chamber back to the IDLE state. If there is a DRAUGHT while the gas is burning, the burning chamber moves to the COLLECTINGGAS state and the whole cycle repeats. If in this state the valve is closed (GASOFF) the burning chamber goes to a substate of VENTILATION state. Depending on concentration value (conditions [ENOUGH] [NOTENOUGH]) this substate is either ENOUGHTOIGN or NOTENOUGHTOIGN. While being in the state ENOUGHTOIGN if SPARK occurs, the collected gas is instantly burnt and the object moves to NOGAS state. An alternative way of arriving in NOGAS state is when the gas concentration drops to zero due to normal ventilation. If while being in VENTILATION the valve is open again the burning chamber returns to an appropriate substate of COLLECTINGGAS, depending on the current value of concentration.

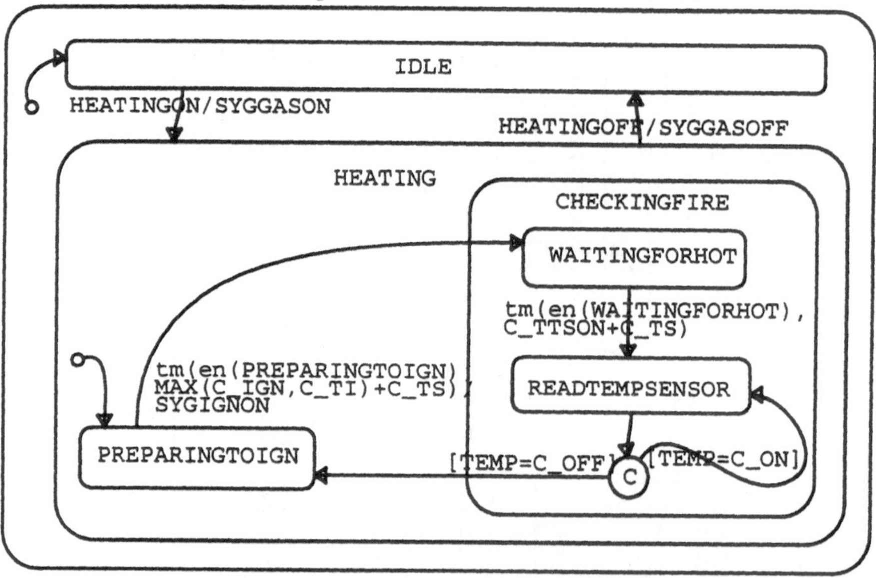

Fig.5. The dynamic model of the control system

Figure 5 presents the control system behaviour. Initially Control is IDLE. In response to the operator-generated event HEATINGON the Control opens the valve (SYGGASON) and moves to the PREPARETOIGN state to enable charging of the ignition device. After the time MAX(C_IGN, C_TI) + C_TS the ignition signal SYGIGNON is sent and Control moves to the state in which it waits to let the sensor get hot and then, after the time C_TTSON the sensor is sampled periodically. If the measured temperature is low the system moves back to the PREPARATOIGN state and the whole cycle repeats. The HEATINGOFF signal moves Control back to the IDLE state.

Verification of the model against the functional requirements confirms that behaviour of the control system is correct in the assumed environment.

4 Safety Analysis

The model presented in the previous section is oriented towards the system mission. However, although this mission is associated with the risk of gas explosion, the model does not identify this possibility in an explicit way. The basic hazard in this system is too high concentration of gas in the burning chamber. To introduce this aspect to our model we start with another model which explicitly distinguishes between safe and unsafe states. The generic form of such model is shown in Fig.6.

Fig.6. The generic model of a hazard

The instance of this model which represents the hazard related to the gas burner system is shown in Fig.7.

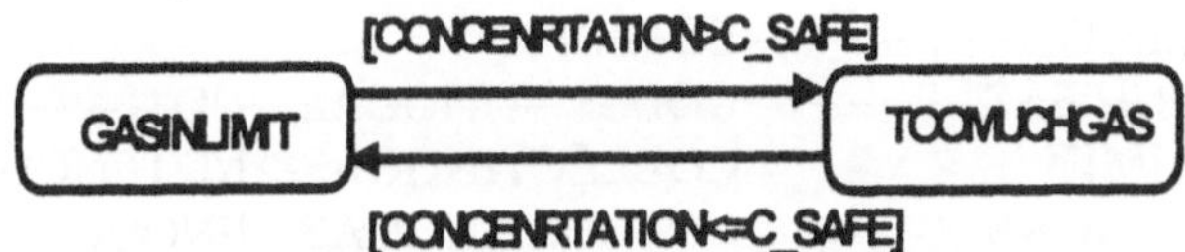

Fig.7. The hazard model of the burning chamber.

The model distinguishes between the state where the concentration is within the safe limits and the state of too high gas concentration. The model is built from the different (orthogonal) point of view comparing to the mission oriented model - the focus is on modelling safety aspects of the system.

At this stage we have two different models of the same system:

- the model focused on the mission aspects, and
- the model focused on the safety aspects (hazard model).

In the next phase we merge the two models into one which leads to the model where the interference between the mission and safety requirements can be studied and analysed. The above is achieved by the following procedure:

Step 1 From the object model we identify *critical objects,* i.e. those objects the hazard definition refers to. In our example we have only one critical object - the BurningChamber.

Step 2 For each critical object, we extend its state diagram with the hazardous states. This is based on the following procedure:

Phase 1 The states of the critical object and the states of the hazard model are "multiplied". The basic idea is shown in Fig.8.

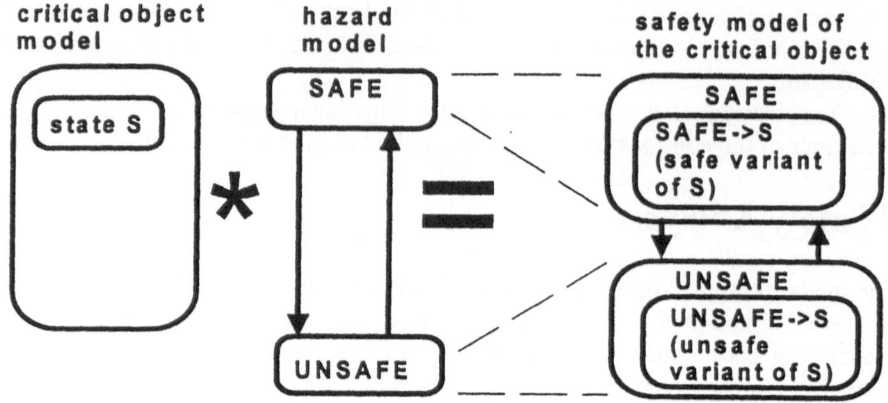

Fig.8. Multiplying the critical object model by the hazard model to obtain the safety model of the critical object.

As the result we obtain the set of *candidate states* of the safety model of the critical object. In our example after multiplying the BurningChamber states (Fig.4.) by the hazardous states (Fig.7.) we get:

~~TOOMUCHGAS - NOGAS~~
GASINLIMIT - NOGAS
~~TOOMUCHGAS - BURNING~~
~~GASINLIMIT - BURNING~~
~~TOOMUCHGAS - COLLECTINGGAS->NOTENOUGHTOIGN~~
GASINLIMIT - COLLECTINGGAS->NOTENOUGHTOIGN
TOOMUCHGAS - COLLECTINGGAS->ENOUGHTOIGN
GASINLIMIT - COLLECTINGGAS->ENOUGHTOIGN
~~TOOMUCHGAS - VENTILATION->NOTENOUGHTOIGN~~
GASINLIMIT - VENTILATION->NOTENOUGHTOIGN
TOOMUCHGAS - VENTILATION->ENOUGHTOIGN
GASINLIMIT - VENTILATION->ENOUGHTOIGN

Phase 2 Each candidate state is validated with respect to the credibility of its occurrence. The incredible candidate states are rejected. In the list above the states which are rejected are crossed out. For example, the state TOOMUCHGAS-NOGAS is obviously impossible. The classification is done as follows (S represents a state of the critical object, SAFE corresponds to GASINLIMIT and UNSAFE corresponds to TOOMUCHGAS):

* SAFE-S is credible and UNSAFE-S is incredible. This situation means that there is no risk associated with the state S (S is inherently safe). In our example, S corresponds to BURNING (burning removes all gas from the burning chamber), NOGAS and NOTENOUGHTOIGN states.
* SAFE-S is incredible and UNSAFE-S is credible. This means that S is inherently unsafe. This situation indicates a direct conflict between the mission and the safety requirements. It is not present in our example.

* both SAFE-S and UNSAFE-S are credible. Here, in some situations the system is unsafe while being in state S, i.e. there might be conflict between mission and safety. This is represented by splitting S into its safe and unsafe variants. In our example this refers to ENOUGHTOIGN substates of the COLLECTINGGAS and VENTILATION states.

Phase 3 The critical object model is enriched with two additional superstates:

- safe part - grouping all inherently safe states and safe variants of other states,
- unsafe part - grouping all inherently unsafe states and unsafe variants of other states.

Candidate states not rejected during the previous phase are clustered into an appropriate superstate. In this phase we concentrate on transitions between states. The transitions are defined as follows:

* all the transitions of the critical object model remain in the safe part of the safety model. If the target state occurs in its safe and unsafe variant, the transition goes to the safe variant of the state.
* for each state split into the safe and unsafe variants, there is a possibility of a transition between those variants. If such a transition actually occurs depends on the validation argument. In our example such a transition occurs from GASINLIMIT->COLLECTINGGAS->ENOUGHTOIGN to TOOMUCHGAS->COLLECTINGGAS (while the gas concentration increases, it first reaches the limit enough to cause normal inflammation and then, after some time, the concentration becomes dangerous) and from TOOMUCHGAS->VENTILATION to GASINLIMIT->VENTILATION->ENOUGHTOIGN (during ventilation, if the concentration was too high, it eventually goes back below the "danger" limit).
* if, in the mission-oriented model, there is a transition between states S and S', then a possibility of transition between the states UNSAFE-S and UNSAFE-S' is considered and validated (this step is performed with respect to those states only which still have their unsafe variants, after Phase 2). In our example the transitions between unsafe states depend on the gas concentration.

In effect of the above steps we obtain a dynamic model of a critical object with explicit distinction of the possibility of hazard occurrence. Such model of the BurningChamber object is presented in Fig. 9.

Step 3 The safety model of the critical object replaces the mission oriented one in the model of the whole system. Then, we analyse the whole model to verify if the system is safe. This is achieved by verifying hazard reachability. In our work we use the STATEMATE tool to support this stage.

The analysis performed for the gas burner system revealed two possible hazard scenarios:

- *the operator repeatedly switches the heating on and off. If the frequency is high enough, the ignition procedure can not be completed and the spark is not generated. Consequently, the gas concentration increases and eventually reaches the dangerous limit.*
- *each time a spark is generated there is a draught which prevents the inflammation. Then again, if this situation repeats many times in short time the gas concentration can become too high.*

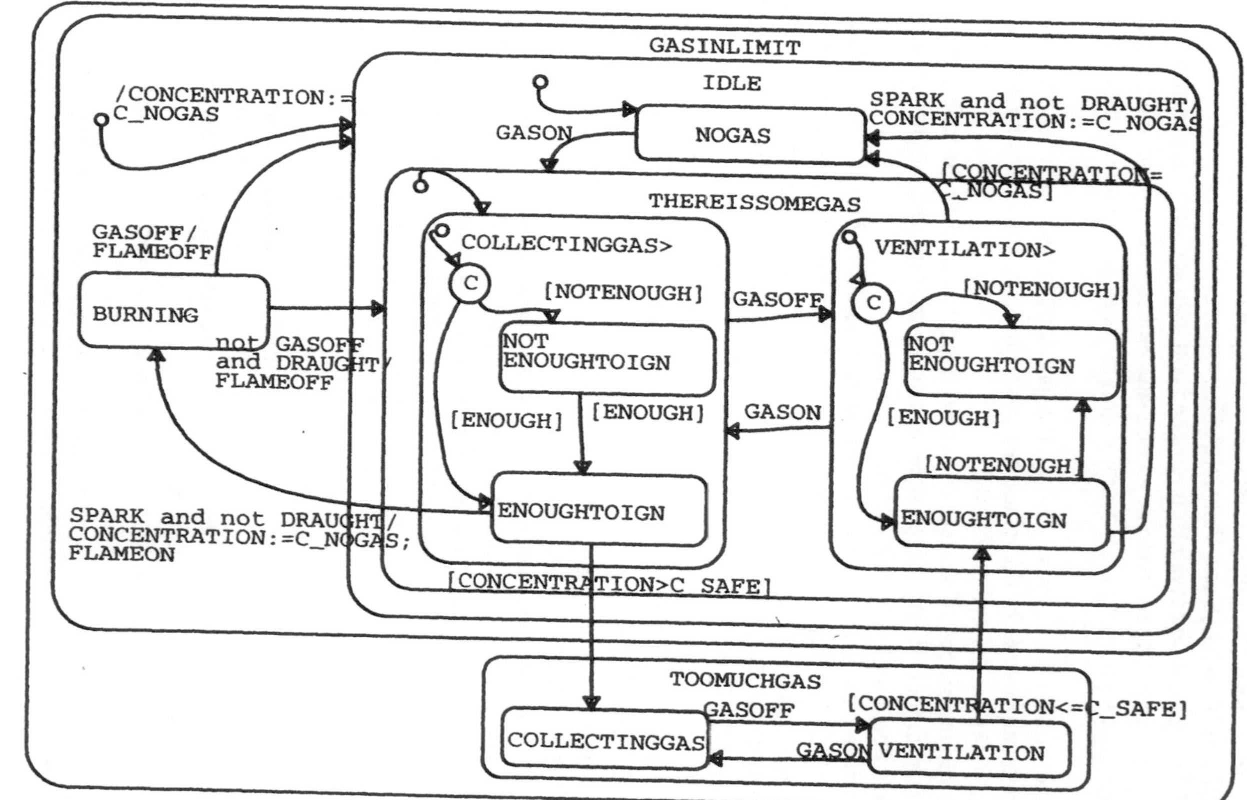

Fig. 9. The dynamic model of the burning chamber taking into account the possibility of hazard occurrence.

After detecting hazard scenarios we look for possible ways of hazard prevention. In Fig. 10. we show a modified version of the Control object which aims at elimination of the second of the above scenarios.

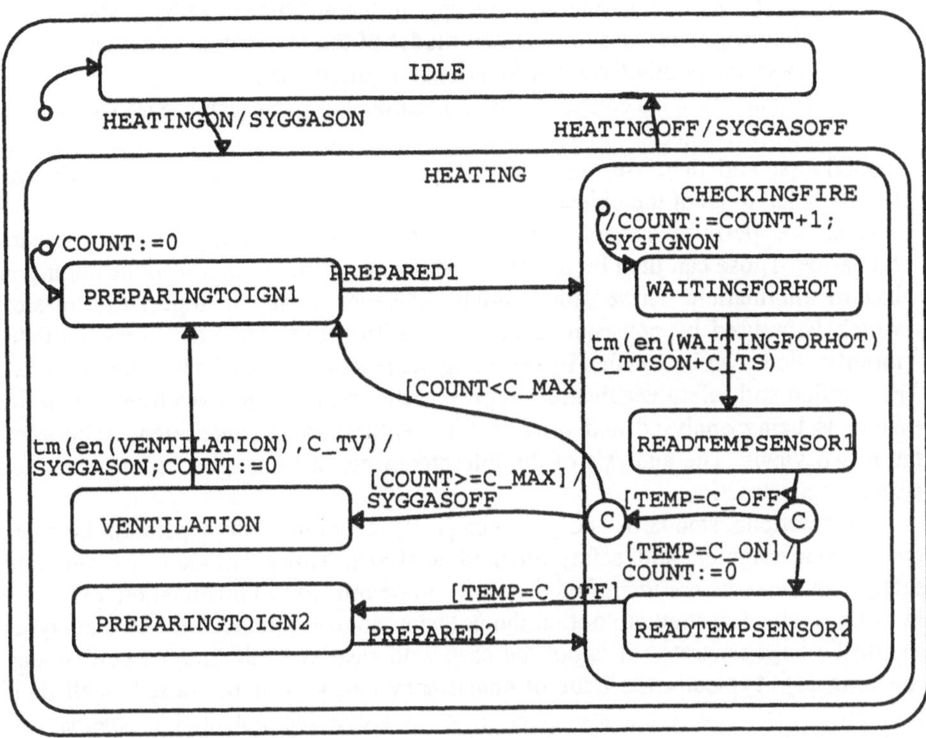

Fig.10. The model of the control system disabling dangerous results of draught.

The changes introduced to the control system (Fig. 10.) include:

- splitting PREPARETOIGN state to two states. The control system distinguishes the situation in which the fire was blown out after the inflammation (PREPARETOIGN2) from the situation in which there was no inflammation at all (PREPARETOIGN1). Time of the burning chamber preparation is different in both cases (events PREPARED1 and PREPARED2).
- a new variable COUNT which counts the number of successive unsuccessful attempts to set fire,
- a new state VENTILATION,

Each time the control system leaves the state PREPARETOIGN1 the COUNT variable is increased. If the fire is set (this is detected by the temperature sensor), the counter is set to zero. If however the counter exceeds the pre-defined limit C_MAX the supply of gas is cut down (SYGASOFF) and the burning chamber is forced to the VENTILATION state. After the time C_TV (enough for a thorough ventilation) the gas is enabled again and the counter is reset to zero. The modified model is subject to mission and safety requirements verification in order to check whether the modifications have restored the safety and have not impacted the mission.

5 Conclusion

The approach presented in this article assumes that safety of the system is analysed using the object-oriented modelling. The approach comprises three basic stages.

1. At the first stage the mission-oriented model of the application is developed and verified from the point of view of functional requirements.
2. At the second stage we consider system hazards. For each such hazard we identify a set of critical objects. For each critical object we develop the model of unsafe behaviours, and then we construct the product model covering both safe and unsafe behaviours of the object.
3. Then we perform hazard reachability analysis to identify possible hazard scenarios. Those can then be used as clues for the system modifications aiming at hazard elimination. These modifications however must not impact the mission which is assured by consecutive mission verifications. The effectiveness of the modifications can be verified by repeating the reachability analysis of the hazards.

The mission and safety verifications represent different perspectives from which the system is being analysed and provide for identification of interferences between these two views. The analysis of the intereferences, in turn, gives some clues for system reconstruction.

International standards and guidance [1, 2] recommend the separation between the mission-oriented and safety-oriented analyses. However, after the software safety requirements are identified they have to be related to functional requirements in order to check if they are not on the collision course. Although the two analyses should be kept separate, it is not the case with respect to the mutual consistency checking [5]. Typically the issue of consistency checking is not raised at all or if addressed, very few hints are given how to solve the potential problems [6]. Problems with consistency checking follows form the fact that it is as complex as safety analysis itself and assuring the completeness of such a check is very difficult. Our approach aims at the maximal reuse of the mission-oriented requirements effort during safety analysis. Thus, consistency between mission and safety is an obvious result of utilising mission-oriented model as the staring point for safety aspects modelling and analysis. Moreover, the process of verification of the mission requirements give some additional arguments for the model validity what is crucial in safety analysis.

Explicit expression of unsafe behaviours in the object model of the system allows to constitute the base for interdisciplinary communication among software engineers, safety engineers and provide for validation by application-domain specialists. In the authors' opinion object-oriented modelling is a viable mean for safety analysis. The whole system, including the target system, plant and environment can be modelled using the same language. The object model provides for the system decomposition. The corresponding dynamic model with its formal semantics provide for precise description of the behavioural properties of objects and the related constraints and is the main vehicle for analysis. Although OMT offers a firm base for system modelling it is not intended for safety analysis and needs some extensions to provide for it. An example of such extension has been presented in this article. In [7] we presented another approach where we concentrate on faults which can be generated by some 'untrustworthy' objects in the system, which can then, through a long chain of intermediate events, contribute to system hazards. In [6] the similar net-based formalism has been used. Applying Petri Nets to safety analysis without structural decomposition results however in an unreadable model even for a relatively simple example.

In our experiments we were supported by the STATEMATE tool which enabled simulation of statecharts - the formalism underlying our dynamic models. Although the approach is not necessarily bound to this tool (nor to the statecharts) the availability of the tool was very helpful while working with larger models and would be necessary in any realistic case study.

The approach presented in this paper is 'holistic' in a sense that we develop a closed model of the whole application which is then subjected to the analysis process. This approach is complementary to the approach presented in e.g. [8, 9] where the analysis is based on a 'partial' model and concentrates on the aspects of faulty behaviour represented by Fault Trees. All those approaches analyse the system from different angels, and in a complementary way contribute to the safety case developed for a given application. The detailed discussion on this subject can be found in [10].

Another interesting problem resulting from the work presented above is *safety monitoring*. This refers to the situation where, for some hazard scenarios, safety can not be restored just by introducing changes to the system. The reason for this can be that the extent of such changes would be too broad and technically infeasible. Typically, we face such problem when hazard depends on human behaviour or on the behaviour of other objects which are beyond the influence of the computer based control system. Then, even if we can not restore safety by changing the design, we can still observe (or *monitor*) the system in order to detect that the system evolves in wrong direction. In such case some emergency actions could be initiated. Two approaches for *safety monitor synthesis* are under investigation. The first results have been reported in [11].

References

[1] International Electrotechnical Commission, *Functional Safety: Safety related Systems IEC 1508*, 1995

[2] Redmill F. J. ed., *Dependability of Critical Computer Systems 1, 2, 3*, Elsevier Applied Science, 1989

[3] J. Rumbaugh, M. Blaha, W. Premerlani, F. Eddy and W. Lorensen, *Object-Oriented Modelling and Design*, Prentice Hall Int., 1991.

[4] i-Logix Inc., Burlington, STATEMATE - Technical Documentation.

[5] Leveson, N.; *Safeware: System Safety and Computers*, Addison-Wesley, 1995

[6] Leveson N. G., Stolzy J. L., Safety Analysis Using Petri Nets, IEEE Transaction on Software Engineering, vol. SE-13, no. 3, March 1987

[7] Górski, J.; Nowicki, B., *Object Oriented Approach to Safety Analysis*, Safety and Reliability of Software Based Systems, Brugge, Belgium, 12-15 September 1995.

[8] Górski, J., Magott, J. and Wardziński, A., *Modeling Fault Trees Using Petri Nets*, Proc. SAFECOMP'95, (G Rabe Ed.) Springer-Verlag, 1995, pp 90-100.

[9] Górski, J.; Wardziński, A., *Deriving Real-Time Requirements for Software from Safety Analysis*, 8th Euromicro Workshop on Real-Time Systems, L'Aquila (Italy), June 12-14, 1996

[10] Górski, J., Nowicki, B. and A. Wardzinski, *Holistic and Partial System Models in Safety Analysis*, Probabilistic Risk Assessment, USA, September 1996

[11] Górski, J., Nowicki, B. *Object Oriented Model Based Safety Monitor Synthesis*, III Conference on Real-time Systems'96, Szklarska Poręba, Poland (in Polish)

ACKNOWLEDGEMENT

The authors would like to acknowledge the support of the EU Copernicus ISAT (Integration of Safety Analysis Techniques) project.

A Processor Architecture Designed to Faciliate the Safety Certification of Hard Real Time Systems

Hans-Peter Meske and Wolfgang A. Halang
FernUniversität Hagen
Faculty of Electrical Engineering
D-58084 Hagen
Germany
{hans-peter.meske | wolfgang.halang}@fernuni-hagen.de
phone: + 49 - 172 - 7 284 283, fax on request

Abstract

The following paper describes efforts to develop a processor architecture that meets the requirements of hard real time computing. The architecture is of the RISC-type with a single, modular CPU. The modules are a Kernel Processor, a Task Processor, a Memory Module and a Controller for internal and external communication. By integrating multiple register files directly accessible by the ALU, the number of main memory accesses decreases and the time for context-switches is reduced considerably. While OS functions, scheduling, time management and interrupt handling are performed by the Kernel Processor, the Task Processor focuses on its primary function, viz., to execute application program code. Assigning the traditionally sequentially performed program-, operating system- and memory-operations to different modules working in parallel results in a significant increase of performance. The reduced instruction set interfacing this architecture allows for a complete and convenient implementation of real time algorithms, especially in distributed systems, without loosing the operational determinism. On the other hand, the simplicity of the design enables the validation in a safety certification process, which was one of the design guidelines.

Keywords
Real-Time, Architecture, Reduced Instruction Set, Predictability, Determinism, Safety Certification

1 Introduction

This paper emphasizes efforts to develop a processor architecture that meets the requirements of hard real time computing, especially for safety sensitive applications. The introduced CPU will be the kernel of the new real-time coprocessor-architecture. The design is kept deliberately simple, because it is less the performance that counts, but rather the specific qualities, that qualify for installation in a real time environment. That doesn't mean, that high performance speed automatically lowers the degree of control over a system, but most of the means to increase speed or the utilization of devices, like caching, virtual addressing or asynchronous communication protocols cause indeterminism up to an extend which is not acceptable for Hard Real Time Systems. Therefore, the processor has to be designed in a fashion that it can allow for a precise a priori prediction of runtime for each application process and can be validated with tolerable expenses. To support the desired properties the design contains, beside a minimal processor kernel, some typical real-time features like a time register or several register banks for fast context switching.

The supporting architecture around the CPU consists of a kernel processor, a memory module, and a controller for internal and external communication. While O.S. functions, scheduling, time management, and interrupt handling are performed by the kernel processor, the task processor focuses on its primary function, viz., to execute application program code, which minimizes the frequency of context switches. Assigning the program, O.S., communication, and memory operations, which are hitherto performed sequentially, to different modules working in parallel results in a significant increase of performance. The paper will show that the architecture is well suited for its purpose by allowing for a complete and convenient implementation of real time algorithms without loosing operational determinism.

Another objective besides the real time abilities is to validate the realized architecture, in other words to formally derive, that the task processor follows its specification and therefore is proven correct. To accomplish this goal, it has to be fully formally specified and designed simple enough to allow for a practicable verification process. By its simplicity the proposed processor is clearly easier to validate than others, e.g. Kershaw's VIPER. Later in this paper we will discuss, that it's not just the verifiable transformation process, but also the proper representation of an idea and the gap between technical design and realization, that have to be taken into account. A proper, full test of the processor is, by any means, impossible due to complexity reasons, as it is the same with every kind of hardware outsides the stage of laboratory prototypes.

The motivation for the design of a fully deterministic architecture to support hard real time applications is derived from the unsatisfying fact that, in spite of all the progress in software development, the hardware aspect of real time research has been neglected. The aptness of the hardware used for real time computing has been assumed implicitly in most cases. The problem is that the timing behavior of conventional platforms is - due to their complexity - hardly or not at all predictable.

The System we are aiming for is a computing system, mostly implemented in a security sensitive environment, that enables - by construction - prediction of its temporal behavior, namely the determination, whether a process can terminate within a given time frame. This motivates the aim of designing a strictly deterministic architecture, which is, by means of a reduced instruction set, kept as simple as possible and able to support all desirable properties of a Hard Real Time System.

2 The Architecture

The major desirable properties of a Real Time System are that it shouldbe timebounded, predictable/deterministic, reliable, fault-tolerant, modular, flexible and simple. The latter is especially important considering the need to be validated with reasonable expense. A concept for an architecture design following the mentioned guidelines was first presented in 1991 by Halang and Stoyenko [5], and developed by Colnariç and Halang [3]. The contemporary version consists essentially of four modules:

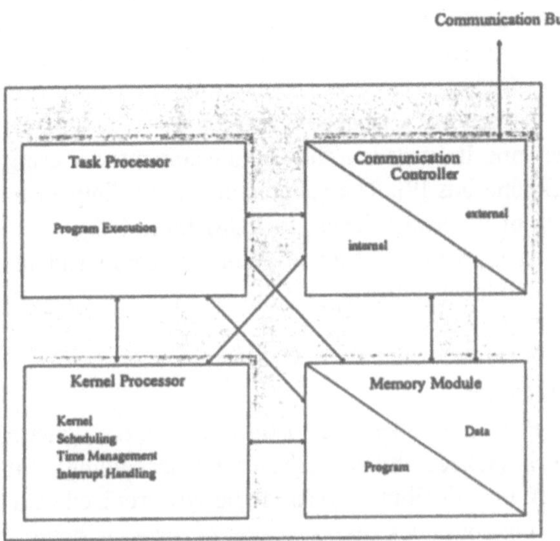

Figure 1 The Architecture

- Memory Module
- Communication Controller
- Kernel Processor
- Task Processor

2.1 Memory Module

The program and data memories are located in the Memory Module, but, in contrast to the von Neumann principle, in separated ranges. The program memory is implemented as a Read Only Memory, to protect it from self-alteration, and is controlled by the Kernel- and Task Processors. The Data Memory, with separated areas for the contents of Variable Files (tagged variables) and Stack Files (return addresses, status words) of pre-empted or terminated tasks, I/O registers for external communication, and a shared (co-operation-) area for the internal communication between tasks running on the same node, is implemented as a Random Access Memory and managed by the Communication Controller and the Task Processor. Data and program memories are implemented as a separate module, because they are the only application-related parts of the architecture.

2.2 Communication Controller

The Communication Controller consists of two submodules. The one for internal communication performs the data exchange between the memory module and the registers of the Task Processor (Variable- and Stack Files) in case of preemption, termination, or re-scheduling of a task. This is done transparently with respect to the running process, using a separate, serial bus. The submodule for external communication performs data exchange with peripherals and other, identical nodes in a distributed network. It works in a fully predictable manner using a token passing protocol on a backbone bus [9]. The submodule itself is implemented in a layered structure consisting of a network layer (ISO/OSI level 1+2), a copy layer to hold received data, a filter layer to select the relevant parameters and an address layer to transform network into local addresses.

2.3 Kernel Processor

The Kernel Processor takes care of operating system functions, as there are process management (schedulability analysis and scheduling), time management (system time register, in a distributed system to be synchronized with the other active nodes), event handling (provides reaction upon changes in status, time conditions and interrupts), and handling of overloads by graceful degradation. The implementation of this administrative co-processor reduces the frequent context switches required to perform O.S.-functions common in single processor systems, and synchronizes asynchronous interrupts by saving them (and their time of arrival) into polling registers. Thus, the task processor can focus on its primary function, resulting in a significant increase of performance without loss of quality in the predictability of temporal behaviour. The risk that a time-critical process will not meet its deadline because of the additional time used for its scheduling can be almost eliminated by this type of architecture.

2.4 Task Processor

The Task Processor will be in the focus of this paper and is the actual 'worker' of the architecture. It executes the process code of applications without the usual operating system-, interrupt-, and input/output-overhead. For the reasons mentioned the design goal was to find in the tense relationship of the aspects programming comfort, complexity, real time support, and verifiability, a simply structured solution, meeting all requirements of an implementable system [2, 6].

The design followed RISC guidelines. That implies the implementation of a reduced instruction set, which prolongs the executable code, but allows for easier validation of the architecture and a more precise prediction of its temporal behaviour. The proposed processor realizes a Harvard-RISC architecture with 16-Bit instruction words, 32-Bit data words, an accumulator (single address instructions, accumulator is first operand and destination register), two bi-directional buses (program and external data) and three variable files for fast context switching.

The modules are:

* an ALU (add, sub, nand)
* some Special Purpose Registers:
  - accumulator
  - program counter (& incrementer)
  - status register (zero, negative, carry, overflow & kernel flags)
  - loop counter (limits number of iterations, & decrementer)
  - loop timer (internal deadline for subprocesses, & decrementer)
  - bank pointer (register bank linked to the current process)
  - timer (external triggered, read only)
* three variable files (64 K words each, lower 2 K directly addressable as index for the upper 62 K)
* an operation decoder (& processor control)
* a program bus (code & address, bi-directional)
* a data bus (data & addresses, bi-directional)

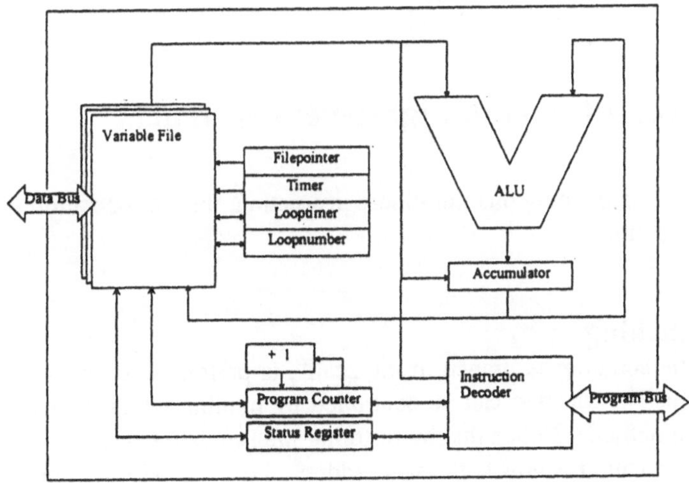

Figure 2 The Task-Processor

The implemented instructions are:

* load (op —> accu)
* store (accu —> op)
* add (accu ::= (accu + op) mod 2^{32})
* sub (accu ::= (accu - op) mod 2^{32})
* nand (accu ::= accu nand op)
* call (if condition then op ::= pc; pc ::= accu; else pc ::= pc+1)
* shift (shifts/rotates on accu)
* immop (load/add little signed constant)

The call-instruction is a hardware supported branch into a subroutine, which is controlled by two special purpose registers to limit the number of loops and their execution time. No 'goto'-constructs are possible, because a subroutine has to end with an unconditional return to the main program.

There are three address modes realized: direct, indirect and external (the latter two using the index registers) in each variable file. There were no tags implemented, though typing is not supported by hardware. The implementation of variable instruction length and floating point arithmetic were avoided to decrease complexity. The introduction of an instruction pipeline, as it is typical for RISC architectures, has to be handled carefully, not to loose the determinism of the behaviour. The analysis, if and how a three stage pipeline possibly can be integrated into the concept, is part of the current research.

3 Representation of high-level constructs

As it is the only one non-linear instruction, the call-operation deserves some special attention.

3.1 Branching

The conditioned branch, using „call", is performed exclusively by calling a subroutine. The function can be controlled by limiting the number of loops and/or giving a timeframe. Notice that by restoring the Program Counter from the stack the program execution returns to the same address, i.e. the call instruction. For control of the number of iterations or the maximum amount of time the call-instruction has to be preceded by a load instruction, which addresses two special purpose registers, the above mentioned Looptime and Loopnumber. The decision to branch to a subroutine is of Boolean nature: "if (condition=true) and (Looptime > 0) and (Loopnumber > 0) then push (PC); PC ::= address; else PC ::= PC+1;".

3.2 Alternatives

The 'if (condition) then ... else ... ' statement is implemented by a load instruction, which loads 1 in the Loopnumber and suspends the Looptime Register (by loading 'FF...F'), followed by two call instructions, one for the 'if'-subroutine, the other one for 'else'. Through a change of the global status, after the termination of the 'if'-subroutine the condition for the following call-instruction, the 'else'-branch, could have become true. The execution of the 'else'-subroutine - a possibly hazardous mistake - is suppressed by the Loopnumber register, which is decreased to 0 by the return.

3.3 Loops and Synchronization

As the decision to execute a subroutine does not only depend on the coded condition, but on the Looptime and Loopnumber registers as well, the call-instruction with a previous 'load' allows to directly implement "for var=...to...do..." constructs by defining the number of loops, suspending the looptime register and asking for an empty condition. Nested loops are supported by pushing the loopnumber register on the stack with every call instruction, and popping it back (decremented by one) with every return.

Giving a certain time frame and a condition which is to be verified in the global Status Register while disabling the loopcounter realizes a "while (condition=true) do..." construct, normally avoided in Real Time Systems, which can be controlled here by the defined maximum time frame.

Finally synchronization between two remote tasks can be accomplished by a timebounded subroutine, which just polls either the assigned interrupt flag in the Status Register or one of the input registers in the Communication Controller.

On the application level timing constraints for a loop are considered only within a task. Constraints for process activation times and deadlines will be part of the responsibility of the kernel processor and stored in a task-header.

4 Remarks about Verification

The increasing complexity of circuits makes them more and more susceptible for faults of all kind. Up to half of all application specific circuits do not work in the desired way, when they are integrated in an environment for the first time [1]. Even the use of sophisticated design tools did not guarantee an error-free product.

In the area of safety-sensitive systems, which, in many cases aren't even be testable as a whole, we cannot afford to apply 'error detection by user' mechanisms. Therefore, we have to think about ways to guarantee the predicted behaviour, either by contruction or through proper analysis.

Let's divide the whole process from the problem to its Hardware-solution into three phases, creation, design and production.

The problem with the creation phase is, that there is no way to formally express it and therefore no serious formal proof exists. It's simply impossible to prove that the verbal expression is a complete and correct representation of the idea or that the idea itself is a proper solution for the problem. In this phase there is no alternative to the experience and the intellectual skills of a human designer.

For the production phase there are several different methods that guarantee up to a certain extent that the product is equivalent to the given layout [10]. Nevertheless, it is impossible to perform a complete test of any chip of practical relevance. Therefore science offers aids, but no real solutions. There is always a certain degree of uncertainty to tolerate.

The design phase is the only one that can be fully formalized. Gajski and Kuhn[4] structured the phase into three domains: behaviour, structure and topology. Each of these domains can be furthermore classified, depending on the level of abstraction. Their „Y-Chart" is a good visualization of this classification.

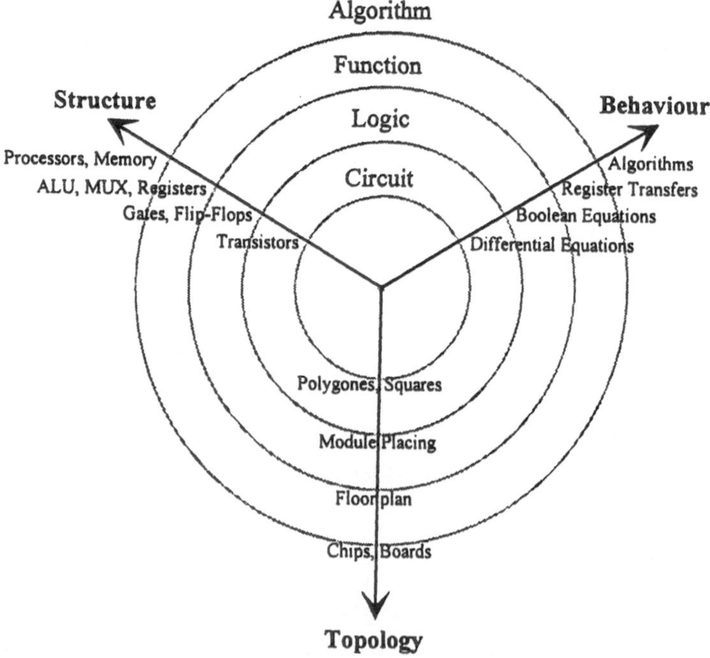

Figure 3 The Y-Chart

Every class in every domain is an entity of the same design. From this point of view, the design process is a sequence of transformations within these entities.

In the typical top-down approach, a behavioural description, mostly given in any VHDL, is synthesized into a structural description, e.g., a register-transfer structure. The second design-step is to synthesize the topology out of the structure.

Beside the synthesizer itself, there are tools, e.g., for analysis, simulation, evaluation or testpattern generation, that support the correct transformation and optimization. In a second step, the whole process can now be verified by means of logic. The most popular tool to accomplish that with a first-degree logical calculus is the Boyer-Moore-Verifier [7], on a higher degree there are theorem-verifiers like HOL, LAMBA or VERITAS.

From all that we can derive, that the design phase is the only one, that enables formal proof. For the creation and the production phase there is no way of a formal verification at all. Therefore we should rather talk about validation instead of verification for the whole process.

With that in mind, complete safety might be an illusion, but there is on the other hand no excuse for not aiming for perfection, even though a label like „100% proven correct" will probably never be serious in this context.

5 CONCLUSION

This paper introduced an architecture, whose lack of complexity makes it suitable for the precise temporal prediction as well as a proper validation of the architecture within the given specification. Additional benefits of a low complexity design are that compilers will be small and the processor will be very fast in simple operations, which make up more than 70 % of standard application programs.

The reduced instruction set, as the interface to the application level, allows internal and external transport of data using a deterministic communication subsystem, arithmetical and logical operations, controlled conditional and unconditional branches, time- and number limited loops, that allow secure handling of time-bounded procedures as well as synchronization with external events. Thus, it is feasible with appropriate architectural support, to satisfy the demand for a practical, minimal instruction set that, in a deterministic way and with predictable timing behaviour, enables the implementation of sufficiently fast and comfortable real time algorithms for distributed environments while observing certain safety aspects.

The global result of the described efforts (in combination with already existing software components) will be the bottom-up design of a complete, homogeneous and deterministic system meeting the requirements of hard real time computing, and, by construction, allowing complete control over all processes. Thereby the systematic transposition of theoretic research and the development of suitable applications will be made possible.

Bibliography

[1.] Anderson, R. (1991). *Design and Test Integration, The Reality Behind Fast Time-to-Market.* Keynote Address at the 14th Annual IEEE Workshop on Design for Testability, Vail, Colorado

[2.] Borst, P. and Meske, H.-P. (1990). *Design and Implementation of a Prolog-RISC-Processor.* Technical Report, University of Karlsruhe, Germany.

[3.] Colnariç, M. and Halang, W.A. (1993). *Architectural Support for Predictability in Hard Real Time Systems.* IFAC Control Eng. Practice, Vol. 1, 1, pp. 51 - 57.

[4.] Gajski, D. and Kuhn, R. (1983) Guest Editors Introduction: *New VLSI Tools*, COMPUTER

[5.] Halang, W.A. and Stoyenko, A.D. (1991). *Constructing Predictable Real Time Systems.* Kluwer Academic Publishers, Boston-Dordrecht-London.

[6.] Heide, J.A. and Halang, W.A. (1991). *Performance Metrics for Real-Time Systems.* Springer: Informatik-Fachberichte, Vol. 295, pp. 121 -- 127.

[7.] Hunt, W.A. (1985) *A Verified Microprocessor*, Ph. D. Dissertation, University of Texas, Austin

[8.] Kershaw, J. (1987) *The VIPER Microprocessor.* Technical Report 87014, Royal Signals and Radar Establishment, Malvern, Worcs. London:HMSO.

[9.] Malcolm, N. and Zhao, W. (1994). *The Timed-Token Protocol for Real-Time Communications.* IEEE COMPUTER, Vol. 27, 1, pp. 35 - 41.

[10.] Wunderlich, H.-J. (1991) *Rechnergestützte Verfahren für den prüfgerechten Entwurf und Test hochintegrierter Schaltungen*, Springer-Verlag

Session 2
Testing, V&V

Formal Verification as a Design Tool -the Transponder Lock Example-

R. Budde, A. Merceron, K.-H. Sylla°
GMD - SET-EES, SchloßBirlinghoven
D-53754 Sankt Augustin
email: {reinhard.budde,merceron,sylla}@gmd.de

Abstract

We describe a methodology for the construction and validation of embedded systems with real-time constraints. Our methodology is based on object-oriented techniques and synchronous programming. This greatly eases the use of formal verification to analyse the system, particularly to support design decisions. We use model checking to verify reactive behaviors and theorem proving to verify datatype behaviors. Our approach has been applied to develop industrial products. It is illustrated here with such a development, a transponder lock.

1 Introduction

Technical systems today comprise a variety of mechanical, electrical, electronic and software components. Software-design has become a crucial task in the design of embedded systems. Such systems have to fulfill strict requirements towards reliability, dependability and real-time properties since they have to react on time to stimuli of their environment.

Our methodology to design embedded systems is based on object-oriented techniques and synchronous programming. Object-oriented techniques allow to achieve the necessary and appropriate flexibility in software design while synchronous programming is well suited to specify real-time constraints.

We use object oriented methods during the whole development process to partition the system into classes. Classes are the units of information hiding and reuse. Every class has a datatype behavior as usual in the object-oriented paradigm. A class is called reactive if it additionally defines a reactive behavior. These two aspects of a reactive class, reactive and datatype behavior, have a sound integration via synchronous programming. Synchronous programming is based on the synchrony hypothesis which stipulates that systems produce their outputs synchronously with their inputs [1]. A consequence is that timing issues may be abstracted in the first stage of the system design, see section 3.

Formal verification (model checking for reactive behavior, and theorem proving for datatype behavior) is used to analyze the system. The partitioning into classes and the synchronous semantics allow to describe and analyze crucial parts of the system in isolation and guarantee that verification results also hold for the complete system. Verification is used not only to prove whether

°This work was partly supported by a grant from the German-Isreali Foundation for Scientific Research and Development (GIF) under contract number G-0301-111.06/93.

programs meet their specification, but also to give feedback about the impact of design decisions on the embedded system. Thus it is used as a design tool. It is this point, which is rarely put forward, that we want to emphasize in this paper.

2 Methodology

Our methodology of system development is closely related to object oriented methods like OOSE [2]. To give a short impression we report typical activities when developing a lock system for contactless transponder keys; in the following we refer to the system as the RFILock.

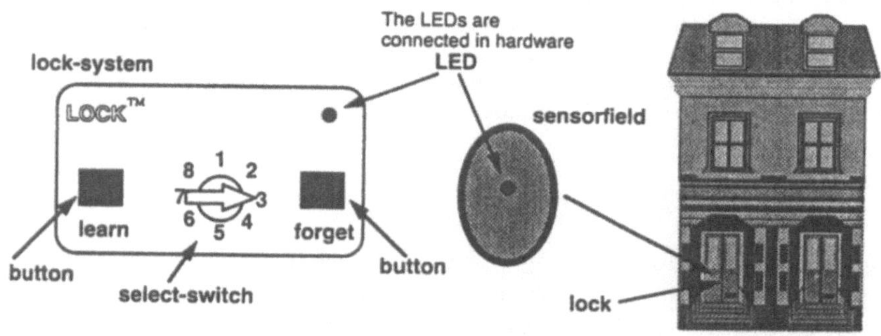

Figure 1: RFIlock overview

We start the development with use cases. Use-cases describe typical ways of using the system, of interaction between the system and the environment and assumptions about the system's and the environment's behavior. Use-cases that describe the human machine interface often are re-written to become a part of the user-manual. Two use-cases are shown in figure 2.

From use-cases objects and classes are identified. Project discussions are stimulated by using CRC-cards [3]. For this method index cards are divided into three regions: the *C*lass name (e.g. Lock), *R*esponsibilities of this class (e.g. open the door for three seconds) and *C*ollaborating classes that help to perform the responsibilities (e.g. class Timer).

The peripheral units (the sensor, the LED, the select switch) are another starting point of the design: they are encapsulated by objects. The corresponding classes are then described by CRC-cards. Thus the design proceeds from known to unknown as opposed to being top-down or bottom-up [3].

After discussions on CRC-cards reach some stability, object- and class-diagrams are drawn, and the so-called reactive objects are identified. Reactive objects are instances of reactive classes and are triggered by input signals and produce synchronously output signals. Usually peripheral objects are reactive. In the object diagram of figure 3 reactive objects are bold. Signals exchanged

use-case **Open the door**: Somebody wants to open the door and holds the transponder key close to the sensor. If the key is recognized as being valid, the LED is switched on and the lock is released. As long as the key remains close to the sensor and additionally 3 seconds after the key is removed from the sensor the door remains unlocked. When these 3 seconds elapse, the door is locked again and the LED is switched off.

use-case **Introduce a key to the lock-system**: The code-switch is set to a number, under which a key should be remembered by the lock-system. This number selects a table-entry to be worked upon. When the learn button is clicked, the system enters a learn mode, which is signaled by slow blinking of the LED. Within the learn time of 2 minutes a key may be hold close to the sensor. If the key is recognized and not yet included elsewhere in the table, it is accepted as a valid key for this lock and stored in the selected entry. The learn mode is stopped and the LED is switched off, if a key is recognized or if 2 minutes have elapsed.

Figure 2: two use-cases

between reactive objects and environment are identified. Often diagrams similar to timeline or message sequence diagrams are drawn to clarify this aspect.

Reactive objects are described in reactive classes. Reactive classes define the input and output signals, called the reactive interface, and the reactive behavior, given either in a graphical Statechart-like notation or in a textual Esterel-like notation, see figure 4. Also the access operations (member functions, methods, features) for both reactive and non-reactive classes are specified. For a reactive class access operations are all private, they can be called only from the reactive behavior. For a data-class the access operations make up the data interface for its clients as usual.

Now access operations may be implemented and simulations can be done. Formal verification is used to prove properties of the system, particularly to validate design decisions and explore design alternatives. An essential feature is that verification is done on the implemented reactive and datatype behaviors and not on a separate model, see section 4. It is checked where dynamic binding –as needed for a powerful object-oriented architecture– can be implemented by static binding for optimization. Cross-compilers are then used to generate code for standard micro-controllers used in industry.

The behavior composed out of the reactive behavior of all reactive objects is used to analyze real-time constraints. The maximal time consumed for reactions in all state transitions is computed to guarantee hard real-time requirements. In the RFIlock worst case reaction must be less than 1/10000 sec. This is mainly due to the physical sensor which is sampled at this rate.

It is an essential insight for the process of developing embedded software, that any activity during development may give feedback to all other activities. The development is incremental and iterative. For instance implementation decisions like implementing floating point operations in software and avoiding a coprocessor can influence architectural design decisions like the signal set exchanged between objects. Thus our methodology carefully supports to re-

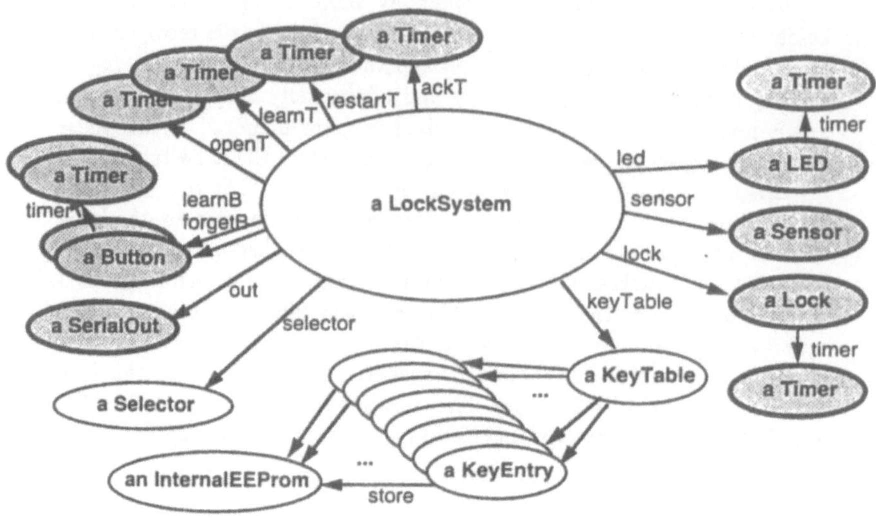

Figure 3: object diagram of the RFIlock

execute proofs in changed designs and tells whether a proof for a proposition still holds if something is modified.

3 Synchrony

Synchrony is the paradigm underlying the reactive behavior of a system and is the basis to integrate reactive and datatype behavior. The synchronous approach makes it possible for the designer to consider a logical time where reactions take place and to concentrate on the functional aspects of the system, physical timing issues being deferred to the implementation stage.

The synchronous algorithm of a reactive system defines the response to stimuli from the system's environment. Remember, that the reactive system is an assembly of reactive objects, for which the following remarks are valid, too. To achieve reaction, the system is thought to be connected with input and output lines to its environment. Signals can flow in and out using these lines. The reaction to input signals is not continuous. Two different phases are distinguished: the system is reacting or it is idle.

When the system is idle, signals on input lines are collected, but not propagated to the system. The change from the idle phase to the reacting phase is effected by an activation (hardware designer would call this a clock-pulse). Whether the activation is periodic, using a timer subsystem of the microcontroller, or depends on environment-conditions is outside the scope of the synchronous system. When the activation takes place, the gathered signals plus a special signal `tick` are provided to the system. Now the system computes the response. Signals are **emit**-ted or **await**-ed, tested whether they are **present**, sub- reactions are computed in parallel (denoted by ||). Access operation are called, both as statements and in expressions. The reaction is complete, if all branches of the (nonsequential) algorithm have committed to **halt**. Then all output signals are made available to the environment and the system is idle again.

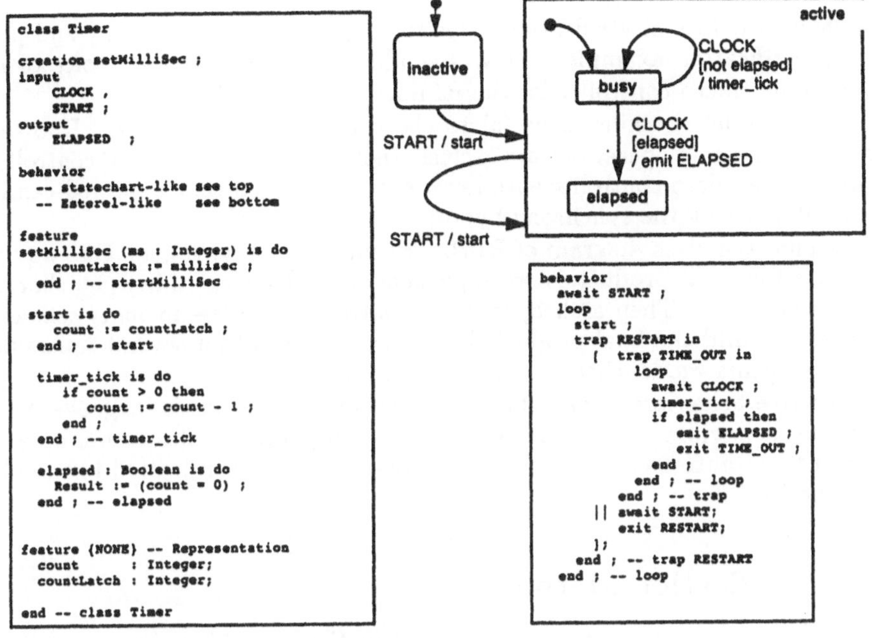

Figure 4: the reactive class Timer

Usually the lines of the output signals are connected to actuators or display-units and effect changes in the environment. These changes may produce input signals, which are collected. Then, if the clock ticks, the next reaction-step is initiated. Each step is called an *instant.* A reacting system is logically disconnected from the environment, i.e. it is impossible to supply input signals during a reaction. There is no rendezvous-like concept in the synchronous model. Emitting signals will never be blocked. This would be the case in asynchronous languages, where a emitter (sender) is blocked if no corresponding process (receiver) awaits the signal. Further, signals are not dedicated to a specific receiver, they are broadcasted to all components of the system, i.e. to all reactive objects. Signals are *not* consumed in an **await**-statement. This facilitates simultaneous reactions to the same signal.

Now perfect synchrony demands, that reactions are instantaneous: the output signals are computed from the input signals without any delay, i.e. in zero time.

At a first glance, this seems to be a simplistic model with no practical use. But from a constructive point of view this claims, that there exists no possibility in the environment to observe the execution-time of a synchronous system.

Synchrony demands, that the system is "quick enough" for the environment. Formal verification helps to validate that perfect synchrony holds. An example is given in section 4.2.1.

4 Formal Verification

We show how formal verification helps to support perfect synchrony and present a verification example that was conducted to support the choice of an appropriate micro-controller for the RFIlock.

First, model checking is useful to check physical timing issues. Knowing whether some data operations are exclusive helps to choose the micro-controller. Sometimes a micro-controller with lower execution speed than expected can be used still meeting the synchrony hypothesis.

Second, the class diagram of RFILock contains several timers. One way to optimize the final product is to implement several logical timers by sharing one `Timer`-object. Then a cheaper micro-controller with less memory cells can be used to build the final product. For that we proved that several timers are exclusive, using model checking and theorem proving.

Reactive behaviors of objects are defined in a synchronous language. Synchronous behaviors are compiled quite efficiently into boolean automata, a kind of finite automata [4]. This fact makes automatic verification like model checking feasible.

4.1 Verification Method

We first express properties to be checked in the branching time logic CTL though we consider using the linear time logic LTL as well [5]. Reactive behaviors contain procedure calls or function calls to the corresponding datatype behaviors. This gives a simple and mechanical way of data abstracting. Procedure calls and function calls are replaced by signals emission and tests for signals presence. The abstract reactive behavior of an object is compiled into a boolean automaton. Proofs are conducted on appropriate objects rather than on the whole system. This feature does not only provide better machine efficiency; it is easier for the human being in charge of the proof to work with an object at a time rather than with the whole system, especially when counter-examples delivered by the model checker have to be understood. The abstraction and the object-wise verification are conservative for the logic ACTL (the logic CTL where only the universal quantifier A is allowed); any property checked as true on an abstract object holds also for the system itself [6]. Model checking an abstract object may lead to formulate assumptions. When assumptions involve signals emitted by the system, they have to be discharged. To prove assumptions, it may be necessary to consider a whole object with data instead of an abstraction. Then we use theorem proving for assumption discharge.

We use the model checker SMV [7] and now start using VIS [8] for checking finite state systems against specifications expressed in CTL. The encoding for either model checker is done automatically. VIS, which accepts Verilog inputs, fits better our boolean automata model –similar to sequential netlists– which leads to more efficiency when automata get big. The theorem prover PVS [9] is built on higher order logic and contains sophisticated decision procedures, which makes it attractive when proofs involving data have to be conducted.

4.2 Verification Example

4.2.1 Supporting Perfect Synchrony: Mutual Exclusion of the Serial Unit and the Sensor

The boolean automaton obtained from the reactive behavior of an object gives information on the instants where data operations are performed. Suppose one wants to check whether data operations `Sensor_Sample` and `Serial_Send` are exclusive. Then it is sufficient to look whether the CTL formula
AG ¬(`Sensor_Start` ∧ `Serial_Start`)
is true, where `Sensor_Start` is the signal emitted when `sensor_sample` has to be performed and similarly for `Serial_Start`. This formula says that always generally *-AG-* signals `Sensor_Start` and `Serial_Start` never appear in the same instant. Doing modular verification, only the objects involving `Sensor_Start` and `Serial_Start` are selected.

This way we proved the exclusion of data operations performed by the serial unit and the sensor of the Lock. These two objects are particularly critical for the actual verification of the synchrony hypothesis since data operations performed by the sensor need 50 micro-seconds, data operations performed by the serial unit need 60 micro-seconds while all the data operations performed by all the other objects put together never exceed 40 micro-seconds. Thus the system meets the specified hard-real time constraint since any reaction will never exceed 100 micro seconds.

4.2.2 Space Optimization: Reducing the Number of Timers

Timers measure time and work as countdown stopwatches. They start measuring time when they receive a `Start` signal and emit an `Elapsed` signal upon termination.

One does not always have to wait for an `Elapsed` signal to be emitted. If, for example, a new key is learned and memorized in less than two minutes, then the learn mode is exited before the emission of the `Elapsed` signal by the `Learn_Timer`. Another frequent situation is when a timer is started and restarted before it elapses. For example, if a new valid key is recognized during the 3 seconds the door stays open, then a new time measure will start at once. The `Open_Timer` is restarted before it finishes its previous measure.

Here we show that the `Open_Timer` and the `Learn_Timer` are exclusive. Both are controlled by the object `RFILock`. When `RFILock` is waiting for an `Elapsed` of the `Learn_Timer`, it emits a `watching_LearnT_Elapsed` signal and, similarly, it emits a `watching_OpenT_Elapsed` while waiting for the `Open_Timer`.

We proved that always generally *-AG-* if `RFILock` starts the `Learn_Timer` while waiting for `Elapsed` from the `Open_Timer`, then it ignores the `open_Timer`: with other words, always at the next instant *-AX-* it does not wait for the termination of the timer anymore. The following CTL formula was checked as true:

AG((`learnT_Start` ∧ `watching_openT_Elapsed`) ⇒
  AX¬`watching_openT_Elapsed`).

Further, if `RFILock` waits for the `Learn_Timer` and not for the `Open_Timer`, then it will always not start the `Open_Timer` until $-A\ldots U-$ the `Learn_Timer` is ignored. In CTL it can be expressed by:

AG((`watching_learnT_Elapsed` $\wedge$ $\neg$`watching_openT_Elapsed`) $\Rightarrow$
$A[\neg$`openT_Start` U $\neg$`watching_learnT_Elapsed`]).

This formula is true with fairness constraint `watching_learnT_Elapsed`, i.e., under the assumption that `RFILock` effectively ignores the `learn_Timer`. `RFILock` ignores the `learn_Timer` either if it starts the `Restart_Timer` or if the `learn_Timer` emits `Elapsed`. Here we show the part of the assumption which concerns the `Learn_Timer`: any time it is started, always in the future $-AF-$ it emits an `Elapsed`. In CTL this is written as:

AG (`learnT_Start` $\Rightarrow$ AF `learnT_Elapsed`).

The dischargement of this assumption involves two steps. The first one is tc check whether, while running, the `Learn_Timer` may be restarted in the module `RFILock`. But this is not possible. Any new `Learn` signal -which indicates that the learn button has been pressed- that occurs while the `Learn_Timer` is already running is just ignored and does not cause any restart of the `Learn_Timer`. The following formula is checked as true on the abstract `RFILock`:

AG (`watching_learnT_Elapsed` $\Rightarrow$
AX (`watching_learnT_Elapsed` $\wedge$ `Learn` $\Rightarrow$ $\neg$`learnT_Start`)).

The second step is to prove that the `Timer`-object works correctly, i.e., if started and not re-started during a measure, it inevitably emits `Elapsed`. To prove it, obviously data have to be taken into account. Data essentially consist of a counter, *count* which is initialized with a positive integer value contained in *countlatch* and decremented till 0. These data are enumerable but large since *countlatch* is 120 000 for the `Learn_Timer`. We proved this second step using the theorem prover PVS. It has the advantage of keeping countlatch as an uninterpreted positive integer, consequently the proof applies to any `Timer` object, not only to the `Learn_Timer`. In the present case, the proof is obvious; it has only to be established that *count* implements a monotonously decreasing function. This is done proving four lemmata given as Hoare triples:

Lemma 1: At any instant s for any emitted signal o:

$\{$`Start`$\}$ o $\{count(s) = countlatch\}$

Any time a timer receives the signal `Start` it initializes its counter.

Lemma 2: At any instant s for any emitted signal o

$\{\neg$`Start`$\}$ o $\{count(s) \geq count(s+1)\}$

where $s+1$ is the instant following s. This means that the value of *count*, except at initialization time, decreases.

Lemma 3: At any instant s

$\{\neg$`Start`$\}$ `timer_tick` $\{count(s) > count(s+1)\}$

At any instant if the signal `Start` is absent, then the value of *count* strictly decreases provided the signal `timer_tick`, which provokes the execution of the procedure `timer_tick`, is emitted.

Lemma 4: At any instant s
$\{count(s) = 0\}$ `timer_tick` {`Elapsed`}
Any time the value of *count* is zero, then the timer emits `Elapsed`.

These four lemmata are quite easy to prove with PVS. The PVS entry is obtained taking the boolean automaton of the `timer` augmented by the operations provided by its datatype behavior. It follows that if the `Learn_Timer` receives a `Start` it will emit an `Elapsed` provided it emits `timer_tick` a sufficient number of times. The emission of `timer_tick` necessitates the input signal `Clock` as the following formula checked on the abstract timer shows:
AG `timer_tick` $\Rightarrow$ `Clock`.

Thus the timer works correctly if the environment provides a infinite sequence of the signal `Clock`.

Considering the whole system, we proved that four timers are sufficient instead of eight in the original design, and we identified the components that can safely share a timer.

5 Conclusion

We used the methodology presented here to develop industrial products, particularly the transponder lock system discussed in this paper and a mass-flow meter [10]. The transponder lock is a relatively small system. It has 16 classes, 7 of them being reactive, and 35 objects, 16 being reactive. It has been realized using a PIC16C86 micro-controller with a 2MHz machine clock. The model checker SMV generated a state space of reachable 2^{18} states for `RFILock` and checking properties took a few minutes on a SUN workstation. The flexibility given by object oriented techniques has proved to be precious to adapt the standard product to individual customers needs.

Our main interest is not to use formal methods only to show that a final implementation meets an initial formal specification. We use an incremental approach in which formal verification supports design decisions. Since synchronous programs are compiled into boolean automata, proofs refer to the actual implementation and not to a separate model. Proofs are performed for individual objects and for object-configurations of a system and their respective classes. In this way they are modularized. This helps to avoid state explosion, a problem that remains even when using sophisticated BDD packages. This helps also the person in charge of the proof, in particular in the process of understanding counter-examples delivered by the model checker in case a property is not checked as true. To help this part of the verification, we develop a tool to simulate counter-examples on the reactive behaviors of objects.

We have shown the use of formal verification to provide a sound argumentation to support design decisions. This does not exclude the use of simulation, prototyping, testing, etc., as further complementary activities, particularly to explore the general behavior of the system.

Currently we finish the development of *embeddedEifel*, an object-oriented design and implementation language that includes synchronous behavioral descriptions into classes. Formal reasoning techniques like model checking and theorem proving are integrated into a compiler and its tool environment.

References

[1] A. Benveniste and G. Berry. The synchronous approach to reactive and real-time systems. *Proceedings of the IEEE*, 79(9), 1991.

[2] Ivar Jacobson, Magnus Christerson, Parik Jonsson, and Gunnar Övergard. *Object-Oriented Software Engineering A Use Case Driven Approach*. Addison-Wesley, 1992.

[3] Kent Beck and Ward Cunningham. A laboratory for teaching object-oriented thinking. Number 24(10) in SIGPLAN Notices, pages 1–6. SIGPLAN, October 1989.

[4] A. Poigné and L. Holenderski. Boolean automata for implementing ESTEREL. Arbeitspapiere der GMD 964, Forschungszentrum Informationstechnik GmbH, December 1995.

[5] E.A. Emerson and J.Y. Halpern. 'sometimes' and 'not never' revisited: On branching versus linear time temporal logic. *Journal Of the Association for Computing Machinery*, 33(1):151–178, 1986.

[6] A. Merceron. Checking synchronous programs using automatic abstraction, modular verification and assumption discharge. Arbeitspapiere der GMD 972, Forschungszentrum Informationstechnik GmbH, January 1996.

[7] K. L. McMillan. *Symbolic Model Checking: An Approach to the State Explosion Problem*. PhD thesis, Carnegie Mellon, 1992.

[8] Vis user's manual. The VIS Group, December 1995. available in http://www-cad.eecs.berkeley.edu/Respep/Research/vis/doc.

[9] S. Owre, J.M. Rushby, and N. Shankar. PVS: A prototype verification system. In D. Kapur, editor, *11th International Conference on Automated Deduction*, volume 607 of *Lecture Notes in Artificial Intelligence*, pages 748–752. Springer-Verlag, 1992.

[10] R. Budde and K.-H. Sylla. Objektorientierte Echtzeit-Anwendungen auf Grundlage perfekter Synchronisation und: Eingebettete Echtzeitsysteme. *OBJEKTspektrum*, (2 and 4):54–60 and 10–16, 1995.

Acceptance Criteria for Critical Software Based on Testability Estimates and Test Results

Antonia Bertolino
Istituto di Elaborazione della Informazione, CNR
Pisa, Italy

Lorenzo Strigini
Centre for Software Reliability, City University
London, U.K.

Abstract

Testability is defined as the probability that a program will fail a test, conditional on the program containing some fault. In this paper, we show that statements about the testability of a program can be more simply described in terms of assumptions on the probability distribution of the failure intensity of the program. We can thus state general acceptance conditions in clear mathematical terms using Bayesian inference. We develop two scenarios, one for software for which the reliability requirements are that the software must be completely fault-free, and another for requirements stated as an upper bound on the acceptable failure probability.

1. Introduction

The only direct method for predicting the operational reliability of a program is inference from "statistical" testing under an operational input profile [1, 2]. For safety-critical software, acceptance requires a long testing campaign with no failures. However, an amount of operational testing sufficient to warrant a high confidence that the software is as reliable as required in some current applications is infeasible [3, 4, 5].

A way forward is to combine the evidence from testing with any other evidence available. This combination can be made rigorous through *Bayesian* methods, in which the assessor can update the *prior* probability of an event, on the basis of new observed data, to produce a *posterior* probability, representing how the strength of belief allowable in the event taking place varies with new evidence. In particular, the use of prior probabilities in Bayesian reasoning explicitly describes the fact that predictions on the basis of statistical inference must also depend on pre-existing information about the events in question.

When judging on the basis of the results of testing, a kind of clearly helpful information is how effective the testing is at discovering faults. One would think that a series of successes in highly effective tests would give the same confidence as a longer series with less effective tests. A measure of test effectiveness that has gained some popularity is *testability*, the probability of a test detecting a failure conditional on the program being faulty, introduced by Voas and co-authors [6, 7, 8, 9, 10] and proposed as a basis for assessing software. The underlying intuition is that a statement about the internal structure of a program (to the effect that any bugs are likely to produce a high failure rate) allows one to draw stronger conclusions from testing than allowed by black-box considerations alone. In [11], we gave a rigorous, Bayesian inference procedure for obtaining the probability that a program is correct, knowing its testability, the test results, and the prior probability of it being correct. However, in that paper we used a *point* estimate of program testability. This amounts to assuming that, if a program does contain faults, it is bound to have a certain, known probability of failure per execution (failure intensity), which is clearly a simplifying, but unrealistic assumption. In reality, we will instead have at most an understanding of which values of the failure intensity are more or less likely. In this paper, we offer two improvements:

1) we describe testability in terms of the prior distribution of the failure intensity of a program. This yields a prediction method which is more applicable in realistic situations, and eliminates the need to reason with a rather abstruse concept like the probability that a program would fail, if it were possible for it to fail;

2) we show how to use this prediction method when the criterion for accepting a program is either the probability that the program is correct (completely fault-free), or the probability that the program has an acceptable failure intensity in operation.

We consider a scenario in which software undergoes a long series of independent test cases, without failure. This is the typical case of interest for safety-critical software, but the mathematics can easily be extended to the case of any number of observed failures. For reasons of space, we only consider a few examples of simple prior distributions, to illustrate some essential facts about reasoning with testability. The other assumptions are that testing takes place with the operational input distribution, and all and only the actual failures of the program under test are detected (*perfect oracle* assumption).

In Section 2, we introduce the notion of failure intensity as a random variable and its probability distribution. Section 3 deals with the representation of assumptions about testability in terms of this distribution. Sections 4 and 5 describe the use of Bayesian inference in judgement about accepting software, according to the acceptance criteria 1) and 2), respectively, and illustrate the method with numerical examples. Section 6 summarises our results and their possible developments and discusses their practical uses.

2. Distributions of the Failure Intensity

In the assessment of software reliability, uncertainty derives from two sources: we do not know which inputs, if any, will cause the software to fail; and we do not know when and whether such inputs will be presented to the software in operation. It is reasonable to describe this uncertainty by stating that, under a given input profile, a program has a certain *probability* of failure when executed once, called a *failure intensity* (often called a "failure rate"). The failure intensity of a program is uncertain because of our limited knowledge about the program: we thus consider it as a random variable, Θ, with a certain *probability distribution*. A way of picturing this is to think about the program to be assessed as having been "extracted at random" from the population of all the programs that *could* have been produced for the same purpose and under the same known conditions: they have different values of Θ, and the distribution describes the frequencies with which different values of Θ appear in this population. Or, the distribution of Θ can be thought of as representing the degrees of one's beliefs ("subjective probabilities", based on whatever evidence is available) that different possible values are the *actual* value of Θ for *this* program. The latter is the Bayesian interpretation. With Bayesian inference, we represent what we expect about the program before testing it via a prior distribution: we can for instance take into account the reliability levels achieved in past products of the same development process. By applying Bayes' rule, we then obtain a (*posterior*) distribution for Θ which also takes account of test results.

Choosing a prior distribution for Θ is a difficult task. It may be appealing to look for a prior distribution that represents "ignorance" about the failure intensity. However, absolute ignorance cannot be uniquely defined. Any representation of "ignorance" embodies a statement about which events are deemed to be equally likely. Many authors (e.g. [3, 4]), represent "ignorance" via the (mathematically convenient) *uniform* prior:

$$P(\Theta = \vartheta) \equiv 1\,, \quad \text{for all } \vartheta \in [0,1]$$

This means that the true value of Θ is as likely to lie within the interval [0.1, 0.2] as is within [0.2, 0.3], [0.3, 0.4], etc. But one can imagine different forms of ignorance, e.g., one might believe that the failure intensity is as likely to fall in [0.1, 1] as in [0.01, 0.1], as in [0.001, 0.01], etc., and this belief would result in a totally different distribution than the uniform prior. This latter way of dividing the event space seems closer to the usual ways of reasoning about software. Our examples will use a variant of this second form of "ignorance", fitting the assumptions of [6, 8, 9]: Θ may only take values in an interval $[\vartheta_1, \vartheta_N]$, $0<\vartheta_1 < \vartheta_N = 1$, *plus* the isolated value $\vartheta_0=0$ (correct software), and the logarithm of Θ has a uniform distribution over the interval $[\log(\vartheta_1), \log(\vartheta_N)]$.

To avoid mathematical complexity and simplify our explanations, we use discrete approximations to our distributions and numerical computations. We thus represent Θ as a discrete random variable, and describe its distribution via a succession

$$a_i = P(\Theta = \vartheta_i),\ \ i=0, 1, \ldots, N;$$

in particular we define $\vartheta_0 = 0$ and $a_0 = P(\Theta = \vartheta_0 = 0)$.

After observing T successful tests, we obtain, by applying Bayes' theorem, a discrete *posterior* distribution $b_0, b_1, \ldots, b_N$ with:

$$(1) \qquad b_i(T) = \frac{a_i(1-\vartheta_i)^T}{\sum_{i=0}^{N} a_i(1-\vartheta_i)^T}$$

where $(1-\vartheta_i)^T$, the probability that no failures happen in T tests *if* $\Theta=\vartheta_i$, is usually called a *likelihood function*.

3. Representing Testability in terms of the Prior Distribution of the Failure Intensity

We refer to the concept of testability introduced by Voas and co-authors, but use our, more precise definition [11]: the testability of a program is the conditional probability that the program fails a test, on an input randomly drawn from a given input profile, *given* a specified oracle and *given that* the program is faulty.

Testability is a rather un-intuitive concept: if the program contained no faults, then it could not fail. Testability describes how likely it would be to fail, *if* it did contain faults, and is then used to decide how likely the program is *not* to contain faults.

If we consider Θ as a random variable, to represent our uncertainty about its actual value, it becomes natural to see testability as a random variable as well. If we consider a point estimate, τ, for this random variable, we can study the relationship between τ and the distribution of Θ. By definition:

(2) τ = P(failure of a test | the program is faulty),

that is

$$(3)\ \tau = \frac{\text{P(failure of a test AND the program is faulty)}}{\text{P(the program is faulty)}} = \frac{\text{P(failure of a test)}}{\text{P(the program is faulty)}}$$

where the last equality is justified since only faulty programs can fail a test, so that the event "failure of a test" is contained in the event "the program is faulty". In our representation of the prior probability distribution of Θ,

$$(4)\qquad \text{P(the program is faulty)} = 1 - a_0 = \sum_{i=1}^{N} a_i$$

and:

$$(5)\qquad \text{P(failure of a test)} = \text{E}(\Theta) = \sum_{i=0}^{N} a_i\,\vartheta_i = \sum_{i=1}^{N} a_i\,\vartheta_i$$

where E(Θ) represents the expected value, or mean, of the random variable Θ. So,

$$(6)\qquad \tau = \frac{\sum_{i=1}^{N} a_i\,\vartheta_i}{\sum_{i=1}^{N} a_i} = \frac{\sum_{i=1}^{N} a_i\,\vartheta_i}{1 - a_0}$$

Voas and co-authors assumed that a lower bound h can be estimated for the testability of a program, i.e., *if* the program is faulty, then its probability of failing a test is at least h. Our chosen prior distribution satisfies this assumption. By increasing ϑ_1, the lower bound on the values of Θ that have non-zero probability, we can thus study the effect of assuming increasing values for testability (i.e., of the failure intensity of faulty programs).

Notice, however, that events with a zero prior probability also have a zero posterior probability. So, stating that a lower bound exists is a very strong statement about the program, as it cannot be changed by any amount of new evidence. By comparison, we observe that the point estimate of testability, τ, will in general change as the probability distribution for Θ changes with the number of successful tests.

The adopted test method affects testability. For instance, test inputs could be taken from an input distribution different from the operational profile. The coverage of the test oracle could be smaller, (or, conceivably, greater) than 1 (in practical terms, failures during testing could go undetected, or vice versa the oracle could, by monitoring internal program variables, detect erroneous behaviour even when this is not propagated to program outputs). In these cases, the probability of a test failure for those programs which are faulty, i.e., testability, will differ from their operational failure intensity. These scenarios can be modelled by a probability distribution for testability, conditional on Θ. For reasons of space, we will not do so in this paper. Simple examples are shown in [12].

4. Acceptance Based on the Probability of Correctness

The reasoning in [8, 9] and similar papers uses the evidence of successful tests to increase confidence that the software under test is defect-free. In Bayesian terms, this means having a prior probability a_0=P(Θ=0)$\neq$0 (if it were a_0=0, no amount of successful test could produce a posterior $b_0\neq$0), and observing how the corresponding posterior probability increases with the number T of successful tests.

From equation (1):

$$(7)\qquad b_0(T) = \frac{a_0}{\sum_{i=0}^{N} a_i (1-\vartheta_i)^T}$$

A non-negligible prior probability (i.e., belief held before observing any testing) that a program is defect-free is implausible in many fields of application of software. However, it may be plausible for simple software developed under very stringent quality criteria, which includes some safety-critical software.

Estimating the probability of the software being correct, rather than its probability of failure or similar reliability measures, has some important advantages. Firstly, it does not depend on testing under the operational profile, which is difficult to derive, but can use any test profile chosen for its effectiveness in revealing faults. This is important because testing in many organisations is organised on the assumption that other test selection criteria are more efficient than operational profiles for revealing faults (the controversy cannot be settled for lack of conclusive evidence; it is also clear that different answers might be true for different organisations and specific situations). Secondly (as we discuss in detail in [13]), a probability of correctness is a lower bound on the probability of correct behaviour over any arbitrary length of time, while an estimated failure intensity or rate implies progressively less favourable predictions with longer times of operation.

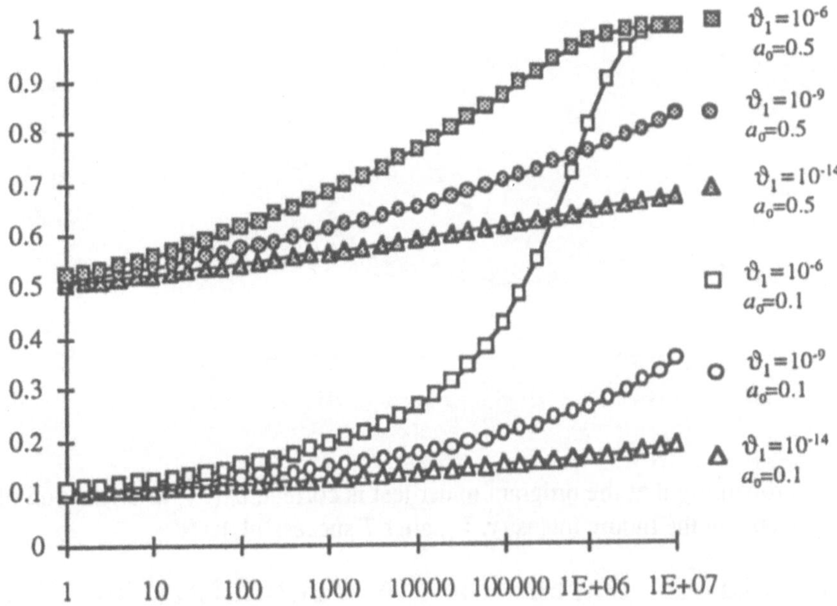

Figure 1 Probability that the program under test is correct, $b_0(T)$, as a function of the number of successful tests, T.

We show in Fig. 1 how the probability of correctness, i.e., the chance that this program is actually one of the perfect ones in our notional population, grows rapidly as we observe successful tests. We show two groups of curves, respectively with a prior probability of correctness equal to 0.1 and 0.5, and within each group we assume different values for ϑ_1, the lower bound on the failure intensity. As expected, the higher the lower bound ϑ_1, the higher the estimated probability of correctness. We can also note that for small numbers of tests the prior probability of correctness greatly affects the posterior; however, as successful tests accumulate, its influence becomes less important: for instance, after 10^6 successful tests, a prior distribution with a_0=0.1 and $\vartheta_1=10^{-6}$ would give a more favourable prediction than a prior starting with a_0=0.5 and $\vartheta_1=10^{-14}$.

In the following figure we also show the posterior probability that the program is correct as a function of ϑ_1, under the assumption that, when we vary ϑ_1, a_0 does not change, and the distribution of Θ when $\Theta \neq 0$ (i.e., for faulty programs) remains uniform between ϑ_1 and 1. We remind the reader that we are assuming that a test input causes the tester to adjudge a test failure if and only if the same input would cause the program to fail in operation. We are thus modelling the effects of the different failure intensities in different programs, not the effects of different degrees of instrumentation of programs to facilitate error detection in testing.

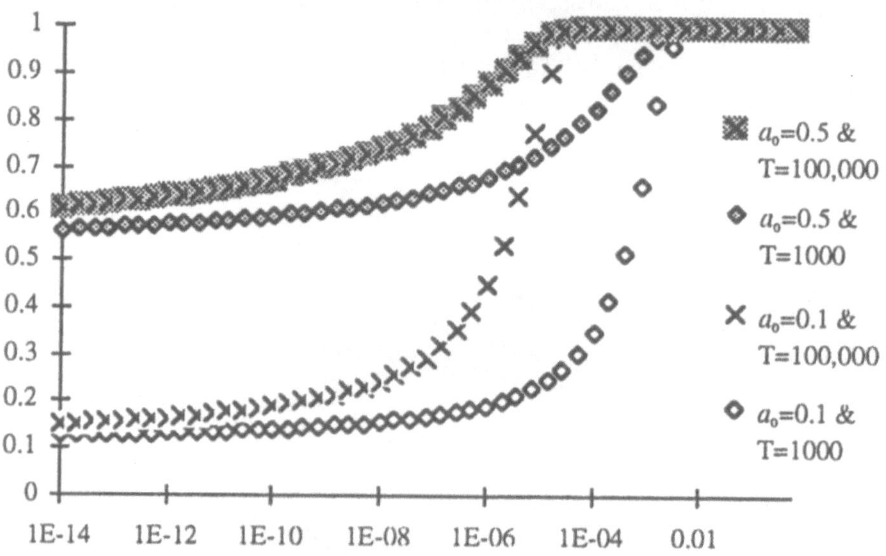

Figure 2 Probability that the program under test is correct, $b_0(T)$, as a function of the lower bound on the failure intensity, ϑ_1, after T successful tests.

5. Acceptance Based on the Probability of Not Exceeding a Given Failure Intensity

Fig. 2 confirms that $b_0(T)$ (for a fixed T) increases with ϑ_1, implying that, out of two programs that pass the same number of tests, the program having a higher testability is more likely to be correct. But this is only one side of the coin. The probability of correctness alone gives us no indication of how likely a program would be to fail, *if* it were faulty. In other words, we may have estimated that our program is very likely to be correct, but if the program happens to be faulty, what risk are we accepting that it is *too* unreliable? In this section, we discuss another acceptance criterion, which seems to apply to most practical situations: a reliability requirement is stated in terms of an allowed upper bound, ϑ_R, on the failure intensity in operation, i.e., it is required that $\Theta \leq \vartheta_R$. So, an appropriate measure on which an acceptance criterion can be based is *the probability that the program satisfies this reliability requirement.* This cannot be 1, since perfect prediction is impossible, but one would reasonably require it to be close to 1 (more or less close depending on the cost of operating unsatisfactory software). Once we have inferred the posterior probability distribution for Θ, the probability of "success", i.e., of the program satisfying its reliability requirement is:

$$(8) \qquad P_{succ}(T) = \sum_{i=0}^{R} b_i(T) = \frac{\sum_{i=0}^{R} a_i (1-\vartheta_i)^T}{\sum_{i=0}^{N} a_i (1-\vartheta_i)^T}$$

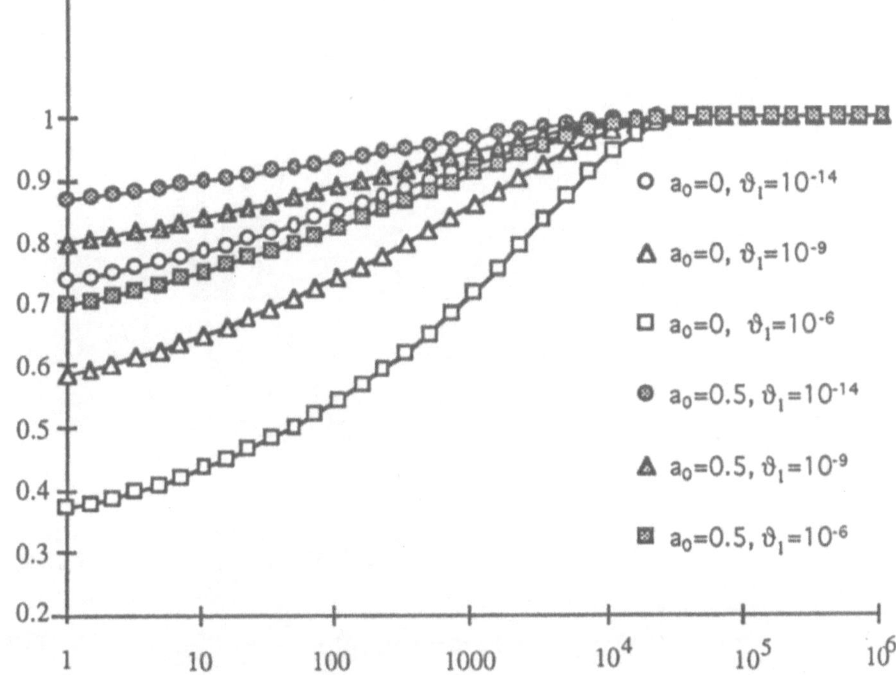

Figure 3. Probability that the failure intensity Θ of the program under test is lower than $\vartheta_R=10^{-4}$, $P_{succ}(T)$, as a function of the number of successful tests, T.

Fig. 3 shows the function $P_{succ}(T)$. Notice that P_{succ} increases with a_0 but decreases with increasing values of the lower bound on Θ, ϑ_1. This is caused by the assumption that the logarithm of Θ is uniformly distributed over $[\log(\vartheta_1), \log(\vartheta_N)]$. Increasing ϑ_1, i.e., assigning a zero probability to a wider range of values of Θ starting from 0, while keeping a_0 constant, means that the remaining possible values have increased probabilities (i.e., the values of the probabilities a_i for $i \geq 1$ are increased). So, increasing ϑ_1, decreases the reliability of the programs and changes the scenario so that the evidence before testing (in particular, the prior probability of "success", $P_{succ}(0)$) is less favourable to the program under test.

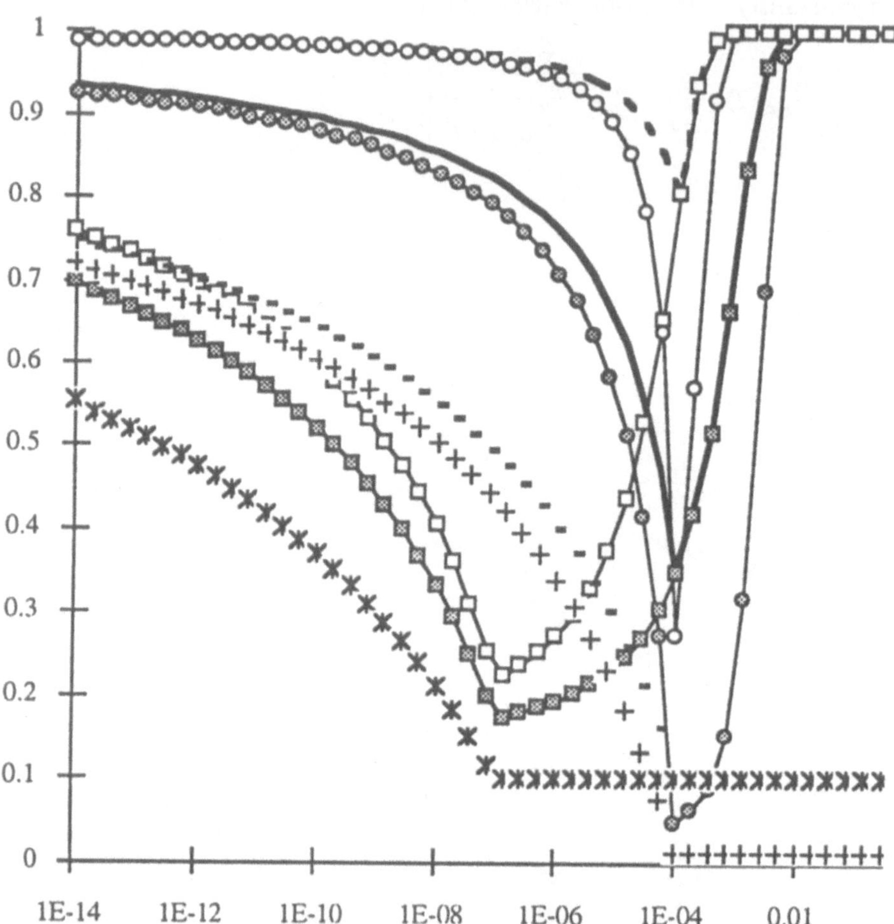

Figure 4 Probability $P_{succ}(T)$ that the failure intensity Θ of the program under test is lower than ϑ_R, as a function of the lower bound on the failure intensity, ϑ_1. The probability of correctness, a_0, is kept constant, and the logarithm of Θ is uniformly distributed between $\log(\vartheta_1)$ and 0. Notice that the curves to the right of the point $\vartheta_1=\vartheta_R$ are just the curves of $b_0=P(\Theta=0)$, since $\vartheta_1>\vartheta_R$ means that all acceptable, non-zero failure intensities have zero probability.

The next two figures offer more insight into the effects of varying ϑ_1. They show $P_{succ}(T)$ as a function of ϑ_1 (as fig. 2 did for $b_0(T)$). However, in Fig 4 the prior distribution of Θ varies with ϑ_1 as we just explained. Fig. 5 studies a completely different scenario: as we increase ϑ_1, we do not change the probability density function for $\Theta>\vartheta_1$, but we increase a_0, the probability that the software is correct, so as to preserve the property $\sum_{i=0}^{N} a_i = 1$. So, Fig. 4 and Fig. 5 represent the reasoning of two assessors with different information. Both know with certainty that there is a lower bound ϑ_1 on the failure intensity of the program, if faulty. However, in Fig. 4 the assessor is considering software for which (s)he has a clear idea of the probability of it being correct, which does not change with different assessments of ϑ_1. The assessor in Fig. 5, on the contrary, knows the distribution of Θ for $\Theta>\vartheta_1$, which does not vary with different assessed values of ϑ_1, and therefore has to change his assessment of a_0 as (s)he varies ϑ_1. In Fig. 4, increasing ϑ_1 causes P_{succ} to plummet, until ϑ_1 exceeds ϑ_R, that is, until one believes that any bug in the software will cause it to be too unreliable. In Fig. 5, increasing ϑ_1 causes P_{succ} to increase slightly ($P_{succ}(0)$ remains constant as ϑ_1 varies), until ϑ_1 reaches the threshold ϑ_R. These are extreme scenarios, of course. In general, we can expect that different situations change all aspects of the distribution of Θ, rather than leaving just some aspects unchanged. However, these figures show how the whole distribution of Θ is relevant, rather than just ϑ_1 or h.

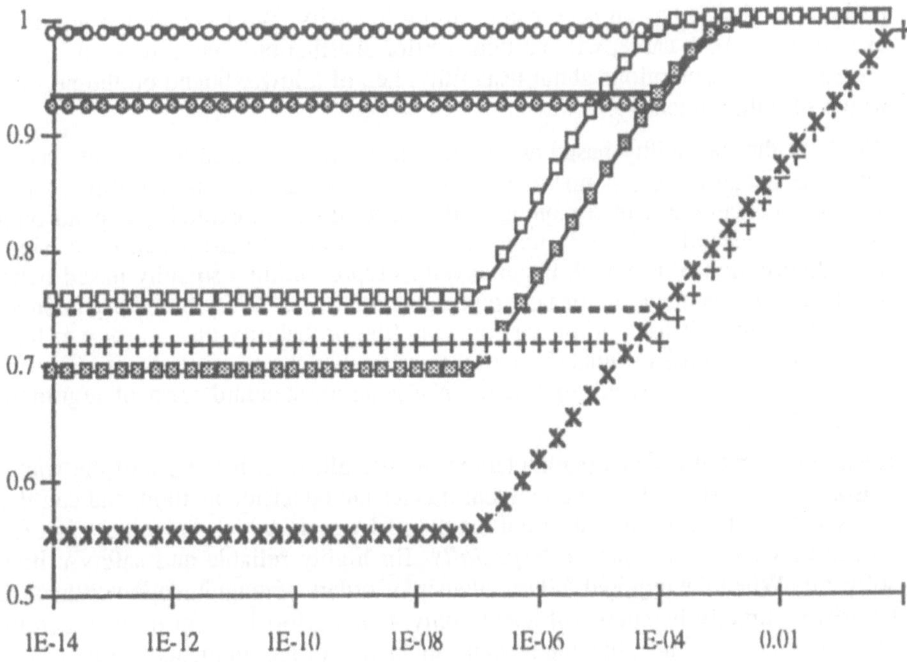

Figure 5 Probability $P_{succ}(T)$ that the program's failure intensity Θ is lower than ϑ_R, as a function of the lower bound on the failure intensity, ϑ_1. The curves are identified by the same symbols as in Fig. 4. However, for each curve, the probability that the software is correct, a_0, starts, for $\vartheta_1=10^{-14}$, at the value shown in the legend of Fig. 4, and then increases as ϑ_1 increases, leaving unchanged the probabilities of the values of Θ greater than ϑ_1.

6. Conclusions

In [11] we reached these conclusions:

- the only ways of increasing testability that are unconditionally beneficial are those that improve one's ability to *detect* undesired behaviour of a program, *without* increasing the likely failure intensity if the program happens to be faulty;
- if program A has higher testability than program B, obtained via a structure that makes faults, if present, more likely to cause failures, and programs A and B pass the same number of tests, this does *not* indicate that program A is a better program than B.

We have now also given precise quantitative expressions of what can be inferred from evidence on testability. In comparison with previous work, we have shown that testability is indeed an interesting measure for an assessor, but is only one aspect of our possible knowledge on a program, and using it without considering other aspects of the prior distribution of Θ may be misleading.

We have instead shown that knowledge about program testability can be integrated in reasoning based on the prior distribution of Θ. In particular, we have shown (equations (7) and (8)) how to compute two specific measures corresponding to reasonable acceptance criteria - the probability that the software is correct, and the probability that it has an acceptable failure intensity. We have shown numerical examples for both cases, with particular prior distributions, and discussed some of the effects of assumptions about testability, i.e., of a lower bound on the non-zero values of failure intensity.

We think that testability-based arguments cannot now be used to evaluate critical software, because there is no trustworthy way of measuring the testability of the programs concerned. Our reasoning in this paper does not explicitly depend on an estimate of testability. Still, in many cases one would be hard pressed to assign a prior distribution of Θ which is perceived as representing a soundly based belief. So, the main advantage of our new representation of testability-based arguments is that it is natural: it uses primitive concepts (the probability that a program has a certain failure intensity) which are easier to grasp than the concept of testability, and subsumes testability-based arguments in a general, standard form of arguments using test results [3, 4, 12].

If estimating the important parameters is so difficult, what is the use of the way of reasoning we described? In the fact that it is a sound decision method, and can thus be used as a check on the informal, haphazard way in which judgement is often passed in software acceptance, *especially* for highly reliable and safety-critical software. When the required failure intensity is orders of magnitude lower than can be proven directly by statistical testing only, much effort is spent in documenting other forms of evidence, like the formal satisfaction of requirements on the form of documentation, the thoroughness of debug testing, and such. All this evidence is then used to reach a decision in an informal, un-auditable, intuitive way, subject to all the common fallacies of intuitive judgement [14, 15]. The method we have described allows checks of the form "After T tests, I wish to be 99 % sure that the software is satisfactory. What kind of prior distribution of Θ would warrant that conclusion? Is this distribution at all plausible, in view of the existing evidence?". Likewise, one can describe a distribution that appears plausible, and then check whether the desired conclusions are sensitive to minor changes in that distribution:

acceptance based on such grounds would appear unsound.

In addition, this kind of reasoning may help with design and project management decisions (e.g., regarding program structure or testing regimes, respectively), because it may show how alternative options, by affecting the knowledge available about the program, would help or hinder the final assessment and acceptance of the program.

Further development of this work are under way in several directions:

- acceptance criteria based on the probability of failure per execution ("failure intensity") can easily be substituted by criteria based on the probability of failure over the operational life of the software, or other periods of operation, as in [16];
- different scenarios can be studied as to the relationship between the probability of failure in operation and the probability of test failure, to take into account factors like imperfect oracle coverage, stress testing, etc. For initial examples of these extensions, the reader is referred to [12].

Acknowledgements

The authors wish to thank Peter Bishop, whose criticism of the use of point estimates for testability prompted them to start this work. This research was funded in part by the European Commission via the "OLOS" research network (Contract CHRX-CT94-0577) and the ESPRIT Long Term Research Project 20072 "DeVa".

References

[1] Musa JD. Operational profiles in software-reliability engineering. IEEE Software 1993; March: 14-32.

[2] Parnas DL, van Schouwen AJ, Kwan SP. Evaluation of safety-critical software. Communications of the ACM 1990; 33: 636-648.

[3] Miller KW, Morell LJ, Noonan RE, et al. Estimating the probability of failure when testing reveals no failures. IEEE Transactions on Software Engineering 1992; 18: 33-43.

[4] Littlewood B, Strigini L. Validation of ultra-high dependability for software-based systems. Communications of the ACM 1993; 36: 69-80.

[5] Butler RW, Finelli GB. The infeasibility of experimental quantification of life-critical software reliability. In Proc. ACM Conference on Software for Critical Systems, in ACM SIGSOFT Software Eng. Notes, Vol. 16 (5). New Orleans, Louisiana, 1991, pp 66-76.

[6] Hamlet D, Voas J. Faults on its sleeve: amplifying software reliability testing. In Proc. 1993 Int. Symposium on Software Testing and Analysis (ISSTA), in ACM SIGSOFT Software Eng. Notes, Vol. 18 (3). Cambridge, Massachusetts, U.S.A., 1993, pp 89-98.

[7] Voas JM, Miller KW. Improving the software development process using testability research. In Proc. of the Third Int. Symposium on Software Reliability Engineering. 1992, pp 114-121.

[8] Voas JM, Michael CC, Miller KW. Confidently assessing a zero probability of software failure. In Proc. SAFECOMP '93 12th International Conference on Computer Safety, Reliability and Security. Poznan-Kiekrz, Poland, 1993, pp 197-206.

[9] Voas JM, Michael CC, Miller KW. Confidently assessing a zero probability of software failure. High Integrity Systems 1995; 1: 269-275.

[10] Voas JM, Miller KW. Software testability: The new verification. IEEE Software 1995; May: 17-28.

[11] Bertolino A, Strigini L. On the use of testability measures for dependability assessment. IEEE Transactions on Software Engineering 1996; 22: 97-108.

[12] Bertolino A, Strigini L. Predicting software reliability from testing taking into account other knowledge about a program. In Proc. Quality Week '96. San Francisco, 1996.

[13] Bertolino A, Strigini L. Is it more convenient to assess a probability of failure or of correctness? Submitted for publication 1996.

[14] Kahnemann D, Slovic P, Tversky A (ed). Judgment under uncertainty: heuristics and biases. Cambridge University Press, 1982.

[15] Strigini L. Engineering judgement in reliability and safety and its limits: what can we learn from research in psychology? SHIP project Technical Report T/030, July, 1994.

[16] Littlewood B, Wright D. On a stopping rule for the operational testing of safety critical software. In Proc. FTCS25 (25th Annual International Symposium on Fault -Tolerant Computing). Pasadena, 1995, pp 444-451.

Developing Dependable Software Using Prototyping and Test-Diversity

W. Kuhn, H. Selami
Austrian Research Centre Seibersdorf
Department of Information Technology
A-2444 Seibersdorf, AUSTRIA

Abstract

Software diversity has been used in many safety-related applications such as flight, nuclear power and railway applications. It has been an important area of research in the recent past too. The benefits and disadvantages of the software diversity are known and have been considered in many publications. But how can we gain the benefits of software diversity, if we want to eliminate the disadvantages of it? In this paper we present a new method of software development and testing, which uses the advantages of software diversity and eliminates the disadvantages of it. This approach can be used to achieve the needed software dependability in safety-related computer systems.

1 Introduction

Dependable software is an essential part of a dependable computer system. The term *Dependability* is defined in [1] as *„the property of a computer system such that reliance can justifiably be placed on the service it delivers"*. This definition is based on the assumption, that the system (and software) requirements are completely and well defined so that the delivered service of it can be evaluated. In the rest of this paper we emphasis on the software aspects only.

One of the most important steps in the software development is to define the user requirements as exactly as possible. After having defined the requirements completely, development can be continued with the aim that the produced software fulfills its required service reliably.

Because of the nature of software faults, which are all permanent development faults (including specification faults), most effort to achieve software dependability should be devoted in the development phases of the software life cycle. The means for dependability are generally described in [1] to be: *fault forecasting, fault prevention, fault removal* and *fault tolerance*.

Among above dependability means the first three (fault forecasting, prevention and removal) can be applied in the development phases of the software life cycle. Software faults should be prevented and removed as soon as possible. Early removal of software faults is very important, because the cost of fault removal increases rapidly in the later phases of the software life cycle. Unremoved software faults are delivered together with the software to the field and exist as dormant faults in the operation phase. Software fault removal during operation is more difficult and expensive.

The 4th dependability mean mentioned above, namely fault tolerance, can be applied to software using **software diversity** [2][3][4]. Software diversity is a known method for detecting software faults during development and operation of a computer system. From the same initial specification, two (or more) software versions are produced independently by different teams in different development environments (different operating systems, programming languages, compilers, development tools, testing methods and tools etc. [5]). These diverse software versions are functionally equivalent according to the same specification. In the field the diverse versions are executed concurrently and the final output values are determined using a decision algorithm based on comparing of parallely generated outputs. Using a voting mechanism with at least 3 diverse versions, software faults can be also tolerated under certain conditions.

The advantage of software diversity is, that low probability of identical software faults existing in diverse versions causes a high rate of error detection in the field. Besides this positive aspect software diversity has several disadvantages too:

- This kind of on-line error detection mechanism implies a run-time system overhead, caused by permanent comparison of results of diverse programs.
- Each decision must be delayed until all outputs are generated. The performance of the whole system is reduced to the slowest one among all diverse versions.
- Specification faults can not be detected, if the same technical specification document is used for all diverse programs.
- Needing several high quality software versions causes high development and maintenance costs.
- The decision mechanism adds more complexity to the system.

But how could we use the benefits of software diversity without having these disadvantages? We discuss the idea of test-diversity in the next section. Section 3 presents the complete software development model using prototyping and test-diversity. Section 4 explains our experiences with the test-diversity. Section 5 summarizes our main results and discusses future work.

2 Diversity for Testing

The effectiveness of „back-to-back" testing of diverse programs against each other during the testing phase of software development has been evaluated and reported in several publications [2][3]. The reported results show generally that this kind of testing has many advantages including high fault detection rate and simplification of testing with large amounts of test-data. As stated in [6] the verification effort of a diverse system may be significantly lower than a single system.

According to this positive results we propose the approach of **diversity for testing**, which tries to use the advantages of software diversity during the test phase of software development. The main idea is to have software diversity in the development phases but not in the operation phase of the software life cycle. *Test-diversity* is the term we use for this kind of software diversity. To do this we implement one main version of the software and one (or more) diverse test version(s) of it. The main version will run in the field and must be of a highest quality. The test versions are used only for test-diversity and can have the quality of a prototype. The outputs of the main and test versions with the same test data are compared during testing. Software faults can be detected in case of inequality of the outputs.

Figures 1 and 2 show the approach of test-diversity with only one test-version. Comparing to the software diversity approach, we see that the parallel execution of diverse software versions are moved from the operation to the development phases of software life cycle. All fault detection and fault removal activities are done in the test phase (Fig. 2), which is more compliant with the nature of software faults as mentioned above. Having just one version running in the field, eliminates the permanent on-line overhead of the software diversity and all other comparison problems related to it (see the disadvantages above).

The task of comparing the results and evaluation of differences is very (time- and safety-) critical, if it is performed during operation of the software. In the case of test-diversity this task is done in the test phase and therefore can be performed more precisely and extensively.

To gain the whole benefits of the test-diversity, the following necessary conditions must be true:

- The guidelines for development of diverse versions must be followed correctly. It means that the main version and the test version(s) must be diverse in any case.
- The task of the test case specification and test data generation and test running must be done systematically and a high level of test coverage must be achieved. We will address this point later in section 3.

As we see in the Fig. 1 the same specification is used for both main and test versions. There is still the disadvantage that certain types of specification faults can not be detected during testing. But if we use different ways for specifying the software functions of the main and the test version, we could be able to detect the specification faults.

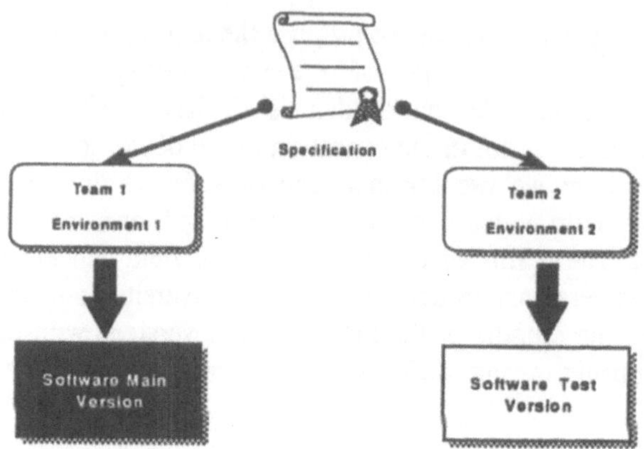

Figure 1: test-diversity, implementation

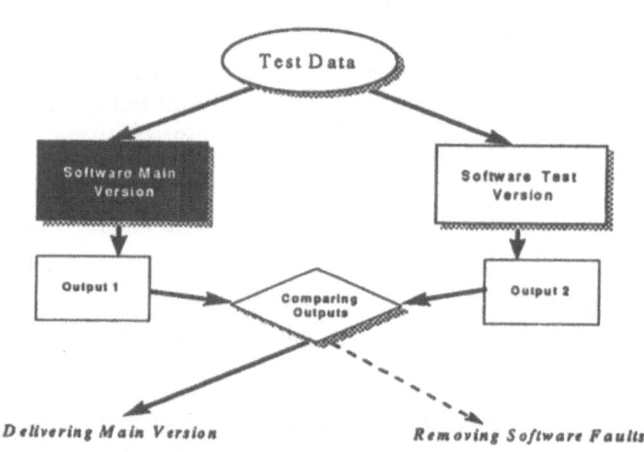

Figure 2: test-diversity, testing

In the next section we present a complete software development model, which combines prototyping and test-diversity to use the benefits of both approaches in a single development model and to resolve the problem of specification faults as described above. However the test-diversity can be used without prototyping at each level of the software structure. This means, test-diversity can be used at procedure-, module- or system-level of the software. It is also possible to use the test-diversity only for some critical parts of the software.

3 The Software Development Model

Figure 3 gives an overview of our software development model. **Prototyping** [7] is used as the very first step of the software development for requirement engineering. Prototyping is a well known requirement engineering method, which gives the user the possibility to experiment with a prototype and helps to fix the requirements completely and clearly.

In the prototyping process three groups of persons are involved:

1. Customer
2. Prototyping team
3. Specification authors

The customer is actively involved to define the requirements. The prototyping team develops the prototype and supports the customer to *play* with the prototype and complete the requirements. The specification authors create the specification document, which is the basis of the development of the main software version. The results of this phase are the specification document and the prototype itself.

The next step is to develop the software main version by the development team according to the specification document.

The prototype and the main version should be developed by different teams using diverse methods. The prototype can therefore be used as the test version for test-diversity in the test phase. After testing and fault removal the main version of the software is delivered to the field.

With this model the advantages of software diversity are used in the development phases of the software life cycle. At the same time, the disadvantages of the software diversity are removed as described in the following:

- No run-time comparison overhead in field: Software-diversity is used only in development phases. In the field there is just one software version. This eliminates the permanent comparison overhead at run-time.
- Detection of specification faults: The specification document is used only for the main version of the software and is not the basis of the implementation of the prototype (test-version).
- Cost effectiveness: The same prototype is used two times in two development phases: requirement engineering and testing.

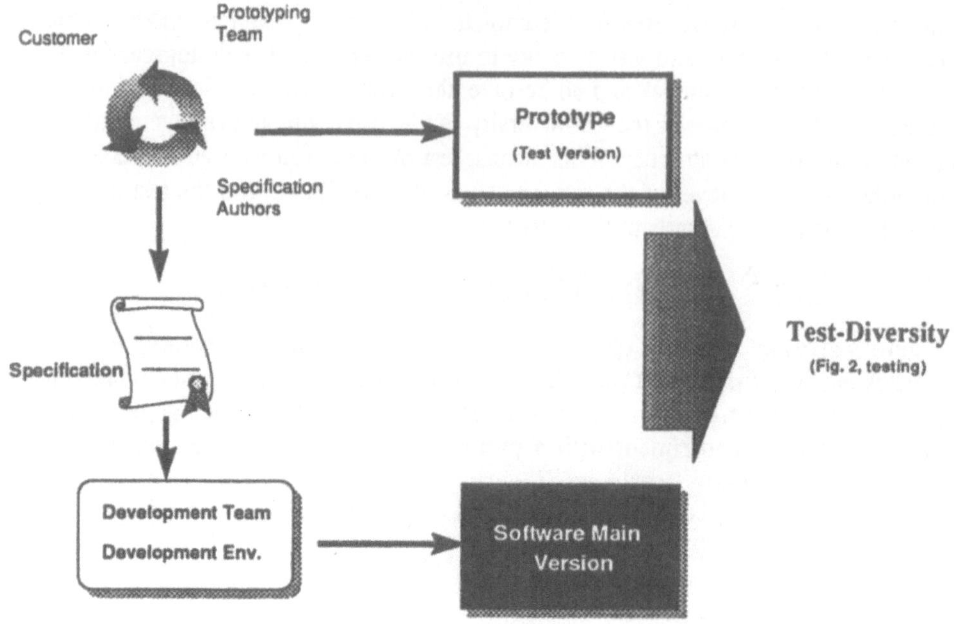

Figure 3 : the software development model

It should be considered that test-diversity is a method for fault prevention and removal and not for fault tolerance. The consequence of this is that no error-masking can be performed during operation of the software. On the other hand, to be able to tolerate one fault at least 3 independent software versions are needed. The high development and maintenance cost of such a computer system makes it necessary to think about alternative solutions in many cases, where the costs of advantages of a diverse system can not be justified.

3.1 Testing

We considered that software faults are all development (including specification) faults, which exist in the software before delivering it to the field. Therefore the very important task of testing is done in the last phase of software development to find as many faults as possible. Test-diversity can help us to do this. During a systematic test several activities should be done:

- test case specification
- test data generation
- expected results prediction

- evaluation of test results

For the first two activities (*test case specification* and *test data generation*) there are methods and tools, which enable us to perform these systematically (the partition testing methods are good examples). However for the rest of the above activities (*expected results prediction* and *evaluation of test results*) there has only been poor methodical support. The test-diversity approach supports exactly these parts of testing. The prediction of expected results is highly simplified using the following rule: „The outputs of the main and the test versions to the test-data must be logically equal".

Testing with a high coverage level usually produces very large amounts of test results, which makes the evaluation of the test results more expensive and complicated. But in case of using test-diversity when we compare the results of the main and the test-versions, we need just to evaluate the differences of them. A difference indicates a possible fault either in the main version or in the test version.

Our experiences show that the differences of the test results, which must be carefully evaluated, occur in only small parts of the whole test results. This simplifies the task of evaluation of the test results.

4 Our Experiences

4.1 Using Test-Diversity in CSS

We have used the test-diversity in our distributed security, alarm and control system PHILIPS-CSS (Scaleable Security System) [8]. Like many other real-time systems, the CSS software consists of a set of processes, which communicate with each other via messages. One of the most important components of CSS is the database-manager which is the only way for all other CSS-Processes for accessing the CSS-database. Figure 4 shows the position of the database-manager in CSS.

To achieve the needed high reliability for such a central component we decided to test the CSS database-manager by applying the test-diversity. We implemented a new main-version database-manager using SQL Module Language [9]. Then we used the original database-manager (with embedded RDML [10]) as the test-version. The diversity of both versions has been achieved through implementation by different persons and different methods. We then compared the results of both the main- and test-versions using the same test data. Most of the differences found led us to find software-faults either in the main-version or in the test-version.

We used the „Classification-Tree Method" [11] for test case specification and test data generation. The test runs of the both versions has been executed sequentially.

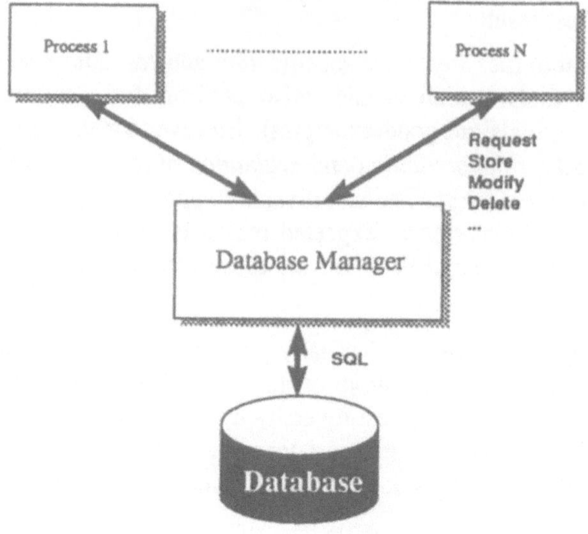

Figure 4 : CSS Database-Manager

For each version there are two kinds of test results:

1. The responses of the database manager to the requesting processes.
2. The contents of the database after test runs.

Both groups of the test results have been saved separately for each version in text-files, each about 40000 lines! After each complete test run the result-files have been compared to find the differences.

It is clear that the evaluation of such a huge volume of test results is impractical in case of a single version. Using the test-diversity and during comparison of result-files only the different parts (together about 500 lines) have been considered and evaluated manually. After finding and removing software faults, the test procedure has been repeated until no differences have been found anymore.

There have been two software faults found in the old database manager, which were really hard-to-detect considering the fact that the old database manager had been running properly in the field for a long time. About 20 faults have been detected in the new (main) version of the database manager through test-diversity.

The new database manager has been running successfully on 9 CSS-installations for several months now (altogether 25 system months).

5 Summary

We considered the nature of software faults and the effectiveness of software diversity for fault detection and proposed the approach of diversity for testing (test-diversity), which is the use of software diversity in the development phases of the software life cycle only. We discussed also the advantages and consequences of having no software diversity in the field. Combining with prototyping we presented a complete software development model, which uses the benefits of both prototyping and test-diversity in a single development model. We explained our experiences with the test-diversity, which show the successful applying of it in the Philips security system CSS.

In the future, the practical aspects of the proposed software development model should be more investigated. More work should be done also on the estimation of remaining faults in the delivered main software version. These and other related questions can be handled within the scope of a research-project, which applies the whole development model from the beginning to the end.

Acknowledgments

We would like to thank our colleagues at the research center Seibersdorf, specially Michael Thuswald and Gert Ernst, for their helpful comments on the earlier drafts of this paper.

References

1. Laprie J.C. Dependability of Computing Systems: Concepts and Terminology, Proc. 25th IEEE International Symposiom on Fault-Tolerant Computing, Pasadena, CA, USA June 27-30 1995
2. Voges U. Software Diversity in Computerized Control Systems ISBN 3-211-82014-0 1988 Springer-Verlag Wien-New York
3. Lyu M. R. Software Fault Tolerance, ISBN 0-471-95068-8 1995 John Wiley & Sons Ltd.
4. Lyu M. R., Avizienis A., Assuring Design Diversity in N-Version Software: A Design Paradigm for N-Version Programming, Dependable Computing for Critical Applications 2, pp. 197-218, ISBN 3-211-82330-1 Springer-Verlag Wien-New York
5. Schoitsch E., The Interaction Between Practical Experience, Standardization and the Application of Standards, SAFECOMP '89, Proceedings of the IFAC/IFIP Workshop, Vienna, Austria, 5-7 December 1989
6. Ehrenberger W., Saglietti F., Architecture and Safety Qualification of Large Software Systems, ESREL '93, Proceedings of the European Safety and Reliability Conference, Munich, Germany, May 10-12 1993
7. Lantz K., The Prototyping Methodology, Prentice-Hall 1986
8. Schoitsch E., Kuhn W., Herzner W., Thuswald M. Experiences in Design and Development of a highly Dependable and Scaleable Distributed Security, Alarm

and Control System, SAFECOMP '91, Proceedings of the IFAC/IFIP/EWICS/SRE Symposium, Trondheim, Norway, 30 Oct. - 1 Nov. 1991

9. DEC Rdb SQL Reference Manual, February 1994
10. DEC Rdb RDML Reference Manual, December 1990
11. Grochtmann M., Grimm K., Classification Trees for Partition Testing, Software testing, Verification and Reliability, Vol 3, PP: 63-82, 1993

Software Reliability Models and Test Coverage.

Bruno Ciciani

University of Rome
"La Sapienza"
Via Salaria 113
00198 Roma, ITALY

Alberto Pasquini

ENEA
(sp 088)
Via Anguillarese 301
00060 Roma, ITALY

Abstract

In recent works some authors proposed to use measures of test coverage in software reliability estimation. They suggested to use these measures either to improve the predictive accuracy of classical reliability growth models or to provide a direct estimation of reliability. This paper provides a short survey of these approaches and describes a case study aimed at investigating the relation between test coverage and reliability growth. Results of the case study are analysed and used to discuss the validity of the proposed approaches.

1 Introduction

Reliability is one of the main attributes characterising the quality of software. Standards of increasing importance [1] emphasise the relevance of this attribute during the evaluation of software and suggest means for its estimation. Software reliability is usually defined as the probability that software will operate without failure, under stated conditions, for a stated amount of time. Then, one of the most direct measure of software reliability is the software failure rate. This can be estimated using reliability growth models. During the testing phase of software the cumulative number of failures occurred till the current time is plotted versus execution time. A reliability growth model is then statistically fitted to the plot. The model curve can be used to estimate the cumulative number of failure at the desired future time.

The present estimation accuracy of these models is sufficient to support project management in keeping under control the software development process [2], they can be useful in detecting project anomalies or to fix stopping rules for the testing phase [2, 3]. But, it is not sufficient to provide the high confidence required to check

---

This work has been partially supported by the European Union under the Programme "Human Capital and Mobility", Network OLOS, Contract: ERBCHRXCT 940577.

if the reliability level specified in the requirements has been reached, or to use the reliability figure during the assessment of software used in safety critical software (that is in application in which a software failure could lead to severe consequences in terms of human life, environmental damages or economical losses). The unsuitability of these models for the reliability estimation of software for critical applications is analysed in detail in [4, 5, 6].

To improve the predictive quality of existing reliability growth models some authors [7, 8], suggested to combine the different predictions obtained from different models. They proposed different criteria to weight the predictions statically or dinamically and in some cases [7] developed tools to automate models selection and combine their predictions. Other authors suggested the use of re-calibration techniques [9, 10] that improve the predictive quality of reliability models by letting them "learn" from their past errors.

These last approaches are similar to those used for VLSI device manufacturing defect level estimation and for hardware fault rate identification proposed in [11, 12, 13]. In particular, in case of hardware fault rate identification the system is supposed to be tested periodically during its operational existence. The value of a test is used not only to determine, with a certain confidence, whether a unit under test is good or not, but it is also used to elicit information about the failure characteristics of that unit. This information, indeed, is used to refine the knowledge of the failure process, i.e., to reduce the variance of the distribution of its parameters. The length of a test is determined starting by an a-priori distribution, which encapsulates the knowledge of the failure characteristics available prior to a particular test. When this next test is completed, and new additional information are available, the prior distribution is modified, and an a-posteriori distribution is derived. This now becomes the a-priori distribution for a next subsequent test. This process can cycle indefinitely. This methodology is based on statistical decision theory referenced in [14, 15].

A completely different approach in software reliability modelling is based on the use of additional information that can be collected during the testing phase. Some authors are suggesting to use test coverage measures either to improve the performances of conventional reliability growth models or to develop new ones. In Section 2, this paper provides a short a survey of these research activities. In Section 3 the paper presents the results of a case study aimed at investigating the relation between test coverage and reliability and discuss the opportunity of using coverage measures in reliability modelling.

2 Coverage Measures in Reliability Models

One of the causes of the low predictive quality of classical reliability growth models is that they consider only few of the information that can be collected during testing and from the code observation. Their predictions are based on: operational profile, time between successive failures or inherent fault density and fault detection rate. Some other such as code complexity and test coverage are not taken into account even if they are known to heavily affect reliability [16, 17]. The reason is that these models use a black box approach to reliability modelling. They only try to simulate the shape of a mathematical function, that is, the shape of the function that represents the reliability growth. They disregard the reason of the phenomenon represented by that function.

Some of the more recent works in the area of software reliability modelling are trying to solve this problem by considering the relation between test coverage and reliability. There is a general agreement on the assumption of a strong correlation between test coverage and reliability, but there is only little quantitative support for this assumption. Veevers and Marshall in [18] tried to define an ideal test coverage metric and proposed to use this metric to measure reliability. Let:

r(c) = reliability of a system after the execution (and the associated fault removal) of a set of test cases that reached the coverage of the program input space denoted by c;

They assume that, under a specific testing regime, an increase in coverage of dc produces, on average, a corresponding increase in reliability proportional to the current amount of unreliability. Then:

(1) $r(c+dc) = r(c)+p \cdot dc \cdot (1-r(c))$

where the proportionality factor $p>0$ indicates the strength of the relationship. They show that (1) leads to:

(2) $r(c) = 1-k \cdot \exp(-p \cdot c)$

where $0 \leq k \leq 1$ is a constant determined by:

(3) $k = 1 - r(c_0)$.

Veevers and Marshall provided also a version of (3) that is applicable in case of discrete input spaces. While their approach seems theoretically sound it works considering the coverage of the input space of software that is extremely hard to measure in practical cases. Then (2) seems to be of little help in real applications.

One of the first research efforts to use coverage measures in reliability modelling was by Jacoby and Masuzawa [19]. They used a classical reliability growth model, namely their Hyper-Geometric Distribution Model (HGDM). This model requires a parameter called "ease of test" function that represents the total number of faults discovered and rediscovered at a test instance. They expressed the "ease to test" function as a function of the test coverage for the software under test and represented the real observed test coverage with a linear function. A linear function does not seem to reflect the real evolution of coverage during testing and corresponds neither to the results of our case study shown in the next Section nor to a more general experiment of a similar type [20]. Jacoby and Masuzawa are now looking at other, more flexible functions.

Piwowarsky and others proposed in [21] a simple reliability growth model that formulates the relationship between increase of statement coverage and decrease of remaining errors in a product. The model was not designed for predicting the reliability of software but rather as a framework for explaining the error removal ratio throughout function test of a real software products at IBM.

Malaiya and others [22] proposed a completely different model, which relates test coverage measures directly with defect coverage and then reliability. They provide the following three parameters model for defect coverage C^0:

(4) $$C^0 = a_0^i \ln[1 + a_1^i \exp(a_2^i C^i) - 1]$$

where C^i is the coverage obtained in terms of a specific coverage metric i and a_0^i, a_1^i, a_2^i are parameters estimated collecting data during part of the testing process. Since the failure intensity is proportional to the remaining number of defects:

(5) $$\lambda = \frac{K}{T_L} N$$

where K is the overall value of fault exposure ratio, T_L is the linear execution time, and N is the number of faults that can be computed as:

(6) $$N = N_0(1 - C^0)$$

then the expected duration between two successive failures can be obtained as:

(7) $$\frac{1}{\lambda} = \frac{T_L}{K} \frac{1}{N_0\{1 - a_0^i \ln[1 + a_1^i(\exp(a_2^i C^i) - 1]\}}$$

where K depends on the operational profile used. Coverage C^i can be measured or estimated as a logarithmic function of the number of test cases executed. Some initial validation of this model have been tried and are described in [22] together with preliminary examples of parameters estimations.

Chen and others [23] suggest to use coverage information to filter the data to be provided as input to reliability growth models. They exclude test cases that do not increase coverage, claiming that this approach eliminates the overestimation of traditional reliability growth models due to the saturation effect of testing strategies.

All the listed approaches consider information concerning test coverage and use them either to improve the predictive accuracy of reliability growth models or to provide a direct estimation of reliability. Unfortunately, none of them has been validated through an extensive field use and both their practical applicability and predictive quality need to be confirmed by real applications. An additional problems is the lack of data concerning the relationship between test coverage and reliability growth. A case study was conducted, as a part of a more general experiment [20], to give some new information about this relation.

3 Case Study

Testing is the execution of a program P in a controlled and systematic way. An operational profile OP is estimated for P. Then, P is exercised with a set of test cases TC generated randomly on the basis of OP. Faults detected during this testing activity are removed. The fault removal process usually continues after the release of P, during its operation ad maintenance when faults are detected through its operative usage. Thus, P evolves, during its lifecycle, toward its correct version P_c. During testing, the interfailure data are recorded and the analysis of these data allow to estimate the reliability of P for the given OP. The actual reliability of P could be measured in terms of probability of failure per execution using the equation [24]:

(8) $$R(P, OP) = 1 - \lim_{n_t \to \infty} \frac{n_f}{n_t}$$

where:

$n_f =$ number of executions of input cases generated randomly on the basis of OP and for which P and P_c had different results

$n_t =$ total number of execution of input cases generated randomly on the basis of OP

During the development process, P_c is not available and therefore the actual reliability can never be measured. But, the current software engineering techniques, and, in particular, the use of configuration management techniques and tools, allow to keep track of the evolution of a software program during its lifecicle and of the faults that have been found during its testing and operational usage.

The approach followed in this case study is to use a program P that has reached the version P_c and for which all the faults have been removed and recorded. We put back in P_c all these faults obtaining P. Then we started a new fault removal process measuring the increase in coverage and having P_c available. Whenever a fault was detected we stopped testing, removed the fault detected and measured the reliability of P using equation (8). This allowed us to measure the actual reliability growth of the software and compare it with the coverage.

The program chosen for case study provides a language-oriented user interface which can allow the user to describe the configuration of array of antennas using a high level language. It was developed for the European Space Agency (ESA) in C language and consists of almost 10.000 lines of code (6.100 executables) divided into several modules. During the integration testing and operative usage of the program, 33 faults were discovered and recorded. We used this program as P_c in our case study.

Tests cases were generated randomly on the basis of OP. The operational profile has been defined in [25] as the set of the occurrence probabilities of the software functions. This case study used the same definition. To estimate an operational profile OP, the possible sub-functions of the program were identified. Transition probabilities between sub-functions were determined by interviewing the program users. The set of test cases TC was obtained generating test cases randomly on the basis of OP. The case study was performed using an automated toolset. This toolset identified and removed the faults from the faulty versions of the program by running the set of test cases TC and using P_c as an oracle. The code coverage was measured using the tool ATAC [26]. This tool provides measures of blocks, decisions, c and p-uses coverage. The execution of TC led to the identification of 28 faults. The others 5 faults remained undiscovered.

Interfailure data were also used to estimate the reliability of the software with classical reliability growth models. We used six between the more common existing models. They are listed in tab. 1 together with some measures of the accuracy and of the goodness-to-fit of their prediction.

The estimation provided by the two more accurate models (Littlewood and Verrall model and Geometric model) are plotted in fig. 1, together with the actual reliability of the program. The models estimation accuracy for the failure data available seems to be strongly related with the number of test cases executed. Models show an acceptable predictive accuracy after the execution of a number of test cases ranging between $5*10^2$ and $5*10^3$. Reliability growth has a characteristic "S" shape and may occasionally decrease during testing. Since no new faults are introduced during our fault removal process, decrease is due to fault masking [20]. Reliability does not increase at the beginning of testing because some faults are overlapping. They affects a large part of the input space of software and the same input space may be affected, contemporaneously, by more than one of them. Reliability does not increase till when all the faults affecting the same input space are removed. This behaviour is strongly dependent on the characteristics and size of software and faults and seems to be not strictly related with coverage growth.

Figure 2 shows the growth of coverage with the execution of test cases. Coverage is measured in terms of Blocks, Decisions and All-uses and is compared with the growth in reliability. It is possible to see that there is not a linear relation between coverage and reliability even if they both tend towards an asymptote after the execution of a large number of test cases. As for the reliability growth models correlation seems to be more clear after the execution of a number of test cases ranging between $5*10^2$ and $5*10^3$.

Model	Accur.	Bias	Noise	Trend	Goodness to fit		
					Kolmo. Distance	χ-square Statistic	Sign. at .05
Geometric model	78.69	.54333	7.2834	.30127	0.09166		no
Littlewood & Verrall quadr.	83.323	.59371	3.687	.31591	0.12478		no
Musa basic time model	N. C.	N. C.	N. C.	N. C.	0.43031		yes
Musa log Poisson model	N. C.	N. C.	N. C.	N. C.	0.13543		no
Brooks & Motley binom.	124.95	N. A.	N. A.	N. A.		26.83	yes
Brooks & Motley Poisson	24.667	N. A.	N. A.	N. A.		1.827	no
N. A. = Statistics not available for interval data analysis							
N. C. = No convergence for these data and model							

Table 1 - Predictive accuracy of six reliability growth models applied to the data obtained from the case study.

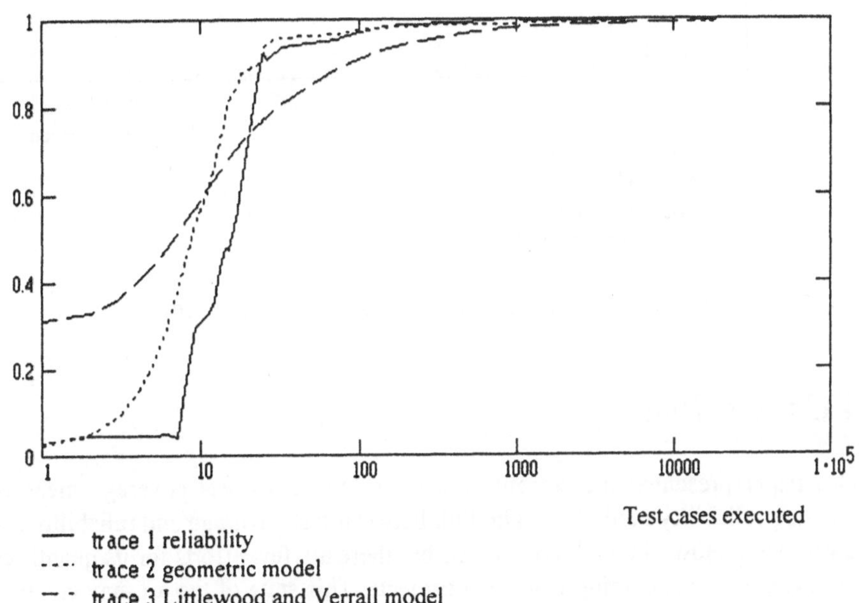

Figure 1 - Estimates provided by the Geometric and Littlewood and Verrall models compared with the actual reliability of software.

The absence of a clear correlation raises some doubts about the possibility of predicting reliability only on the basis of coverage measures, or of using a linear function to model the evolution of coverage with the execution of test cases. No clear answer are possible for the approaches based on filtering the data to be provided as input to reliability growth models or on integrating coverage measures in classical reliability growth models. However, these observations are based on the results of a single case study that is influenced by several parameters such as: code size, type of application, type of faults. For this reason results cannot be easily generalised and much more data are needed to reach a general conclusion.

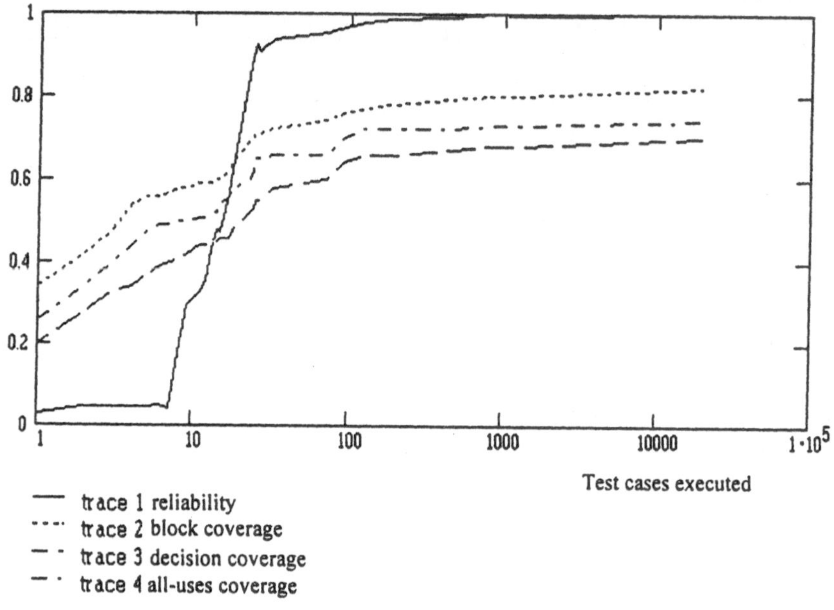

Fig. 2 - Test coverage compared with the actual reliability of software.

4 Conclusions

This paper presented the current approaches for using test coverage measures in software reliability modelling. The link between test coverage and reliability growth is very well known in qualitative terms but there are few efforts for its quantification and even less data coming from real projects. This state of the art may create some problems for the validation of the reliability models that use coverage measures. As a matter of fact the existing ones have not yet been adequately validated. This paper reports the results of a case study that shows a clear positive correlation between increase in coverage and growth in reliability, but, evidences that this relation is not linear and that it may be affected by the characteristics and the size of the software and of the faults under examination.

References

[1] IEEE 982.1 - 1988, Standard Dictionary of Measures to Produce Reliable Software.

[2] Musa J. D., Iannino A., Okumoto K., Software Reliability - Measurement, Prediction, Application, McGraw-Hill, 1987.

[3] Bishop P. (ed.), Prediction and measurement of software reliability, Dependability of Critical Computer Systems 3, Elsevier Science Publ., 1990.

[4] Butler R. W., Finelli G. B., The unfeasibility of Experimental Quantification of Life-Critical Software Reliability, Proc. of the ACM SIGSOFT '91, New Orleans, Louisiana, ACM Press, 1991.

[5] Keiller P. A., Miller D. R., On the use and the performance of software reliability growth models, Reliability Eng. and System Safety, 95-117, 1991.

[6] Littlewood B., Strigini L., Validation of ultra-high dependability for software-based systems, Communication of the ACM, Vol. 36, No. 11, Nov. 1993.

[7] Lyu M. R., Nikora A., Applying Reliability Models More Effectively, IEEE Software, Vol. 9, no. 4, July 1992.

[8] Lu M., Brocklehurst S., Littlewood B., Combination of Predictions Obtained from Different Software Reliability Growth Models, Proc. of the Tenth Software Reliability Symposium, Storage Tek, 1992.

[9] Brocklehurst S., Littlewood B., New Ways to Get Accurate Reliability Measures, IEEE Software, Vol. 9, no. 4, July 1992.

[10] Brocklehurst S., Chan P. Y., Littlewood B., Recalibrating Software Reliability Models, IEEE Trans. on Software Eng., Vol. SE-16, no. 4, 1990.

[11] Jone W. B., Defect Level Estimation of Circuit Testing Using Sequential Statistical Analysis, IEEE Trans. on Computer Aided Design of Integrated Circuits and Systems, Vol. 12, No. 2, February 1993, pp. 336-348.

[12] Yu P. S., Krishna C. M., Lee Y. H., An Adaptive Optimisation Model with Applications to Testing, Computer Performance and Reliability, Iazeolla G., Courtois P. J., Boxma O. J. (ed.), Elsevier Science Publishers, 1988, pp. 503-515.

[13] Yu P. S., Krishna C. M., Lee Y. H., Optimal Design and Sequential Testing of VLSI Testing Strategy, IEEE Trans. on Computer, Vol. 37, No. 3, March 1988, pp. 339-347.

[14] De Grood M. H., Optimal Statistical Decisions, McGraw-Hill, 1970.

[15] Fu K. S., Sequential Methods in Pattern Recognition and Machine Learning, Academic, New York, 1968.

[16] Hennell M. A., Testing for the Achievement of Software Reliability, Reliability Engineering and System Safety, Vol. 32, pp. 119-134, 1991.

[17] Munson J. C., Khoshgoftaar T. M., The Use of Software Complexity Metrics in Software Reliability Modelling, Proc. of the ISSRE '91, IEEE Computer Society, 1991.

[18] Veevers A., Marshall A. C., A Relationship Between Software Coverage Metrics and Reliability, Software Testing, Verification and Reliability, Vol. 4, 3-8, 1994.

[19] Jacoby R., Masuzawa K., Test Coverage Dependent Software Reliability Estimation by the HGD Model, Proc. of the ISSRE '92, IEEE Computer Society, 1992.

[20] Del Frate F., Garg P., Mathur A. P., Pasquini A., On the Correlation between Code Coverage and Software Reliability, Proc. of the ISSRE '95, IEEE Computer Society, 1995.

[21] Piwowarski P., Ohba M., Caruso J., Coverage Measurement Experience During Function Test, Proc. of the ICSE '93, IEEE Computer Society, 1993.

[22] Malaiya Y. K., Li N., Bieman J., Karcich R., Skibbe B., The Relationship between Test Coverage and Reliability, Proc. of the ISSRE '94, IEEE Computer Society, 1994.

[23] Chen M., Horgan J., Mathur A. P., Rego V., A time/structure based model for estimating software reliability, SERC-TR-117-P, Purdue University, Dec. 1992.

[24] Nelson E., Estimating Software Reliability from Test Data, Microelectronics and Reliability, Vol. 17, Pergamon Press, New York, 1978.

[25] Musa J. D., Operational Profiles in Software Reliability Engineering, IEEE Software, Vol. 10, no. 2, March 1993.

[26] Horgan J. R., Mathur A. P., Assessing Tools in Research and Education, IEEE Software, May 192, pp.61-69.

Defining the Unit Testing Level of Synchronous Data Flow Programs

Pascale Thévenod-Fosse[1], Christine Mazuet[2] and Yves Crouzet[1]

[1] LAAS-CNRS
7, avenue du Colonel Roche
31077 Toulouse Cedex, FRANCE

[2] Schneider Electric
SES/ST/ETS
38050 Grenoble Cedex 9, FRANCE

1 Introduction

Synchronous data flow languages are languages designed for programming reactive and real-time systems, and especially those having safety-critical requirements [1-3]. Software testing is a gradual process which begins with unit testing, followed by integration testing and system testing. Through the example of the language Lustre [4], the paper focuses on unit testing of synchronous data flow programs. To define a unit testing approach, two issues must be addressed: (i) how to design unit test cases and, (ii) given a software how to define the unit level i.e., the set of modules (called nodes in Lustre) which have to be the target of unit testing. In previous work concentrated on the first issue, a testing strategy has been defined and experimented with [5]. Based on these results, in this paper interest is focused on the second issue.

The proposed approach follows a bottom-up exploration of the program call graph. It aims to minimize the test effort by: (i) gathering as many nodes as possible to test them all together and, (ii) avoiding redundant testing of nodes called by different nodes. Based on both guidelines, we define a three-step algorithm and exemplified it on a case study (21 nodes). The associated test patterns have been generated and mutation analysis [6] was conducted to assess their efficiency.

Section 2 briefly recalls the method for determining test patterns previously defined for Lustre nodes. Based on this background, Section 3 presents the algorithm for defining the unit testing level. Section 4 focuses on the experimental results. The basic principles underlying the extension of the method to deal with the definition of integration testing levels are described in Section 5.

2 Background

In [5], the criterion proposed as a guide for determining test cases is the coverage of the transitions of the finite state automaton produced by the Lustre compiler (Section 2.1). Then, the method for generating test inputs proceeds according to two complementary principles: probabilistic and deterministic (Section 2.2).

2.1 The Lustre Automaton

Detailed presentations of the language Lustre have been published elsewhere (e.g., [3, 4]). Only the elements necessary for understanding the paper are quoted below.

A Lustre program has a cyclic behavior, and that cycle defines a sequence of times. Any variable or expression denotes a flow which is a pair made of a possibly infinite sequence of values, and a sequence of times related to the cyclic behavior of the program. A program is structured into nodes (subroutines). A node contains a set of equations and any variable which is not an input parameter has to be defined by a single equation, e.g., the equation "X = Y+Z;" means that at each time t, $X_t = Y_t + Z_t$. In addition to classical operators (arithmetic, boolean, relational, conditional), temporal operators are available, e.g., "pre" (delay) and "->" (initialization): "X = pre(Y)" means that at each time $t \neq 0$, $X_t = Y_{t-1}$ and X_0 has an undefined value; "X = Z -> Y" means that $X_0 = Z_0$ and $X_t = Y_t$ for any $t > 0$. For example, the equation "Nat = 0 -> pre(Nat)+1" defines the series of natural integers.

The Lustre compiler produces sequential code. First, it expands recursively all the nodes called by the program so that the code generation step starts with a node which does not call any node. Then it may perform an automaton-like compilation, based on an analysis of the internal states of the program. For this, a set of state variables of boolean type, whose values induce the code to be executed in future cycles, is chosen: this set identifies the program state, and the code to be executed during a cycle is specified for each possible value of the state. The result is a finite state automaton in which: (i) each state is characterized by a combination of values of state variables; (ii) each transition is labelled by a combination of input or internal values which defines the condition that triggers it; a transition consists in executing a linear piece of code and corresponds to an elementary reaction of the program. It is worth noting that the compiler may introduce an initial state in the automaton: this state corresponds to the very first execution of the program, during which the variables are set to their initial values defined by the operator "->".

This automaton is expressed in a format code common with other synchronous languages (e.g., Signal, Esterel) [2]. A code generator takes this format as input and produces a program in the C language.

2.2 Design of Statistical and Deterministic Test Inputs

2.2.1 Principle

The testing strategy previously defined [5] combines statistical testing with deterministic testing of special values. Statistical testing is based on an unusual definition of random testing [7, 8]: it involves exercising a program with inputs that are generated randomly, but not naively according to a uniform distribution over the input domain (which is a more conventional random testing approach [9]). Indeed, the aim of statistical testing is to provide a "balanced" coverage of a model of the target software, no part of the model being seldom or never exercised during testing; and for this, the input distribution used to generate the test patterns is deduced from a probabilistic analysis of a software model. In cases of Lustre programs, the model retained is the finite state automaton produced by the compiler: the statistical test inputs thus defined allow us to provide a stringent coverage of the automaton during testing (Section 2.2.2), yet without tracking any special values inferred from the automaton. As a result, additional test inputs are deterministically selected to ensure the coverage of special values which involve (i) the automaton initial state — if any

— and its output transitions and, (ii) boundary values related to the conditions that trigger the automaton transitions. First experimental results showed the high fault revealing power of the mixed strategy which makes use of the complementarity between statistical and deterministic test cases [5]. Due to the lack of space, we just outline below the method for designing statistical test patterns.

2.2.2 *Statistical Test Patterns*

To provide a balanced coverage of a software model, the general method for generating statistical test patterns takes advantage of the information provided by a test criterion C, which defines a set S_C of elements (of the target model) to be exercised during testing. The test patterns are then defined by two parameters, which are determined according to the criterion retained: (i) the test profile, or input probability distribution, from which the patterns are randomly drawn and, (ii) the test size, or equivalently the number of input patterns (i.e., of program executions) that must be generated.

Search for a test profile. In order to decrease the test size N it is necessary to ensure that each element of S_C will be rapidly exercised when finally a test set is generated from the test profile. Therefore, the probability of exercising the least likely element of S_C must be as high as possible; the optimum profile to minimize N would be the one where all elements have the same probability of being exercised, but it may not always be possible to find that optimum profile. In practice, the search for a proper test profile may proceed either analytically or empirically [8].

Assessment of the test size N. The test size N must fit an objective of test quality with respect to (wrt) the elements of S_C, the test quality (denoted q_N) being the probability that the least likely element of S_C be exercised at least once during N program executions with random inputs [7]. Let P be the probability that a random input activates the least likely element under the test profile retained. N and q_N are linked by the relation: $(1\text{-}P)^N \geq 1\text{-}q_N$. Hence, relation (1) gives the minimum test size required to reach a target test quality q_N:

$$N_{min} = \ln(1-q_N) / \ln(1-P) \qquad (1)$$

The result of this is that, on average, each element is exercised several times. For example, $q_N = 0.9999$ means that, during N_{min} executions: (i) the probability that the least likely element be never exercised is only 10^{-4} and, (ii) on average this element will be exercised 9 times.

Application to Lustre programs. The criterion retained is the coverage of the Lustre automaton transitions [5]. Then, the influence of the input distribution on the transition probabilities can be studied on the stochastic graph obtained by replacing the input conditions that label the transitions with their occurrence probabilities in the distribution. But, relation (1) used to assess N_{min} assumes that the probabilistic analysis is conducted on a strongly connected graph; and Lustre automata do not always depict strongly connected graphs because of the initial state that may be introduced by the compiler (Section 2.1). As a result, statistical testing focuses on all the program states and transitions, except the initial state and its output transitions that will be tested in a deterministic way (Section 2.2.1): the probabilistic analysis is then conducted on the stochastic graph obtained after deletion of the initial state; this graph, denoted S-graph, is strongly connected. Starting from S_C = {S-graph transitions}, the test profile is analytically derived: for each transition k, the expression of its occurrence probability p_k in steady-state conditions, as function of the input probabilities, is deduced from the S-graph. Then, the test profile is determined from the set of equations $\{p_k\}$ in order to maximize $P = \min\{p_k\}$.

3 Definition of the Unit Testing Level

The method for selecting the nodes which have to be the target of unit testing aims to minimize the test effort by: (i) gathering as many nodes as possible to test them all together and, (ii) avoiding redundant testing of nodes called by different nodes. Point (i) induces the bottom-up exploration of the call graph from its terminal nodes, and the exploration can go on until the complexity of the observed node is no longer tractable wrt the probabilistic analysis conducted for designing statistical test patterns. Here, a node complexity will then be expressed in terms of the transition number of its S-graph. In practice, when a S-graph possesses more than about 30 transitions, the node will not belong to the unit level and the exploration of the corresponding branch in the call graph will be stopped. Beyond this order of complexity, the assessment of the transition probabilities becomes difficult and, even if a test profile is derived, a prohibitive test size could be required to reach a high test quality due to the large number of transitions to be activated under a single distribution. Point (ii) will require a specific algorithm for multiple occurrence nodes.

Let UNIT be the set of nodes which have to be the target of unit testing. The algorithm for defining UNIT is based on the program call graph completed with the complexity measures of the observed nodes. It consists of three steps: 1) initialization, 2) treatment of multiple occurrence nodes and, 3) treatment of single occurrence nodes. These steps are described below and exemplified on a case study. The algorithm assumes that all the terminal nodes are tractable wrt the unit testing strategy; if this hypothesis is not verified, the probabilistic analysis cannot be conducted on some S-graphs and the associated nodes have then to be tested according to another criterion (this problem being out of the scope of the paper).

3.1 Step 1: Initialization

The program call graph is built and completed with the numbers of transitions of the S-graphs associated with the terminal nodes. A partition of the terminal nodes into two subsets T0 and T1 is defined: T0 contains the terminal nodes that occur more than once in the call graph, and T1 contains the other ones. Initially, UNIT is empty (ϕ).

Example: the program SWITCH. The program is extracted from a fault-tolerant monitoring system which regulates the outputs of eight gates or burners in a thermal power station. It belongs to that part of the system which controls analogical signals coming from two Programmable Logic Controllers (PLC) in standby redundancy. Each PLC provides 8 analogical signals, one specific to each gate or burner. The program SWITCH processes one pair of signals related to a same gate or burner. Its informal specification is detailed in [10]. It is made of 21 Lustre nodes. Figure 1a shows its call graph completed with the transition numbers of the S-graphs of the 11 terminal nodes: calc is the only terminal node with multiple occurrences.

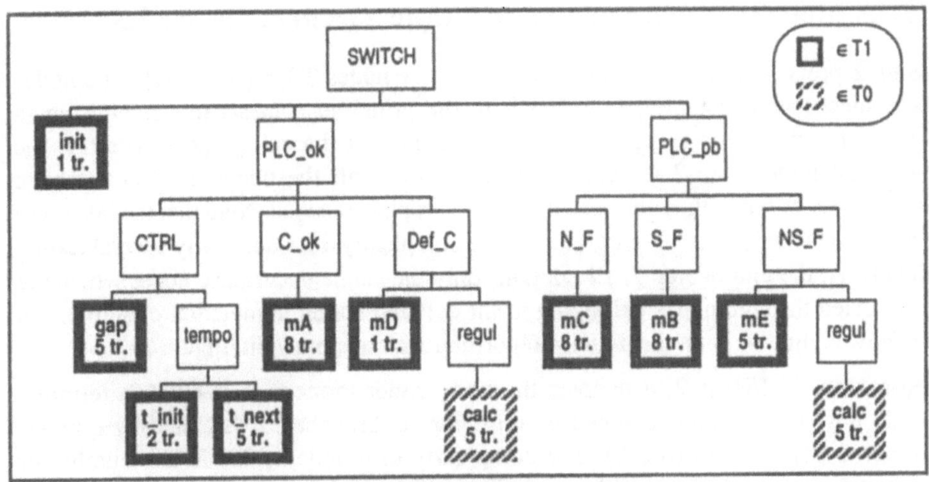

(a) After completion of Step 1: UNIT = ϕ

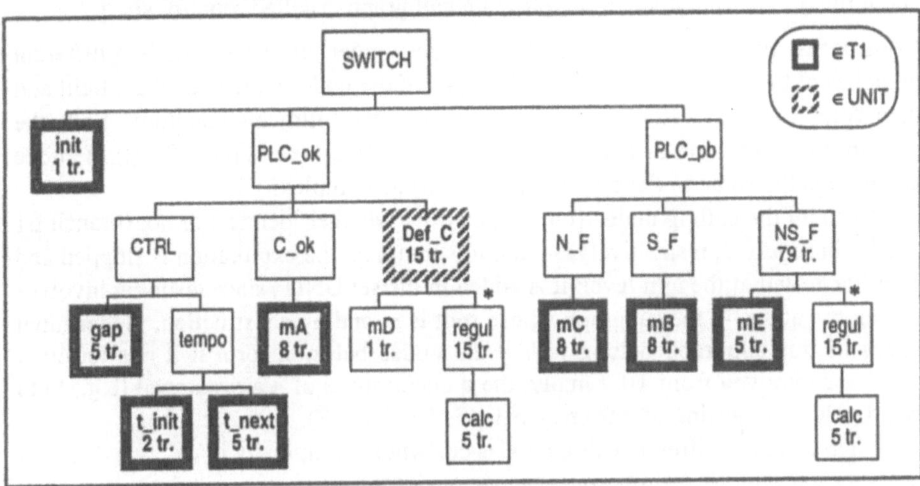

(b) After completion of Step 2: T0 = ϕ

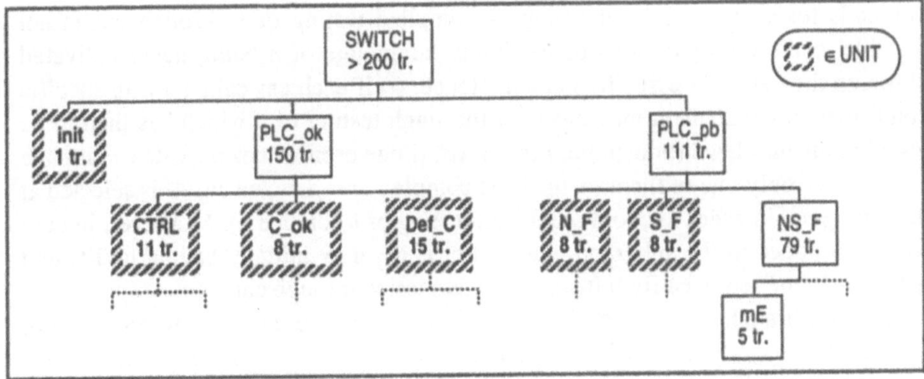

(c) After completion of Step 3: T1 = ϕ

Figure 1. Call graph of the program SWITCH (tr.: transitions)

3.2 Step 2: Treatment of Multiple Occurrence Nodes

Step 2 processes the set of multiple occurrence nodes $T0 = \{t_1, ..., t_p\}$. The nodes are successively examined: for each t_i, the principle consists in climbing up in parallel and as high as possible all the branches of the call graph whose t_i is the terminal node. Step 2 is completed either when all the nodes of T0 have been processed, or when the root node of the call graph is reached. Note that the later case corresponds only to small programs whose complexity does not justify several testing levels. At the end of step 2, T1 contains only the single occurrence nodes which are not tested through nodes belonging to the set UNIT after completion of step 2. Let us now comment upon the detailed algorithm which is drawn in Figure 2.

Notation. In Figure 2, *n* denotes the node under process: it is either a terminal node $t_i \in T0$, or a node reached by climbing up branches. It is called by q nodes denoted $cn_1, ..., cn_q$ ($q \geq 2$). The complexity of a node x, that is, the number of transitions of its S-graph, is denoted *C*(x): it is 'ok' when its value does not exceed about thirty. The root node of the program call graph is called root, for short.

Comments. By construction, a node *n* (≠ root) under process occurs in q different branches of the call graph. First, the S-graphs of the nodes which call *n* are built and their complexity are calculated. Then, while the two following conditions hold, the bottom-up exploration of the q branches goes on (loop α in Figure 2): (i) the node calling *n* is the same in the q branches and, (ii) its complexity is 'ok'.

When all the calling nodes $\{cn_1, ..., cn_q\}$ — whether identical or not (branch β1 and β2 in Figure 2, respectively) — are too complex, the exploration is stopped and *n* will be tested at the unit level: it is added in the set UNIT; since testing *n* involves testing the nodes in the subgraph whose root is *n*, and by construction, the terminal nodes of this subgraph belong to T0, the nodes below *n* (or *n* if it is a terminal node) are removed from T0. Finally, the q occurrences of *n* are stamped (e.g., *) to avoid redundant testing of *n* from step 3 (see Section 3.3).

When there are different calling nodes cn_k whose complexity is 'ok', two different cases are possible depending on whether or not the nodes cn_k call for *n* in special conditions that exclude some points of the input domain of *n*. Indeed, when a node cn_k calls for *n* in special conditions, a complete testing of *n* through cn_k is not feasible, some configurations of the input parameters of *n* being never activated although they are processed by node *n*. Hence: (i) if each cn_k calls for *n* in specific conditions, none of them may ensure a thorough testing of *n* which has then to be tested at the unit level (branch β3, Fig. 2); (ii) if one or more cn_k provide a complete testing of *n*, only one of them — the least complex one, denoted cn_j — is selected at the unit level, in order to avoid redundant testing of *n* (branch δ). Moreover, in case (ii) some nodes in the subgraph whose root is cn_j may initially belong to T1, and have now to be removed from it since they are tested through cn_j.

Finally, branches ε1 and ε2 in Figure 2 correspond to the cases where the root node of the program is tractable at the unit level. As a result, all the program nodes are tested through the root and step 3 of the algorithm becomes useless ($T1 = \phi$).

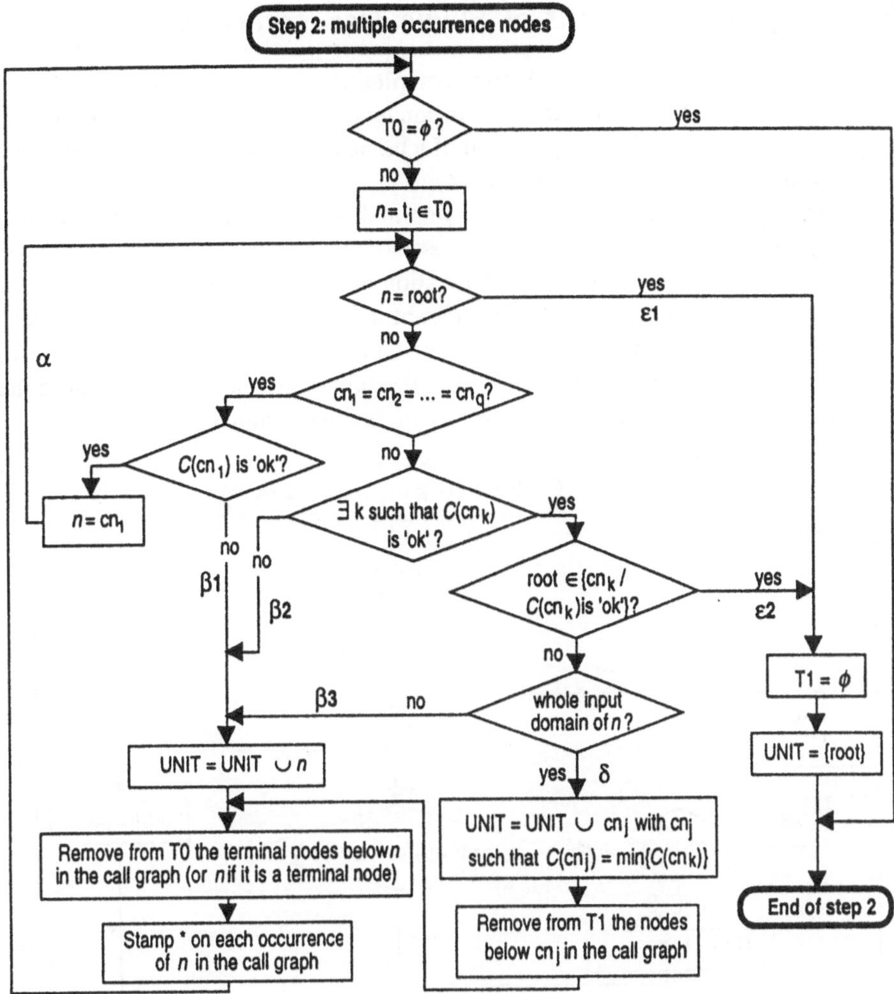

Figure 2. Processing of the terminal nodes with multiple occurrence (node set T0)

Example: the program SWITCH. The bottom-up exploration is performed in parallel in the two branches whose terminal node is calc. The algorithm works as follows (see Fig. 1a, 1b):

- n = calc; both occurrences of n are called by the same node regul, whose S-graph possesses only 15 transitions. The exploration goes on (loop α in Figure 2).
- n = regul; 2 nodes call n: Def_C (15 transitions) and NS_F (79 transitions). Only C(Def_C) is 'ok' and Def_C calls for n with input parameters that do not exclude any point of the input domain of n. Hence, Def_C is retained at the unit level (branch δ, Fig. 2); mD which belongs to the subgraph whose root is cn_j = Def_C, is removed from T1; calc which belongs to the subgraph whose root is n = regul, is removed from T0; both occurrences of regul are labelled * to point out that they are tested (through Def_C).
- T0 = ϕ; Step 2 ends with UNIT = {Def_C}, T1 = {init, gap, t_init, t_next, mA, mC, mB, mE} as shown in Figure 1b.

3.3 Step 3: Treatment of Single Occurrence Nodes

Step 3 processes the nodes ∈ T1 after completion of step 2. Figure 3 gives the algorithm. The bottom-up exploration is conducted on a single branch of the call graph at the same time. The exploration of a branch goes on (loop α, Fig. 3) until:

- either the root node is reached (branch ε);
- or, the subgraph whose root is the calling node cn contains a node X labelled * during step 2 (branch δ); then, cn is not retained at the unit level to avoid redundant testing of X; this means that the verification of the subgraph whose root is *n* is postponed to integration testing since part of it is already tested at the unit level;
- or, the calling node cn is too complex (branch β): *n* will be tested at the unit level (*n* ∈ UNIT); since nodes which belong to the subgraph whose root is *n* are tested through *n*, terminal nodes below *n* (or *n*) are removed from T1.

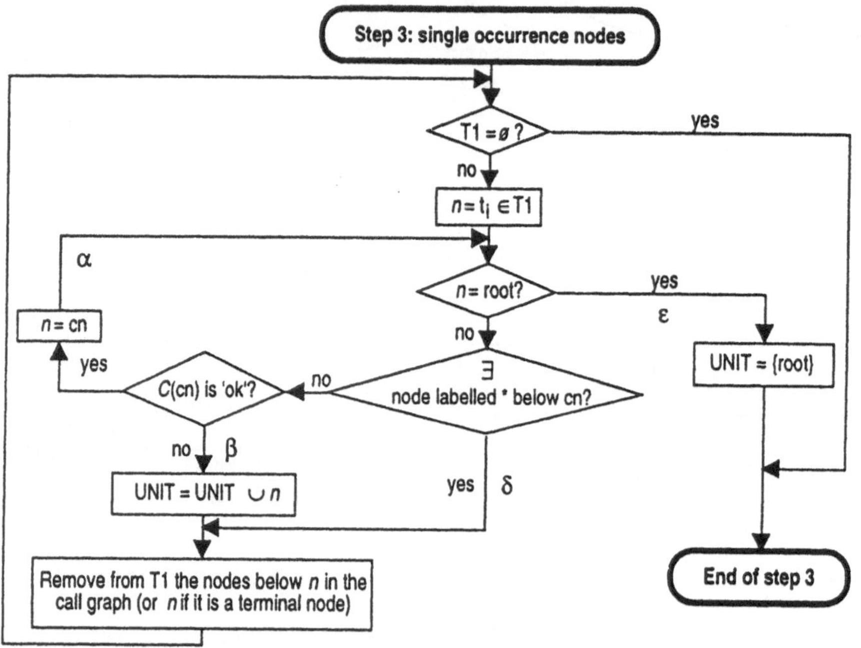

Figure 3. Processing of the terminal nodes with single occurrence (node set T1)

Example: the program SWITCH. Step 3 proceeds as follows (Fig. 1b, 1c):

- *n* = init; *C*(cn = SWITCH) > 200 ⇒ β: init ∈ UNIT and init is removed from T1.
- *n* = gap; *C*(cn = CTRL) = 11 ⇒ α: *n* = CTRL, *C*(cn = PLC_ok) = 150 ⇒ β: CTRL ∈ UNIT and {gap, t_init, t_next} are removed from T1.
- *n* = mA; *C*(cn = C_ok) = 8 ⇒ α: *n* = C_ok, *C*(cn = PLC_ok) = 150 ⇒ β: C_ok ∈ UNIT and mA is removed from T1.
- *n* = mC; *C*(cn = N_F) = 8 ⇒ α: *n* = N_F, *C*(cn = PLC_pb) = 111 ⇒ β: N_F ∈ UNIT and mC is removed from T1.
- in the same way, *n* = mB leads to: S_F ∈ UNIT and mB is removed from T1.

- n = mE; the subgraph whose root is cn = NS_F contains the node regul which is labelled * (tested through Def_C from step 2) ⇒ δ: mE is removed from T1, i.e. the processing of mE is postponed to the integration phase where it could be possible to test it through NS_F (see Section 5).
- T1 = ϕ; Step 3 ends with UNIT = {init, CTRL, C_ok, Def_C, N_F, S_F} as shown in Figure 1c; the subgraphs which are no more drawn in Figure 1c are tested through the nodes ∈ UNIT; five nodes are not tested at the unit level, namely: {mE, NS_F, PLC_pb, PLC_ok, SWITCH}.

Based on these results, unit testing experiments involving mutation analysis, have been conducted to assess the efficiency of test patterns generated according to the mixed strategy outlined in Section 2.2. They are described below.

4 Experimental Study

4.1 Mutation Analysis

A mutation is a single-point, syntactically correct change, introduced in the program P to be tested [6]. A mutant of P consists of a program which differ from P in containing one mutation from a list of faults representative of the most likely faults introduced by programmers using the language of P. A mutant is killed (i.e., distinguished from P) by a set T of test patterns if its output history differs from that of the original program P. A mutation score is the ratio of the non-equivalent mutants of P (i.e., those which are distinguishable from P under at least one pattern from the input domain) which are killed by T. A high score indicates that T is efficient for P wrt mutation fault exposure, thus providing a measure of the efficiency of T.

A specific mutant generator is required for each programming language, in order to produce syntactically correct changes representative of the programmer faults. We have developed a proper generator that introduces changes that take into account the peculiarities of Lustre [5]. As regards the nodes involved in the experiments reported here, the generator has produced a total of 2317 non-equivalent mutants.

4.2 Test Patterns and Results

Due to the simplicity of the node init, testing experiments were not conducted on it. Indeed, this node is called only at the very first execution cycle of the whole program SWITCH to set the output variables to initial values, and its S-graph contains a single transition (Fig. 1). As a result, any test case allows to execute the transition and, in practice, init will always be tested through the root node SWITCH in a later integration phase. This example leads us to a more general remark: when the algorithm for building the set UNIT retains a node n whose S-graph is so simple that its unit testing is obvious, it would be better to postpone the verification of n to integration testing, in order to minimize the test effort.

For each node ∈ UNIT (except init), a test profile was searched and the associated test size N for a target test quality of 0.9999 was calculated. Then 5 different sets of N statistical test patterns were randomly generated based on each test profile in order to expose eventual disparities, that is, in order to assess the ability of the input patterns to reveal faults repeatedly whatever the input values drawn from a given profile; they are denoted Sk. Additional deterministic test patterns related to special values were identified for only two nodes: CTRL and Def_C; they are denoted D.

The experimental results are tabulated in Figure 4 which gives for each node:

- the number of non-equivalent mutants produced by the generator;
- the size N of the test sets Sk, and of the sets D for CTRL and Def_C;
- the lowest and the highest numbers of mutants remaining alive after completion of (i) each of the 5 statistical sets Sk and, (ii) the mixed strategy got by applying the deterministic test patterns D in addition to each of the statistical sets;
- the values of the corresponding mutation scores.

Node	# Mutants	Test sets	N	# Mutants alive	Mutation score
CTRL	311	Sk Sk + D	146 146 + 17	4 0	0.987 1
C_ok	503	Sk	69	0	1
Def_C	493	Sk Sk + D	223 223 + 39	3 - 8 0 - 2	0.984 - 0.994 0.996 - 1
N_F	506	Sk	69	0 - 1	0.998 - 1
S_F	504	Sk	69	0 - 1	0.998 - 1

Figure 4. Unit testing of the program SWITCH: results of mutation analysis

4.3 Comments

The sets Sk supply high mutation scores. For C_ok, the score is perfect (= 1): each of the 5 sets kills the 503 mutants. As regards CTRL, 4 mutations are never revealed by the sets Sk; but they are typical cases of special values, and thus, they are revealed by the deterministic patterns D (see [5] for more details on CTRL results). For Def_C, from 3 to 8 mutants are left alive by the sets Sk; their analysis shows that:

- 3 mutants are never killed whatever the set Sk; they are cases of boundary values which are the target of the deterministic test set D;
- 7 mutants are killed only by some (at least 2) of the 5 sets Sk: they correspond to the insertion of the operator "pre" which turns the memoryless behavior of the program into a history-sensitive behavior. Revealing these faults requires that the mutants be supplied with two successive input patterns in a specific order. Nevertheless, 5 of the 7 mutants are killed by the deterministic patterns D.

More generally, some mutations, such as the insertion of the operator "pre", can induce a faulty behavior that is difficult to reveal. The analysis of the results related to N_F and S_F corroborates this conclusion: in both cases, 4 of the 5 sets Sk reach the perfect score; only one mutant is not revealed by one set and it corresponds to the insertion of the operator "pre".

These results support the high fault revealing power – even wrt subtle faults – of the statistical test sets designed according to the transition coverage criterion, and show how deterministic testing of special values deduced from the Lustre automaton provides an additional efficiency.

5 Conclusion

As regards unit testing of Lustre programs, we have defined a complete and rigorous approach, that is (i) a method for determining complementary statistical and deterministic test patterns from Lustre automata and, (ii) an algorithm for defining the unit nodes with the aim of minimizing the test effort. The definition of the unit level is based on the bottom-up exploration of the call graph, the automata complexity being a stopping criterion of the exploration. The experimental investigation conducted on the program SWITCH has shown the feasibility of the algorithm, and the results are in favor of a high efficiency of the test patterns thus derived.

Testing of the nodes whose structure is no longer tractable wrt the probabilistic analysis for designing statistical patterns, has been postponed to the integration phase. In essence, integration testing disregards the structure of nodes previously tested to focus on their interactions. Then, this reducing principle allow us to progressively decrease the automata complexity by removing the parts corresponding to previously tested nodes. As a result, the method defined in the paper can be generalized to deal with the definition of integration levels by referring to simplified automata [10].

Finally, it is worth noting that the application field of this work should be much larger than Lustre programs. Indeed, the whole testing approach is based on the analysis of Lustre automata, and these automata are expressed by the compiler in a format code common with other synchronous languages (e.g., Signal, Esterel [2]); hence, the results should apply to other synchronous languages.

References

1. Benveniste A, Berry G. The synchronous approach to reactive and real-time systems. Proceedings of the IEEE 1991; 79(9):1270-1282.
2. Benveniste A, Gauthier T, Le Guernic P et al.: Synchronous technology for real-time systems. In: Proc. Real Time Systems Conference (RTS), Paris, 1994, pp 104-122.
3. Halbwachs N. Synchronous programming of reactive systems. Kluwer Academic Pub., 1993.
4. Halbwachs N, Caspi P, Raymond P, Pilaud D. The synchronous data flow programming language Lustre. Proceedings of the IEEE 1991; 79(9):1305-1320.
5. Thévenod-Fosse P, Mazuet C, Crouzet Y: On statistical structural testing of synchronous data flow programs. In: Proc. 1st European Dependable Computing Conference (EDCC-1), Berlin, 1994, pp 250-267.
6. DeMillo RA, Lipton RJ, Sayward FG. Hints on test data selection: help for the practicing programmer. IEEE Computer Magazine 1978; 11(4):34-41.
7. Thévenod-Fosse P: Software validation by means of statistical testing - Retrospect and future direction. In: Dependable Computing and Fault-Tolerant Systems, vol. 4, Springer-Verlag, 1991, pp 23-50.
8. Thévenod-Fosse P, Waeselynck W, Crouzet Y: Software statistical testing. In: Predictable Dependable Computing Systems, Esprit Basic Research Series, Springer-Verlag, 1995, pp. 253-272.
9. Duran JW, Ntafos SC. An evaluation of random testing. IEEE Transactions on Software Engineering 1984; SE-10(4):438-444.
10. Mazuet C. Stratégies de test pour des programmes synchrones - Application au langage Lustre. Doctoral Dissertation, INPT, Toulouse, 1994.

Tolerant Software Interfaces: Can COTS-based Systems be Trusted Without Them?

J. Voas, F. Charron
Reliable Software Technologies Corporation
Sterling, VA 20166 USA

K. Miller
Department of Computer Science, Sangamon State University
Springfield, IL USA

Abstract

We have investigated an assessment technique for studying the failure-tolerance of large-scale *component-based* information systems. Our technique assesses the tolerance of the interfaces between component objects in order to predict how the software will behave if anomalous failures exit certain components and enter others. (Note that we are not talking about graphical user interfaces, but rather the mechanisms that link software components together.) These failures can originate from incorrect code, bad input data from a failed hardware devices, or bad input data from human operators. Our approach is applicable to systems for which source code is available, as well as systems for which no source code is known (e.g., systems composed from executable Commercial Off-The-Shelf (COTS) components), and addresses several of the larger problems associated with software maintenance.

1 Introduction

Large-scale information systems are made up of networks of closely related subsystems that may include distributed hardware and distributed software components. Even if a system is composed of correct components, the *interactive* complexity and coupling between components can cause unexpected errors to emerge from component interfaces. In our view, a system is "failure-tolerant" if it is capable of continuing its mission regardless of any *potential* anomalies that might possibly occur against it. Our solution for assessing the failure tolerance of large-scale systems is to assess the tolerance of the interfaces between components in order to predict how the software system will behave when actual anomalies arise, meaning when corrupt information enters into the *state* of an executing program and is passed between components. (Note that we are not talking about graphical user interfaces, but rather the mechanisms that link software components together.)

Suppose that you knew that the components of a proposed system were known to be failure-tolerant. The natural question that arises when combining these components into a larger system is: *how can we use our confidence in the failure tolerance of individual components to come up with a measure of failure tolerance for the entire system?* Three key concerns regarding the quality of the composite involve: (1) the failure tolerance of the interfaces between components, (2) determining whether the failure of an individual component is likely to cause the failure of another component or, worse, the failure of the entire composite, and (3) the manner by which the components are composed (for example, does A call B or does B call A, and what difference does it make). The technique that we will report on here is able to analyze concerns (1) and (2). Analyzing (3) is out of the scope of this paper, however the approach that this effort describes might be extendible to this problem.

2 Approach

Information systems are the most complex artifacts in human history. Today's information systems include distributed hardware and distributed software components, and interactions with other machines and humans, all of which directly or indirectly comprise "the information system." The number of different failure modes for only one of these components can be intractable, and when taken over all components in combination, the number of different ways that system failure can occur becomes "effectively infinite." For instance, can you enumerate all of the failure modes of a human operator of a electrical power plant when 5 different warnings occur simultaneously? And which of these failure modes will exacerbate the situation, and which are harmless? Today, we build systems that can tolerate certain anomalies, but we often fail to determine which other anomalies cannot be toleranted. These other anomalies constitute the weakest links of the system, and we need better methods for assessing the strength of the more brittle parts of the system. To make matters worse, the weakest links can quickly vary with only the slightest deviation in how the system is configured and used.

The problem is further exacerbated by the complexity imposed by concepts such as object-oriented programming and Web-based development. These technologies engender the possibility of programs no longer being monolithic, but rather being thousands or millions of little parts distributed globally, executing whenever called, yet still being part of one or many systems. These distributed technologies create a scalability problem for virtually all software assessment metrics.

The greatest impetus for using component-based technology for rapid development is the resulting gain in productivity. Today, organizations are touting future product cycle times of 3-4 months that were formerly 12-18 months. To accomplish this requires enormous success with reuse and componentization of fundamental objects that get reused in almost all developments. To make this happen gracefully, a development manager must be confident that fundamental objects constructed elsewhere are compatible and reliable for his or her application. The industry standard is that 50% of bugs are detected after component integration, not during component development and testing. Hence technologies for decreasing the problems of component integration are imperative.

In Leveson's book, "Safeware" [1], on pg. 36, she acknowledges the necessary role of interfaces in complex system design:

"One way to deal with complexity is to break the complex object into pieces or modules. For very large programs, separating the program into modules can reduce individual component complexity. However, the large number of interfaces created introduce uncontrollable complexity into the design: The more small components there are, the more complex the interface becomes. Errors occur because the human mind is unable to comprehend the many conditions that can arise through interactions of these components. An interface between two programs is comprised of all the assumptions that the programs make about each other... When changes are made, the entire structure collapses."

Later on pg. 410 of [1], Leveson succinctly describes the importance of thwarting the propagation of failures through the many interfaces that complex systems employ:

> "A tightly coupled system is highly interdependent: Each part is linked to many other parts, so that a failure or unplanned behaviour in one can rapidly affect the status of others. A malfunctioning part in a tightly coupled system cannot be easily isolated, either because there is insufficient time to close it off or because its failures affects too many other parts, even if the failure does not happen quickly. ..... Accidents in tightly coupled systems are a result of unplanned interactions. These interactions can cause a domino effects that eventually lead to a hazardous system state. Coupling exacerbates these problems because of the increased number of interfaces and potential interactions: Small failures can propagate unexpectedly."

Recognition of the fact that the quality of today's systems is highly dependent on tolerant interfaces has led us to develop an appropriate fault-injection methodology for software interfaces. The purpose of such a method is to assess the strength (failure tolerance) of the interfaces with respect to component failures and problems that enter in the system from external sources. Our method can be applied to source code interfaces or interfaces between executable components; however the thrust of this research pertains to interfaces between executable components. Interfaces are the mechanisms by which information is passed between components during processing, with the *final interface* being the mechanism through which the output passes. Our approach observes how interfaces behave when we intentionally corrupt data passing through them; the tolerance of an interface under simulated stress gives valuable information about the likelihood that the interface will produce acceptable results when the stress is real (i.e., after release).

Software components can be combined in series and parallel just as mechanical systems are. The most common way to combine components, whether physical or software, is to connect them in a series. When physical components are connected in a series, the quality of the whole is less than that of the worst component. However, for software, this may not be true, as another faulty component may actually make the system more reliable (by cancelling the effect of the other fault). Parallel component construction of physical systems is employed as a mechanism for increasing overall quality through redundancy. Once again, when this paradigm is applied to software, the expected increase in quality through redundancy may not occur. Hence the quality-based arguments for designing from components that are used for engineering physical systems do not necessarily hold for software systems.

Because software components in series or parallel are not analogous to hardware components in series or parallel, it is prudent to assume that hardware engineering reliability theories are not applicable to component-based failure tolerance models. Our approach side-steps the series and parallel concerns, and instead focuses on the quality of a system's interfaces.[1] By not looking at failure tolerance solely as a dependability problem, we are able to *explicitly* ignore how the components are interconnected, and place our attention on the interfaces that do the connecting. The results of our analysis technique (which will execute the code to give the information that we want to collect) are *implicitly* dependent on how the components are interconnected.

2.1 How the Technique Works

Our technique employs a modification of the fault-injection methods described for measuring software attributes such as software testability, safety, and vulnerability in [4, 3, 2]. Unlike the fault-injection employed in these applications, the new method described in this paper does not require access to component source-code. A major advantage concentrating on fault-injection on interfaces is that the analysis requires considerably less computing than was required when fault-injection was embodied in software instrumentation that is injected into the source code.

In this technique, there are two key events whose probabilities are *indirectly* measured after fault-injection is performed: "propagation across" and "propagation from." Propagation across refers to the probability that an anomalous event that is input to a component will still be observable at the exit interface of the component, i.e., if the component gets bad input, will the component produce bad output? We need to know this information before we can assess how the failure of one component may affect another.

"Propagation across" a component will be quantified by first applying "perturbation functions" to interfaces feeding components their input data and then sniffing component exit interfaces to see whether the corruption propagated.[2] This process will be repreated n times, and the frequency of times that sniffing detects propagation will be used as the estimate of the probability. *Perturbation functions* are the fault-injection instrumentation mechanisms that we use to force incoming interface data to be corrupted. Assessing the likelihood of "propagation from" occurring will be harder, particularly if we are talking about "propagation from" an executable component. "Propagation from" is the case where something inside the component fails that then forces the component itself fails. If a particular component is very unreliable, then it has very high "propagation from." "Propagation from" is very straightforward to assess via our previous fault-injection commercialized methods when access to source code is a non-issue [5]. But if source is not available, then fault-injection methods for executable components (that allow for corruptions to be injected via

[1] Global parameters are considered as a part of the interface in our model.

[2] We are not only concerned with whether the exact same corruption that we injected propagated, but instead with whether any information coming out of the component is corrupt. After all, the original corruption may have caused a completely different corruption, which if observable at the exit point, signals that propagation across the component occurred.

instrumentation that is written in the executable language) will be necessary. Currently, we know of no ability to measure "propagation from" in this case.

2.2 flipbit

Information passed between interfaces is simply millions of 0s and 1s. To corrupt this information, we use a software procedure that inverts bits, **flipbit**. *flipbit* and its many derivatives (that cannot all be detailed here) are the perturbation functions that we employ.

The first argument to function **flipbit** is the original value that we wish to corrupt, and the second argument is the bit to be flipped (we assume little-endian notation). The function **flipbit** is then written in C as follows and linked with the executable. NOTE: The ^ represents the **XOR** operation in C and the << operator represents a **SHIFT-LEFT** of **y** positions. Also, ~ represents the negation operator.

```
void flipBit(int *var, int y)
{
   *var = *var ^ (1 << y));
}
```

flipBit can be used to model various kinds of program state corruptions, including:

- n Random bits flipped, ($n \geq 1$).

```
void flipNbits(int *var, int n)
{
        int bits = 0;
        int bitPos = 1;
        int i,j,k;
        int xbit;

        for (i = 0; i < n; i++)

        {
           bits |= bitPos;
           bitPos <<= 1;
        }

        for (j = 0; j < sizeof(int) * 8; j++)
        {
           xbit = lrand48()

           if ((!!(bits & (1 << xbit))) !=
               (!!(bits & (1 << j))))
           {
              flipBit(&bits, xbit);
              flipBit(&bits, j);
           }
```

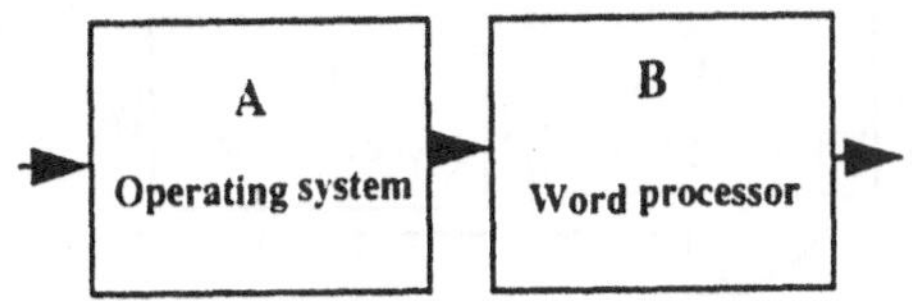

Figure 1: **System S made of two COTS components in a series**

```
        }

        for (k = 0; k < sizeof(int) * 8; k++)
           if (bits & (1 << k))
              flipBit(var, k);
}
```

2.3 Executable Code Fault-injection

Assessing the impact of components for which very little may be known is a conundrum. The expense of customized, one-of-a-kind systems has become so prohibitive that the practice of depending on someone else's code is ubiquitous. And third party code will almost certainly be delivered in a format that precludes any other party from seeing it. Without seeing such code, how can its impact be adequately assessed before it is added to a system?

To explain our approach to this problem, consider the system in Figure 1; Figure 1 represents a legacy system, S, comprised of two COTS components in a series. To implement the information-flow shown in Figure 1, suppose that there exists a main program that is composed of a Lisp-like series of calls where A returns a result to B.

```
Program
    B(A);
End.
```

Suppose however that a new executable component, D, is needed in S between components A and B, as shown in Figure 2. S is now:

```
Program
    B(D(A));
End.
```

We want to know whether D will seamlessly integrate into S. To know this, we will find answers to the following questions:

Question 1 If D is faulty, will its incorrectness affect the functionality of S?

Question 2 If D is correct, but the interface between D and A is faulty, will that affect the functionality of S?, and

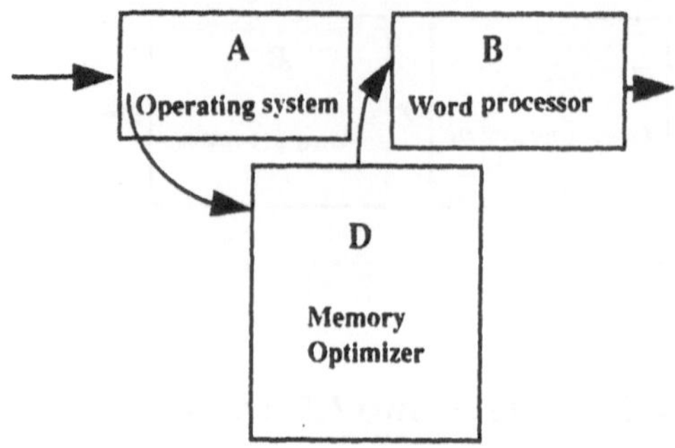

Figure 2: **Integrating Component D into** S

Question 3 If D is correct, but the interface between D and B is faulty, will that affect the functionality of S?

Question 4 Might D exacerbate any problems in A or B that have so far not noticeably degraded S's functionality?

To answer Question 1, we will inject artificial faults (representing possible faults in D) into the out-going parameters from D, by `B(fault-inject`$_1$`(D(A)))`. To answer Questions 2 and 3, fault-injection will simulate incorrect interfaces between D and A and also between D and B, which will include problems such as incorrect parameter orderings and incorrect parameters, via `B(D(fault-inject (A)))` and `B(fault-inject`$_3$`(D(A)))`. Note that the fault injection functions which answer Question 1 and Question 3 differ in which parameters are injected. To answer Question 4, we can try combinations of events, such as `B(fault-inject(D( fault-inject(A))))`. Once this fault-injection has been performed, we gain insight into how likely it is that our modified S will continue to perform satisfactorily after the addition of component D.

2.3.1 *UNIX Examples*

This section discusses four examples of how applying fault-injection perturbations to the output of operating system calls can measure the impact of propagation across the calling programs. Software with embedded UNIX system commands could encounter problems, especially if it uses the output returned by the system commands in subsequent statements in the program. The output from a system call could get corrupted for a variety of reasons. Bugs in system calls have been well-documented and could cause such calls to return unexpected output [6]. Corruptions can occur within the stack of the program if memory resources have been exhausted. Physical failures in the computer could be responsible for noisy returned values as well. These are all examples of the impact that corruptions in the return values of calls to the operating system could potentially have on the calling application programs.

As a first example, the usage of the **vmstat** command is explored. A program may contain a **vmstat** system call to determine the amount of real and virtual memory currently available. Such a program may attempt to allocate a large block of memory on condition that a sufficient amount of memory is indeed available. If the output from the **vmstat** command is perturbed numerically, a value indicating a larger amount of available memory than currently exists could potentially be returned to this program. The result could be disastrous for the program, which would fail to successfully allocate the needed amount of memory.

A second example involves the usage of the **mkdir** system command, which attempts to create a directory with a given name if possible. The **mkdir** command is representative of a large number of UNIX commands that output only a newline character if successful, and an appropriate error message if unsuccessful. For the **mkdir** command, error messages begin with the string "**mkdir:**." A program that has not been carefully designed might test the resulting output assuming that output can be returned in only these forms. For instance, it could test whether the output from a **mkdir** call begins with the "**mkdir**" string, and determine whether the attempt to create a new directory failed or not based on the result. A corruption of a single character in the "**mkdir**" string that begins a resulting output error message would be misinterpreted as an indication that the **mkdir** command succeeded. On the other hand, a test based on the expectation that all successes will be indicated with a single newline character would encounter similar problems for the same class of perturbation. A single character perturbation causing a newline character to be at the beginning of a **mkdir** error message would result in the incorrect determination that the **mkdir** command succeeded, when in fact it failed.

A program that uses the **printenv** system command to determine an environment variable value would be susceptible to corruptions to the returned output. The result from a **printenv** command might be the path name for a directory in which to locate certain configuration files for that program. If the **printenv** output was perturbed such that extraneous characters appeared at the end of the returned string, the returned directory path name would be invalid.

Another case involves the usage of the **find** system command. Imagine a program that uses the **find** command to determine the full path name of a file it wishes to delete. If the file is located, and its full path name has been corrupted to the name of another file that exists, the wrong file would be deleted by the program. For example, a file called **critical_program.OLD** is to be located using **find** and then deleted. Suppose that it is located in the file system, having the path name **/home/apps/critical_program.OLD**. If the output from the **find** becomes corrupted in such a way that it is truncated to the path name **/home/apps/critical_program**, the program will attempt to delete the file named **critical_program** in the same directory. In the event that **critical_program** exists, the program will inadvertently delete it.

A commercial fault-injection tool exists that allows the user to define functions that perturb data values during execution of the program. The tool can be applied to software that makes UNIX system calls in order to perturb the result of those calls in order to determine the impact that a failure of UNIX will have on the software. To do this, the source code can be instrumented such that the data states associated with the returned output from the system calls are perturbed. The tool supports hardware-fault perturbations, standard random numerical perturbations for both discrete (integer) and real (floating-point) values, and assorted character string perturbations.

3 Summary

Our approach is applicable to both source code and executable Commercial Off-The-Shelf (COTS) components. Today it is commonplace for system developers to face the possibility that many of the key components of their systems are "black boxes." They have no control of the quality in these boxes. A related benefit of our approach is that it can *assess* the failure tolerance of legacy systems. which even if the source exists, may be so unreadable and poorly documented that it might as well be written in an executable format. This measurement methodology provides assessments of system-level failure tolerance for any system that contains software whose quality is suspect.

Once it is decided that a component will be integrated into an existing system, it is prudent to explore the aforementioned questions about the possible affects of the integration. Fault-injection applied to interfaces is applicable to these questions. The benefits of applying fault-injection to component interfaces are that it:

1. Provides a measure of system-level *component-to-component interface tolerance*,

2. Provides a plausible measure of the likelihood of common failure modes between parallel components (whose function is supposed to be identical) as well as a measure of component independence for components in series, and

3. Provides a measure of the impact of a new COTS component on the quality of a legacy system in which the new component is substituted or otherwise placed.

In summary, predictions of component-to-component interface propagation provide a portion of the information necessary to predict failure tolerance. Interfaces are the means by which components communicate; interfaces are the glue for today's information systems that are object-based. Failure-tolerant interfaces are imperative if we are to build failure-tolerant information systems.

Contact Address

Voas may be reached at RST Corporation, Suite 250, 21515 Ridgetop Circle, Sterling, VA 20166, USA, phone: (703) 404-9293, jmvoas@rstcorp.com.

Acknowledgements

This work has been partially supported by DARPA Contract F30602-95-C-0282 and NIST Advanced Technology Program Cooperative Agreement No. 70NANB5H1160.

References

[1] N.G. Leveson. *Safeware: System Safety and Computers*. Addison-Wesley, 1995.

[2] J. VOAS, G. MCGRAW, A. GHOSH, F. CHARRON, AND K. MILLER. Defining an adaptive software security metric from a dynamic software failure tolerance measure. In *Proc. of Ninth Annual Conference on Computer Assurance*, National Institute of Standards and Technology, Gaithersburg, MD, June 1996.

[3] J. VOAS AND K. MILLER. Dynamic testability analysis for assessing fault tolerance. *High Integrity Systems Journal*, 1(2):171–178, 1994.

[4] J. VOAS AND K. MILLER. Software Testability: The New Verification. *IEEE Software*, 12(3):17–28, May 1995.

[5] Reliable Software Technologies Corporation. *PiSCES Software Analysis Toolkit (R) User's Manual*, 1996. Version 2.0.

[6] B. P. MILLER, L. FREDRIKSON, AND B. SO. An empirical study of the reliability of unix utilities. *Communications of the ACM*, 33(12):32–44, December 1990.

Modeling Software Dependability Growth under Input Partition Testing

Yinong Chen*
Dept. of Computer Science
University of the Witwatersrand
Johannesburg, SOUTH AFRICA
yinong@cs.wits.ac.za

Jean Arlat
LAAS-CNRS
7 Avenue du Colonel Roche
31077 Toulouse, FRANCE
arlat@laas.fr

Abstract

One problem of the input domain-based software reliability models is the large number of test cases required to obtain a high confidence in reliability estimation, because testing has to be restarted from beginning after any fault correction is performed. This paper intends to overcome this problem by considering relations between the programs before and after fault corrections and therefore making use of the testing data collected from the previous testing stages. For this purpose we propose an input domain-based reliability growth model. Both partition and random testing can be used to generate input cases for test runs. It is generally considered in the model that input generation, fault detection and fault correction are all imperfect. It will be shown that some existing reliability models can be viewed as the special cases of our model.

Keywords: reliability modeling, software testing, software dependability, testing coverage

1 Introduction

An important quality measure of a system is its dependability attributes, e.g., reliability during application, the fault coverage during testing. *Software reliability* has been defined as the probability that no failure occurs in a specified environment during a specified (continuous) exposure period. The exposure period can be the calendar time, or *application runs* corresponding to a selection of *input cases* from the *input case domain* (*ICD*) of the programs [1]. For the evaluation of software reliability there exist different kinds of reliability models. [2] classifies software reliability models according to the development phases of software life-cycle, while [3] divides them according to the nature of failure process into times-between-failures models, failure-count models, fault seeding models, and input domain-based models. Since fault corrections are necessary in testing and debugging phases of life-cycle, reliability growth models which take fault corrections into account are mainly used in this phase. The models that do not account for fault corrections, say input domain-based models, can only be applied in this phase by treating the program after

* This work was partly sponsored by the fellowship of the European Commission Human Capital and Mobility Programme, CaberNet, at LAAS-CNRS, Toulouse.

each correction as a new program. This causes the problem that a very large number of test cases have to be generated after each fault corrections.

Reliability growth models usually define the software reliability as the probability

$$R(T) = Prob\{\text{no failure within time period } [0, T]\}$$

where T is the exposure period whose time unit is the calendar or CPU time. On the other hand, input domain-based models [4,5,6], define software reliability as the probability

$$R(N) = Prob\{\text{no failure over } N \text{ application runs}\}$$

where N is the exposure period whose time unit is the number of application runs. Assuming that input cases are selected independently, $R(N)$ can be expressed by $R(N) = R^N$, where R is the *reliability per application run*. Since $R(N)$ can be simply calculated from R, R becomes the main objective to be estimated by input domain-based models, and R is thus called *reliability* for short.

Variants of input domain-based models are studied in [7], where weights can be added to different input cases. However, no input domain-based models so far take the fault corrections into consideration.

This paper studies an input domain-based reliability growth model, in which fault corrections are considered and the failure data collected before error corrections are also used for the reliability evaluation. This approach reduces the problem of input domain-based models where a very large number of failure data have to be collected after each fault correction.

The remaining part of the paper is organized as follows. §2 defines testing strategies, a fault model, fault distributions, the fault detection and the fault correction. Collection of failure data and the parameter estimation are discussed in §3. The input domain-based reliability growth model and the relation with some other models are studied in §4. Finally, §5 concludes the paper.

2 Testing Strategies and the Fault Model

This section introduces some important concepts and defines the basic models needed for the proposed model.

2.1 Partition and Random Testing Strategies

Def.1 A partition of the input case domain *ICD* is defined by a set $\{C_1, C_2, ..., C_m\}$, where, $C_i \subseteq ICD$, $\bigcup_{i=1}^{m} C_i = ICD$, and $C_i \cap C_j = \phi$ (empty set) for $i, j = 1, 2, ..., m$ and $i \neq j$.

Def.2 **Partition testing** consists of (test) rounds of test runs. A **test run** is an execution of the program under test by applying an input case randomly selected from a class. If the *ICD* of the program is partitioned into m classes, a **test round** consists of exactly m test runs, aimed at **covering** (generating input cases from) all the m classes of the partition. If partition testing covers some classes more than once in a test round, some other classes will be **uncovered**. Let u be the expected number of uncovered classes in a test round, the portion of covered classes, $(m-u)/m$, is called **partition testing**

coverage. If $u = 0$, $(m-u)/m = 1$, the coverage is perfect, otherwise, it is imperfect.

Def.3 **Random testing** is a special case of partition testing when the number of classes of the partition is one.

Note, in both def.2 and def.3, the random selection is assumed to be uniform in the analysis of this paper. A further extension of the work could be done by using a statistical distribution to characterize the input selection probability, as defined in [8].

2.2 Fault Model

In this section, we turn to discuss the faults in programs under test.

Def.4 An incorrect output of a program is a **failure**. A program contains **faults**, if it will give a failure when certain input cases are applied. An input case is a **failure-causing input case** related to a fault if its application can detect the fault by causing a failure. The set of failure-causing input cases related to a fault f_i is called a **fault region** G_i. The ratio $\Theta = |G_i|/|ICD|$ is called the fault size of fault f_i or fault region G_i. The set of all failure-causing input cases related to all faults is the **detection set** (*DS*) of the program.

In other words, a failure is a manifestation of a fault in the output domain, while a fault region is the complete manifestation of a fault in the input domain.

Evidently, if all input cases are equally likely during application, the fault size of a fault is the probability that a failure-causing input case is generated from its fault region, or the fault is detected in one run (application or test run). Thus, $|DS|/|ICD|$ is the probability that any fault is detected in a test run or a failure occurs in an application run. $|DS|/|ICD|$ also called **failure probability**. According to the definition of software reliability for input domain-based model, the reliability per application run of a program is then

$$R = 1 - |DS|/|ICD| = 1 - \sum_{i=1}^{n} \Theta_i, \quad \text{where } n \text{ is the number of faults.}$$

Def.5 Assuming that there exist n faults $f_1, f_2, \ldots, f_n$, in a program and $C_1, C_2, \ldots, C_m$ are classes of an *ICD* partition of the program. If the fault region corresponding to fault f_k is distributed in d_k classes, $1 \le d_k \le m$, then the d_k classes are called **fault classes** of fault f_k. The parameter set (m, n, d_k, Θ_k) for $k = 1, 2, \ldots, n$, forms a **fault distribution**.

Remark: Fault, fault region, fault class and fault size are related but different. A fault region G_k is the set of failure-causing input cases related to a fault f_k, and the fault size Θ_k of G_k is a probability, instead of a set. A fault class is a class of the partition that contains failure-causing inputs of one or more faults. Fault and fault region can be used interchangeably in some cases, for example, the fault size of a fault or a fault region, and failure-causing input of a fault or a fault region.

As an example Fig.1 shows the structure tree of a program with 10 paths. If the structure tree is used as the criterion to partition the *ICD*, we can obtain a partition with 10 classes. If there are three fault f_1, f_2 and f_3 in the program, the corresponding fault regions are shown in the classes. Classes C_1, C_2, C_3 and C_4 are the fault classes related to f_1. Classes C_8 and C_9 are the fault classes related to f_2. Class C_9 is the fault class related to f_3.

A fault correction is a behavior aimed at localizing and removing the faults in the program which are detected. The effect of a fault correction can be so large that the program after the correction has no relation at all with the program before correction. In this case, neither the failure history nor the fault correction history of the old program is useful for any reliability growth model.

Thus, in our reliability growth model we only considered fault corrections which ensure that the partition after the corrections still satisfies the given partitioning criterion, so that we can continue to test the program without having to make a new partition.

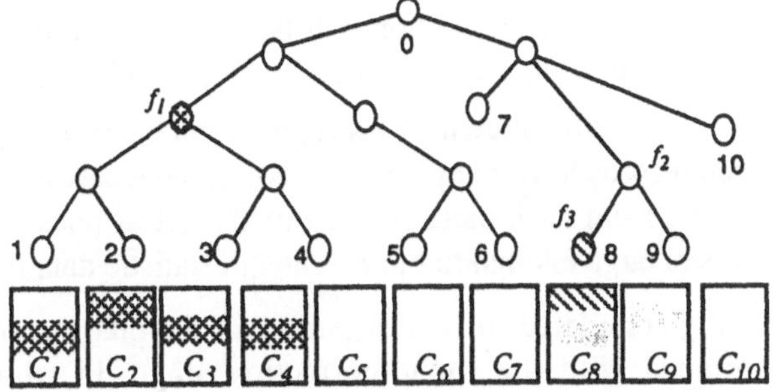

Fig.1 A partition with a fault distribution

Def.6 A **fault correction** is a change (decreasing, increasing, or none) of the fault sizes of some classes of a given partition.

According to the number of corrections needed to remove a fault (change its fault size to zero) we can informally distinguish two kinds of faults: **easy faults**, the fault sizes of the faults are changed to zero after a small number of corrections; and **difficult faults**, the fault sizes of the faults are not changed to zero after a large number of corrections, because they are either difficult to be detected or difficult to be corrected. The easy faults are usually removed after a period of testing. The remaining faults are all the difficult faults, which are the main problem to be studied in testing and evaluation of software reliability.

With fault correction in consideration, the fault sizes and the fault distributions will be changing with fault corrections.

Def.7 Let $(m, n_{j-1}, d_{k(j-1)}, \Theta_{k(j-1)})$ be the fault distribution before the jth fault correction, and it is expected to be changed into $(m, n_j, d_{kj}, \Theta_{kj})$ after the jth fault correction, where, $\sum_{k=1}^{n_j} \Theta_{kj} = \rho \cdot \sum_{k=1}^{n_{j-1}} \Theta_{k(j-1)}$.

Parameter ρ is called the **fault size change rate** and $1-\rho$ is called the **correction rate.**

3 Parameter Estimation

So far we have characterized testing strategies, fault distribution, and fault correction by a set of parameters u, n_j, d_{kj}, Θ_{kj}, ρ. Now we discuss how to organize the testing process and how to use the failure data collected during testing to estimate the values of these parameters. To simplify the process, this section only considers the cases that there are no overlaps of fault regions with a class. For this purpose, we divide partition testing into two stages: a phase-correction stage and a round-correction stage.

- The **phase-correction stage** consists of q **test phases**, and each test phase consists of p test rounds. If faults are detected during testing, testing will not be interrupted for fault correction. After the p rounds of testing, all the faults detected in the phase are corrected together. The fault correction is viewed as a part of a test phase.
- The **round-correction stage** consists of t test rounds. If any failure is detected, fault correction will be performed at the end of a round. The fault correction is viewed as a part of a test round.

A **failure table**, shown in Fig.2, is maintained during the phase-correction stage, which records how many failure-causing input cases have been generated (number of failures) from each class in each of the q test phases. Of course, p and q must be numbers large enough to obtain enough statistic data for parameter estimation.

Having completed the q test phases in the phase-correction stage, testing will be continued with the round-correction stage, in which fault correction will be performed after each round (if any fault is detected). The final parameter values estimated in the first stage will be used as the initial parameter values of the second stage. Changes of parameter values caused in the second stage will be predicted according to certain probability distributions.

In the sequel of this section, we will briefly study the phase-correction stage and the parameter estimation. The round-correction stage will be studied in full details in §4.

Class	no. failures / no. test runs in test phase 1		no. failures / no. test runs in test phase 2		no. failures / no. test runs in test phase 3		. . .	no. failures / no. test runs in test phase q	
C_1	w_{11}	s_{11}	w_{12}	s_{12}	w_{13}	s_{13}	...	w_{1q}	s_{1q}
C_2	w_{21}	s_{21}	w_{22}	s_{22}	w_{23}	s_{23}	...	w_{2q}	s_{2q}
...	...	...	...	...	...	...	...	...	...
C_i	w_{i1}	s_{i1}	w_{i2}	s_{i2}	w_{i3}	s_{i3}	...	w_{iq}	s_{iq}
...	...	...	...	...	...	...	...	...	...
C_m	w_{m1}	s_{m1}	w_{m2}	s_{m2}	w_{m3}	s_{m3}	...	w_{mq}	s_{mq}

Fig.2 The failure frequency table

Let s_{ij} be the total number of test runs which have selected input cases from C_i in the jth test phase, in which w_{ij} failures are observed. For $i = 1, 2, ..., m$ and $j = 1, 2, ..., q$, w_{ij} and s_{ij} are recorded in Fig.2. Based on these data and the specification and/or implementation of the program under test, the values of the parameters u, n_j, d_{kj}, Θ_{kj}, p can be estimated as follows.

Since p test rounds are performed in each test phase, each classes should be tested exactly p times, i.e., $s_{ij} = p$, if the testing coverage is perfect. Otherwise, $s_{ij} \neq p$.

If $s_{ij} \geq p$, class C_i can be considered to be tested at least once in average in each test round. If $s_{ij} < p$, class C_i is not tested in at least $p - s_{ij}$ rounds, which reflects the imperfect coverage of this test phase. Then the parameter u can be estimated by the average number of classes which are tested less than p times in the test phase: $u_j = \frac{1}{m}\sum_{i=1, s_{ij}<p}^{m}(p - s_{ij})$.

The above estimation is only based on the statistic data in the jth test phase. Since the number of uncovered classes is a global parameter independent of the test phases, we can further build the average among all the value $u_1, u_2, ..., u_q$ to get the final estimate of u:

$$u = \frac{1}{q}\sum_{j=1}^{q} u_j = \frac{1}{q}\sum_{j=1}^{q}\Big(\frac{1}{m}\sum_{\substack{i=1 \\ s_{ij}<p}}^{m}(p - s_{ij})\Big) = \frac{1}{q \cdot m}\sum_{j=1}^{q}\sum_{i=1, s_{ij}<p}^{m}(p - s_{ij})$$

Now let us briefly discuss how to find the regions of faults, so that the classes which are affected by a single fault belong to a region. For a given partition, there are usually some relations among the classes. The relationship is determined by the specification or implementation of the program. As an example, the structure tree in Fig.1 can be used to explain the relationship among fault classes. If the classes corresponding to paths 0→1, 0→2, 0→3, and 0→4 have non-zero fault sizes, it is more likely that a common point of these paths contains a single fault than that the private parts of these paths contain four independent faults. Thus, classes C_1, C_2, C_3 and C_4 should be viewed as a fault region. The classes corresponding to paths 0→8 and 0→9 have non-zero fault sizes, they can be considered as another fault region. The fault f_3 has an overlap with fault f_2 in the region. The determination of overlapping faults is beyond the scope of this paper.

How to determine the fault regions will not be further discussed in this paper. It will be assumed that after the jth test phase, we can obtain n_j fault regions, $G_{1j}, G_{2j}, ..., G_{n_j j}$, and the number d_{kj} of fault classes in each fault region according to Fig.2 and the specification and/or implementation of the program.

Based on the principle of input domain-based models, the fault size Θ_{kj} of fault region G_{kj} after the jth test phase can be estimated by the average ratio of the number of failures and the number of test runs related to the classes in G_{kj}, that is,

$$\Theta_{kj} = \sum_{C_i \subseteq G_{kj}} \frac{w_{ij}}{p \cdot m} = \frac{1}{p \cdot m} \cdot \sum_{C_i \subseteq G_{kj}} w_{ij}, \text{ where } k = 1, 2, ..., n_j,\ j = 1, 2, ..., q.$$

According to def. 7 and the estimate of Θ_{kj}, the imperfect correction rate $1 - \rho_j$ of the jth correction in the phase-correction stage can be estimated as:

$$1 - \rho_j =$$

$$1 - \sum_{k=1}^{n_j}\Theta_{kj} \Big/ \sum_{k=1}^{n_{j-1}}\Theta_{k(j-1)} = 1 - \sum_{k=1}^{n_j}\sum_{C_i \in G_{kj}} w_{ij} \Big/ \sum_{k=1}^{n_{j-1}}\sum_{C_i \in G_{k(j-1)}} w_{i(j-1)}$$

Further build the average value among all $\rho_1, \rho_2, ..., \rho_q$, we obtain the final estimation of the fault correction rate:

$$1-\rho=1-\frac{1}{q}\sum_{j=1}^{q}\rho_j=1-\frac{1}{q}\sum_{j=1}^{q}\left(\sum_{k=1}^{n_j}\sum_{C_i\in G_{kj}}w_{ij}\Big/\sum_{k=1}^{n_{j-1}}\sum_{C_i\in G_{k(j-1)}}w_{i(j-1)}\right)$$

The final parameter values, u, n_q, d_{kq}, Θ_{kq}, ρ, estimated in this stage will be used as initial values in the next stage.

4 The Proposed Model

In §3, the values of the set of necessary parameters are estimated. As the result we obtain a set of initial values of parameters: m, u, n_q, d_{kq}, Θ_{kq}, ρ, $k = 1, 2, ..., n$.

Since the round-correction stage will use the same partition and the same testing algorithms, m and u will remain unchanged in this stage. Moreover, since test phases have been repeated many times in the phase-correction stage, most easy faults have been removed. In this testing stage we mainly consider the difficult faults which cannot be removed even after a large number of corrections. This means that the number of fault regions n_q and the number of classes d_{kq} in fault region G_k should remain unchanged, only the Θ_{kq} and the ρ will change with increasing number of test rounds and fault corrections. As constants of this testing stage we will rename n_q and d_{kq} as n and d_k, and rename Θ_{kq} and ρ as Θ_{k0} and ρ_0. The remaining fault size of ICD from the last stage (or initial fault size for this stage) is then $\sum_{k=1}^{n}\Theta_{k0}$. The values of Θ_{k0} and ρ_0 after r^{th} test round in this stage will be denoted as Θ_{kr} and ρ_r. Beside theses parameters, we need some extra parameters to characterize the increasing difficulty of fault detection and correction in the new testing stage, which will be discussed when they are introduced.

In §4.1, the reliability growth model under partition testing will be studied. In §4.2, the relations of the model to the other models will be discussed.

4.1 Dependability Growth under Partition Testing

Let G_k be a fault region with d_k classes and its fault size be $\Theta_{k(r-1)}$ after the $(r-1)^{th}$ test round. During the rth test round, one input case will be generated exactly from each class of G_k. Since the input cases from the $m-d_k$ classes irrelevant to G_k will not contribute to the detection of G_k, we need only consider the other d_k input cases. In average, the probability that anyone of the d_k input cases is a failure-causing input case of fault region G_k is $\frac{|G_k|}{d_k\cdot|C|}=\frac{m\cdot|G_k|}{d_k\cdot|ICD|}=\frac{m\cdot\Theta_{k(r-1)}}{d_k}$.

Since there are d_k classes in fault region G_k, d_k input cases will be selected from G_k. Thus, the probability that the fault corresponding to G_k is detected is $1-\left(1-m\cdot\Theta_{k(r-1)}/d_k\right)^{d_k}$. In this expression it is assumed that the testing coverage is perfect, i.e., $u=0$. If $u\geq 0$, the probability will be changed into $1-\left(1-m\cdot\Theta_{k(r-1)}/d_k\right)^{a_k}$, where $a_k=d_k\cdot(m-u)/m$ is the average number of input cases which are generated from the fault region G_k. In the probability expression, fault detection probability is assumed to be independent of the number of test rounds performed. This assumption is not accurate in most cases. According to the human learning curves [10,11], we can use an exponential function $e^{-\lambda(r-1)}$ to reflect the fact that the detection probability per run may decrease with increasing number of test rounds, because easier faults are more likely to be detected and removed in the earlier rounds. Decreasing of detection probability is controlled by parameter λ. If $\lambda = 0$, $e^{-\lambda(r-1)} = 1$, the detection probability will not suffer from decreasing with test rounds' increasing. This case refers to perfect fault detection. Taking this factor into account, the detection probability is generalized as follows:

$$1-\left(1-(m\cdot\Theta_{k(r-1)}/d_k)\cdot e^{-\lambda(r-1)}\right)^{a_k} \tag{1}$$

Similar to fault detection, we first assume that fault correction is perfect (the entire fault size related to a fault will be removed if the fault is detected), the fault size $\Theta_{k(r-1)}$ of G_k will be completely removed if the fault is detected. Therefore we have the fault size removed from fault region G_k by the rth test round:

$$\left(1-\left(1-(m\cdot\Theta_{k(r-1)}/d_k)\cdot e^{-\lambda(r-1)}\right)^{a_k}\right)\cdot\Theta_{k(r-1)}$$

If fault correction is imperfect, only a portion of $\Theta_{k(r-1)}$ can be removed. According to the definition of fault correction rate, the removed portion is $\Theta_{k(r-1)}\cdot(1-\rho_{r-1})$, which reflects the fact that the fault correction rate will decrease with the increasing number of test rounds, because easier faults are more likely to be removed in the earlier rounds. The decreasing strength is controlled by parameter μ: the fault correction rate $1-\rho_r$ can decrease very fast or very slow, depending on the selection of μ values.

Then, we have a more general form of the fault size removed from fault region G_k by the rth test round: $\left(1-\left(1-(m\cdot\Theta_{k(r-1)}/d_k)\cdot e^{-\lambda(r-1)}\right)^{a_k}\right)\cdot\Theta_{k(r-1)}(1-\rho_{r-1})$

Since there are n fault regions, and if we have totally completed t test rounds, then the fault size removed after t test rounds, $FSR(t)$, is

$$FSR(t) = \sum_{r=1}^{t}\sum_{k=1}^{n}\left(\left(1-\left(1-(m\cdot\Theta_{k(r-1)}/d_k)\cdot e^{-\lambda(r-1)}\right)^{a_k}\right)\cdot\Theta_{k(r-1)}(1-\rho_{r-1})\right) \tag{2}$$

After the rth correction, the fault size, $\Theta_{k(r-1)}$, of G_k will be changed. Considering all imperfect factors, the new fault size can be calculated as follows:

$$\Theta_{kr} = \Theta_{k(r-1)} - \left(1-\left(1-m\cdot\Theta_{k(r-1)}/d_k\right)^{a_k}\right)\cdot\Theta_{k(r-1)}\cdot(1-\rho_{r-1}) \tag{3}$$

According to the definition of reliability, the initial fault size and the fault size removed during t rounds of testing $FSR(t)$ in equ. (2) we have the final *reliability growth model for partition testing*:

$$R(t) = 1-\left(\sum_{k=1}^{n}\Theta_{k0} - \sum_{r=1}^{t}\sum_{k=1}^{n}\left(\left(1-\left(1-\frac{m\cdot\Theta_{k(r-1)}}{d_k}\cdot e^{-\lambda(r-1)}\right)^{a_k}\right)\cdot\Theta_{k(r-1)}\left(1-\rho_{r-1}\right)\right)\right) \quad (4)$$

where, $\sum_{k=1}^{n}\Theta_{k0}$ is the initial fault size of the program before the round-correction stage and $\sum_{k=1}^{n}\Theta_{k0} - ESR(t)$ is the remaining fault size. Thus, $1-\left(\sum_{k=1}^{n}\Theta_{k0} - ESR(t)\right)$ is the reliability of the program after the tth test round.

4.2 Relation with the Other Models

This subsection shows that some existing software reliability models [4,5,6,7,12] can be viewed special cases of our model in Equ. (7).

Special case 1: Model under general random testing

Random testing is a special case of partition testing when the number of classes $m = 1$, which implies the number of classes in each fault region $d_k = 1$ and $a_k = 1$. Taking $m = 1$, $d_k = 1$ and $a_k = 1$ into equ.4, we have

$$R(t) = 1-\left(\sum_{k=1}^{n}\Theta_{k0} - \sum_{r=1}^{t}\sum_{k=1}^{n}\left((\Theta_{k(r-1)})^2\cdot e^{-\lambda(r-1)}\cdot\left(1-\rho_{r-1}\right)\right)\right) \quad (5)$$

Special case 2: Models under perfect random testing

If the fault detection and correction are perfect, that is, $\lambda = 0$ and $1-\rho_{r-1} = 1$, we obtain the reliability growth model under perfect random testing:

$$R(t) = 1-\left(\sum_{k=1}^{n}\Theta_{k0} - \sum_{r=1}^{t}\sum_{k=1}^{n}(\Theta_{k(r-1)})^2\right) \quad (6)$$

Moreover, if there is only a single fault $(n = 1)$ and only one test round is carried out, then we have

$$R(1) = 1-\left(\Theta_{10}-(\Theta_{10})^2\right) = 1-\Theta_{10}(1-\Theta_{10}) \quad (7)$$

Special case 3: Models under perfect testing, detection and correction

If the testing coverage is perfect, we have $u = 0$ and $a_k = d_k$. If the fault detection and correction are perfect, we have $\lambda = 0$ and $1-\rho_{r-1} = 1$. Taking these conditions into equ.4 we obtain

$$R(t) = 1-\sum_{k=1}^{n}\Theta_{k0} + \sum_{r=1}^{t}\sum_{k=1}^{n}\left(1-m\cdot\Theta_{k(r-1)}/d_k\right)^{d_k}\cdot\Theta_{k(r-1)} \quad (8)$$

Furthermore, if there is only a single fault and only one test round is carried out:

$$R(1) = 1-\Theta_{10} + \left(1-(1-m\cdot\Theta_{10})^{d_1}\right)\cdot\Theta_{10} = 1-(1-m\cdot\Theta_{10})^{d_1}\cdot\Theta_{10} \quad (9)$$

Comparing the equ. (6) with (8) and equ. (7) with (9), we can see the elementary difference between the reliability growth models for partition and random testing.

5 Conclusion

The main contribution of this paper is the proposal of an input domain-based reliability growth model which takes fault correction history into account. This model makes it possible to apply the input-domain-based reliability model in the testing-debugging phase, without having to ignore all the failure statistics collected in the previous stages of fault corrections. A wide range of measurable parameters are included in the model to fit various kinds of testing strategies with different testing coverages, detection probabilities and correction rates. Some existing models can be viewed as special cases of this model. The model has been applied to compare the efficiency of testing strategies, through which some new factors which affect the efficiency of testing strategies were therefore discovered [13,14].

References

1 J. D. Musa, A. Lannino, K. Okumoto, Software reliability: measurements, prediction, application, McGraw-Hill Book Company, New York, 1987.

2 C. V. Ramamoorthy, F. B. Bastani, Software reliability — status and perspectives, IEEE, Trans. Soft. Eng., SE-8, No. 4, 1982, pp. 354 - 371.

3 A. L. Goel, Software reliability models: assumptions, limitations, and applicability, IEEE, Trans. Soft. Eng., SE-11, No. 12, Dec. 1985, pp. 1411 - 1423.

4 W.H. MacWilliams, Reliability of large real-time software, in Proc. IEEE Symp. Computer Software Reliability, New York, May 1973, pp. 1 - 6.

5 J.R. Brown and M. Lipow, Testing of software reliability, in Proc. Int. Conf. Reliable Software, Los Angeles, CA, April 1975, pp. 518 - 527.

6 E. Nelson, Estimating Software reliability from test data, Micro-electron. Rel., vol. 17, 1978, pp. 67 - 74.

7 M.Z. Tsoukalas, J.W. Duran, S.C. Ntafos, On some reliability estimation problems in random and partition testing, IEEE, Trans. Soft. Eng., SE-19, No. 7, July 1993, pp. 687 - 697.

8 P. Thévenod-Fosse and H. Waeselynck, "An Investigation of Statistical Software Testing," Journal of Software Testing Verification and Reliability, vol. 1, no. 2, 1991, pp.5-25.

10 Y. Tohma, et al, The estimation of parameters of the hypergeometric distribution and its application to the software reliability growth model, IEEE Trans. Soft. Eng., SE-17, No.5, May 1991, pp. 483 - 489.

11 J.-C. Laprie, K. Kanoun, C. Beounes, K. Kaaniche, The Knowledge-Action-Transformation approach to the modeling and evaluation of reliability and availability Growth, IEEE Trans. Soft. Eng., SE-17, No. 4, 1991, pp. 307 - 382.

12 S.N. Weiss, E.J. Weyuker, An extended domain-based model of software reliability, IEEE, Trans. Soft. Eng., SE-14, No. 10, 1988, pp. 465- 470.

13 Y. Chen, J. Arlat, An Input Domain-Based Reliability Growth Model and Its Applications in Comparing Software Testing Strategies, LAAS Report 95105, April 1995.

14 Y. Chen, J. Arlat, Comparing software testing strategies using reliability growth, in R. Mittal, et al (ed), Fault-Tolerant Systems and Software, Narosa Publishing House, Madras, December 1995, pp. 48 - 54.

Round Table

An Holistic Approach to Dependability ?

Moderator: A. Pasquini, ENEA, Rome, ITALY
Participants: W. Goerke, Karlsruhe University, GERMANY
K. Kanoun, LIS-LAAS, Toulouse, FRANCE
A. Rizzo, University of Siena, ITALY

Abstract

This panel will address the problem of methods for dependability analysis and evaluation of computer system, where the human, hardware and software components are considered by an integrated approach. The state of the art consists of well developed methods in particular domains, but dependability is a total system concept. However practitioners have come to accept that their approaches address only some aspects of this total system view. This panel will move from the point of view of the specific participant competencies to discuss: the extent to which the individual approaches synergise (and the state of the art in doing that) or are antagonistic; the difficulties hindering an integrated approach to dependability.

Introduction

Alberto PASQUINI - ENEA
Email: pasquini_a@casaccia.enea.it

Safety-critical systems require an assessment activity to verify that they are able to perform their functions in specified use environments. This activity would benefit from evaluation methodologies that consider these systems as a whole and not as the simple sum of their parts. Indeed, analysis of accidents involving such systems as shown that they are rarely due to the simple failure of one of their components. Accidents are the outcome of a composite causal scenario where human, software and hardware failures combine in a complex pattern.

Well known examples include: space applications such as when the Phobos I flight control system and the ground control caused the failure of the space mission; medicine when a combination of an architectural flaw with a software fault and operator misbehaviour in the Therac 25 radiation therapy machine caused the over radiation and death of some cancer patients; nuclear power when a failure of the Crystal River process control system and the operator caused a radioactive water release. These examples are all drawn from large scale, low volume systems. The same kind of problem will also surface in increasingly volume produced items that will incorporate programmable components involving hardware, software and interaction with operators. An obvious area is in the automotive industry whenever there are strong pressures to decrease costs and increase functionality through the use of programmable elements.

On the contrary, dependability analysis and evaluation of safety critical systems involving computers are based on techniques and methodologies that concern human, software and hardware components separately. Most integration efforts are limited to hardware and software components with the questionable assumption that their evaluation can be performed independently and then combined, for example using the traditional reliability graphs when evaluating reliability. Therefore the assessors of these systems have the difficult task of integrating the results of completely different and not compatible methodologies at different stages of advancement.

Furthermore, the research activity of experts and scientists usually addresses only one among the hardware, software and human components. There is an evident lack of communication between the researchers of these disciplines that have separate conferences, scientific journals and research networks.

An innovative approach is needed for the dependability analysis and evaluation of safety-critical systems. Instead of regarding the human, hardware and software components as effectively independent, this approach must take a holistic view seeking to identify the component inter-dependencies and incorporate the evaluation within a common framework. This aim requires the joint effort of experts and scientists from different disciplines: engineers, computer scientists, mathematicians, psychologists, and cognitive scientists.

Human as a Back-up for System Safety

Winfried GOERKE - Karlsruhe University
Email: goerke@ira.uka.de

Simplified the development of system dependability can be characterised in the following way: The classic view addresses functional limits of equipment, especially if reliability is concerned, hence back-up procedures in case of failure are required, originally by human operator or fail-safe design. More recent developments take increased levels of automation everywhere into account, such that a substitution of human functions also with respect to intelligent decisions can be observed, hence computer control of complex systems forms a basic solution approach.

As a consequence also dependability has shifted to systems as a whole: the user does not care for the reason for the denial of service, he simply requires an organisation masking all the detailed effects which might lead to failures of certain sub-components. He requires a design incorporating all possible deficiencies of functional details, being able to deliver the specified service in spite of possibly detailed malfunctions. Society has become used to the availability of sophisticated technical systems and depends upon them to a very large degree. But occasionally their limits in capability can be noticed: supply systems break down and disrupt service (power systems, telephone systems, railways), but service can be restored after more or less short times. Limited safety, however, may lead to loss of human lives or cause damage to health or property which cannot be accepted, even if they seem to be unavoidable.

How can we increase safety or better decrease risk? From a hardware point of view fault tolerance and redundancy indicate possible solutions, but only to the limited extent of the physical device behaviour. A component breakdown can be masked to tolerate its occurrence as far as the system is concerned. But there are system aspects where this approach cannot be used, in particular the overall system design. As we have learned by bitter experience from the limited dependability achieved so far in e.g. air transportation and space flight, it is overwhelmingly the human part which will finally be blamed if an accident has occurred. This leads to the following questions:

- Has the hardware system design been sufficient as far as fault tolerance is concerned?
- Are there software components demonstrating design flaws or bugs robust to removal?
- Was there an operator mistake as far as back-up by human control is concerned? How was training organised, in particular with respect to the system state calling for back-up?
- What about system integration? Should not the influence of human errors be excluded from possibly causing a fatal system state because it is well known how error prone humans will react? How can this be achieved?
- Are our methods insufficient as far as human interaction is concerned?

Software Reliability Evaluation

Karama KANOUN
E-mail: kanoun@laas.fr

Software faults are design faults in the broad sense of the term, including faults injected from the requirements to operation and maintenance. Usually, software reliability evaluation consists in applying reliability growth models to the failure data collected on the software under consideration. This means that the field of software reliability evaluation addresses systems that are not life critical but that could be "money" critical. Since the obtained results are not accurate for all systems under any circumstances, several methods have been developed to improve the estimations obtained from these models. However, even for accurate estimations, the relevance of the measures estimated varies with the considered life-cycle phase. For instance, the most relevant estimations are obtained for software systems in operation, on multiple installations, since we have usually a sample of failures that are representative of system behaviour.

In the software reliability evaluation field, emphasis has mainly been placed on the development of a) reliability growth models and their related staff (such as model validation criteria) considering the software as a black box, b) methods to improve the estimation accuracy, c) some structural models, taking into account the structure of the software and the transfer of control between the various components, mainly in stable reliability, and d) more recently and still in an infancy phase, methods incorporating past experience on previous similar software systems.

Most of these methods are applicable to hardware field; one of the questions to be addressed is: could they be used or, at least, adapted to human error collection and/or data processing ? Indeed, data collection for quantification of human behaviour is a growing concern now. Could the human reliability field benefit from the software reliability field ? Does the human behaviour exhibit a trend during system operation (more experience can improve the knowledge about the systems and their environment and hence the number of human errors can be reduced). Is the notion of reliability growth meaningful when dealing with human behaviour ?

Indeed, a first step towards a holistic approach has been jumped: a few models for system dependability evaluation, considering hardware and software faults together with their interactions (but excluding human errors) have been developed since several years (based on Petri Nets and Markov Chains). But how could the human behaviour be incorporated into these models ? More precisely, how could models integrating from the beginning the three aspects can be derived ? Is such an aim realistic ? These questions are of prime importance for those systems in which the safety depends heavily on the cooperation between the "technical system" and the human operator. An example of such systems is the Air Traffic Control System where the role of the technical system is only to process and provide information to the controller who has to take the final decisions (the technical system has not direct impact on the controlled environment).

Human Error and Distributed Cognition

Antonio RIZZO - University of Siena
Email: rizzo@unisi.it

System Reliability Analysis has made noticeable progress in assessing technical component reliability, but the introduction of Human Reliability Analysis is still a very difficult task.

Human interaction with objects and artefacts in both physical and social systems is frequently characterised by errors. However, "errors" are very often best attempts at accomplishing desired and sensible goals, and should not be necessarily attributed to incompetence. People may not have any way of foreseeing the unintended and deleterious consequences of their decisions and actions, and when negative modifications occur in the world, they might regret having taken a certain action or made a particular decision.

The recent user-centred design approach of man-machine systems considers the error as the product of breakdowns in communication between the humans and the physical artefacts. These breakdowns may play a fundamental role in the design process by revealing and creating space for problems and opportunities. Moreover, as these authors have observed, a structured analysis of the processes that cause breakdowns and of the processes that allow to handle breakdowns should be useful to the designers. The inevitability of human error have led system designers to make a great effort to minimise the incidence of errors, to maximise their discovery, and to make easier their recovery or the repair of their consequences. However, despite the claims for user-centred systems, most of the available guidelines and design principles for error recovery are concerned with errors occurring at the syntactic level of the human-computer interaction. That is, error concerning users performances that violate rules in the grammar of the interaction. Guidelines and principles are not concerned with higher semantic and pragmatic levels of the interaction. How can a more general approach based on distributed cognition be identified and can it represent a sound support for a global analysis of system dependability ?

About the Panel Participants

Alberto Pasquini - He is with the Italian Agency for New Technology, Energy and the Environment (ENEA) where he works in European research projects in the area of software engineering and, more specifically, software quality and reliability. He was member of the Italian licensing authority for nuclear power plant where he was involved in the dependability evaluation of computer systems used to control nuclear power plants. His current interest are in software reliability and reliability evaluation of systems involving human components. He is the scientific coordinator of the OLOS network.

Winfried Goerke - He received his diploma in electrical engineering and graduated as Dr.-Ing. at the University of Karlsruhe, Germany. Being a full professor in the computer science department of this university his main research interests are related to fault-tolerant computing systems. Among his scientific publications there are textbooks on reliability engineering, fault diagnosis of switching circuits, microcomputers and fault-tolerant computing systems. He is a member of GI (German Computer Society) and ITG/VDE (society of information techniques within the association of German electronic engineers).

Karama Kanoun - She is currently Charge de Recherche at LAAS-CNRS. She joined LAAS in 1977 as a member of the "Fault-Tolerance and Dependable

Computing" group. Her current research interests include modeling and evaluation of computer system dependability considering hardware as well as software. She has authored and co-authored more than eighty papers. She has conducted several research contracts and she has been a consultant for some French companies and for the International Union of Telecommunications. Dr Kanoun is a member of the working group of the European Workshop on Industrial Computer Systems (EWICS): "Technical Committee 7 - Reliability, Safety and Security", a member of the AFCET working group "Dependability of Computing Systems". She acts as a referee for several international conferences and journals. Besides serving on program committees of international conferences, she served as a program committee co-chair of the international Symposium on Software Reliability Engineering (ISSRE'94) and she served as general chair of ISSRE'95.

Antonio Rizzo - He teaches Human-Computer Interaction at the Communication Science Department of the University of Siena. He worked at the Human Factors Unit of the National Research Council - Institute of Psychology in Rome. He has been involved in research on cognitive ergonomics since mid-eighties by working in several European founded projects and as consultant of national industries. His current interest are in the field of Multimedia Design and Distributed Cognition. In particular, he is concerned with the design of a Multimedia System for supporting disabled people involved in tourism activity; and with the development of an integrated approach for designing safety critical applications. He is the University of Siena co-ordinator for the Apple Design Project '96.

About the OLOS Network

OLOS is a network funded by the European Community under the programme Human Capital and Mobility, Contract: ERBCHRXCT 940577. Partners of the network are European Universities, Research Centres and Industries active in the dependability analysis and evaluation of computer systems. The interdisciplinary competencies that are present in the network ensure the resources needed for a holistic approach to dependability. Goals of the network are:

- To develop interdisciplinary competencies, especially among young researchers, concerning "global system dependability";
- To define and develop the concept of "global system dependability" in order that various dependability and reliability notions and methodologies can be seen to make a contribution to overall dependability;
- To promote the development of an integrated set of methodologies to be used for the dependability analysis and evaluation of those critical systems that require the combination of hardware, software and human resources.

More detailed information concerning the OLOS network can be obtained through the authors or visiting the World Wide Web site: http://rep1.iei.pi.cnr.it/OLOS.

Invited Paper

Software-based critical systems

Jean-Claude Laprie

LAAS-CNRS and *LIS*, Toulouse, France

Abstract

Computerization of critical systems has given rise to much debate. This paper is aimed at giving recommendations and directions in order to undertake the design of software-based critical systems in an effective and cohesive way.

The current situation is portrayed in the first section, based on statistics on system failures and on the procurement cost of critical software.

This current situation, together with technological trends lead to identify three driving forces which are commented upon the second section: i) cost-effective highly dependable systems via re-use, ii) evolution towards integration (vs. federation), iii) fault evolution.

The discussion of the driving forces lead to state four recommendations and directions in the third section: i) supplement off-line validation with on-line protection via fault tolerance, ii) extend the applicability of dependability measures for dependability prediction, iii) establish a theory of composability of dependability properties, iv) build dependability-explicit development processes.

1- Current situation

Our society has become increasingly dependent on computing systems and this dependency is especially felt upon the occurrence of failures. Figure 1 gives examples of disastrous (or potentially disastrous) failures over the past fifteen years. Significant trends can be drawn from these examples: a) failures become more and more distributed (which is no surprise with ever-increasing networked applications), and two of the examples are relative to nation-wide failures (AT&T outage in the USA and credit card denial of authorization in France), b) although some failures are due to hardware misfunction (physical faults), the sources of failures are mostly design faults, possibly in combination with operational misuses (interaction faults), be the latter without or with malicious intention. In terms of consequences, the failures generally relate primarily to economics; however, some can lead to endangering human lives as second order effects, or even directly as those failures

---

The work reported in this paper has been partially supported by the ESPRIT Long Term Research Project DeVa (Design for Validation, Project no. 20072).

	Faults			Failures		Availability/ Reliability	Safety	Confidentiality
	Physical	Design	Interaction	Localized	Distributed			
June 1980: False alerts at the North American Air Defense (NORAD)	✔			✔		✔		
April 1981: First launch of the Space Shuttle postponed		✔		✔		✔		
June 1985 - January 1987: Excessive radiotherapy doses (Therac-25)		✔		✔			✔	
August 1986 - 1987: the "wily hacker" penetrates several tens of sensitive computing facilities		✔	✔	✔				✔
November 1988: Internet worm		✔	✔		✔	✔		
15 January 1990: 9 hours outage of the long-distance phone in the USA		✔			✔	✔		
February 1991: Scud missed by a Patriot (Dhahran, Gulf War)		✔	✔	✔		✔	✔	
November 1992: Crash of the communication system of the London ambulance service		✔	✔		✔	✔	✔	
26 and 27 June 1993: Authorization denial of credit card operations in France	✔	✔			✔	✔		
4 June 1996: Flight 501 failure of Ariane 5		✔		✔		✔		

Figure 1 - Examples of computer-related failures

labeled as having safety consequences. Finally, failures can impact one or several attributes of dependability, such as availability or reliability, safety, or confidentiality, as indicated on figure 1.

Our dependency on computing systems is amplified by the fact that computerization of activities has enabled a complexity such that manual backup is simply not possible, be it for safety-critical applications such as airliner fly-by-wire control, or for commercial applications such as airline reservation and pricing.

Software, and thus design faults, are generally recognized as being the current bottleneck for dependability in critical applications, be they money- or life-critical. A simple reason is that the computer systems involved in such applications are tolerant to physical faults.

An examination of statistics for largely deployed systems confirms this state of affairs, as exemplified by figure 2 which compares failure data for non-fault-tolerant and fault-tolerant traditional systems (i.e. whose architecture is typically a central computing system, which is accessed through a network of terminals), and for large scale client-server networks (several thousands of workstations), that is a type of architecture which tends to replace the traditional architectures.

A close examination of those statistics show that fault-tolerance provides for traditional systems an improvement of two orders of magnitudes in terms of time to failure, and that, on the average, fault-tolerant computing systems become obsolete before system failure. Although the data for client-server networks relate to non-

Traditional systems		Client-server networks (non fault-tolerant)
Non-fault-tolerant	**Fault-tolerant**	
(Japan, 1383 organizations [Siewiorek & Swarz 92, ch. 2]; USA, 450 companies [FIND/SVP 93])	(Tandem Computers [Gray 90]; Bell Northern Research [Cramp et al. 92])	(Tandem Computers [Wood 94])
Mean Time To Failure: 6 to 12 weeks Average outage duration after failure: 1 to 4 hours	Mean Time To Failure: 21 years (Tandem)	Average availability: 98%
Failure sources Hardware 50% Software 25% Communications-Environment 15% Operations - Procedures 10%	Failure sources Software 65% Operations - Procedures 10% Hardware 8% Environment 7%	Failure sources Design 60% Operations 24% Physical 16%

Figure 2 - Failure sources and frequencies for computing systems

fault-tolerant systems, the dominance of design faults come from the fact that, in such architectures, a user usually exploits a large number of various software, located on a similarly large number of servers.

Interaction faults becoming the second source of system failure is not due to the fact that human operators become more error-prone, but is simply due to our progressive mastering of physical faults, and to the permutation among failure sources which ensue. This permutation effect has been noticeable for a long time in fault-tolerant systems (see figure 2, and, e.g., [Toy 78, Davis & Giloth 81]), and goes beyond computing systems, as indicated by the statistics of figure 3 [Ruegger 90] which show that, in commercial flights, human errors have become the first source of accidents, in spite of having significantly decreased in absolute terms over the years.

	Accidents per million takeoffs			
	1970-1978		1979-1986	
Technical defects	1.49	(45%)	0.43	(33%)
Weather	0.82	(25%)	0.33	(26%)
Human error	1.03	(30%)	0.53	(41%)
Total	3.34		1.29	

Figure 3 - Primary cause of accident of domestic commercial flights in the USA

However, a significant proportion of interaction faults can actually be traced to design faults [Norman 83, Leveson 95], be they due to either a) a poor design of man-machine interfaces or interaction procedures, or b) to a lack of assistance from the system to its human operators in tasks where human reasoning and judgment is ultimately necessary, such as facing situations of multiple system faults. The current trend in the so-called human factors domain towards a better distribution of tasks between humans and machines, performed at the initial system requirements stage (see e.g. [Sheridan 94]), just strengthens the above mention of the traceability of the majority of interaction faults to design faults.

Computerization of safety-critical systems has given rise to much debate. In short, proponents advocate that programmed systems enable, thanks to more elaborate functionalities, a better control of the total system, and thus an improved safety ; opponents reply that more elaborate functionalities mean increased complexity, and thus a decrease in safety. In [Ichiyen & Joannou 91], it is reported that the failure rate after 288 year of cumulated operation for a programmed nuclear monitoring system is inferior to the failure rate for the conventional (i.e., electro-mechanical and harwdwired electronics) part of the system, and that the programmed part of the system exhibited no unsafe failure, whereas the one fourth of the failures of conventional part were potentially unsafe. It has to be noted that the programmed part of the system is rather simple (10 000 lines of code). Quantitative estimates of failure rates of critical software seldom appear in the literature. Two exceptions are the following data:

- in [Laryd 94], processing of the data relative to 11000 years of cumulative operation of nuclear control and monitoring system led to an estimate of 5.10^{-8}/h,
- in [Shooman 96], the exploitation of large database of airworthiness directives (consecutive to critical operational problems) led to estimate that current avionics software failure rate is comprised between 10^{-7}/h and 10^{-8}/h.

A consequence of the recognition that probabilistic assessment of software reliability, prior to operation, to levels commensurate with safety requirements (e.g. 10^{-9}/h or 10^{-5} per demand) is currently out of reach, has led to highly labor intensive approaches for the development and validation of operational life-critical software. Be they undertaken via traditional software engineering approaches or via mathematically formal approaches, orders of magnitudes of effort dedicated to the development and validation of such software are in the range of 10 man.years per 1000 lines of code, for software ranging from a few thousands to a few tens of thousands lines of code [Craigen et al. 93; OFTA 94, ch. 13]. Verification and validation activities typically amount to 75% of the total development costs for such critical software and the real benefit in terms of reliability improvement to be gained from any increase in those costs is doubtful. Furthermore, those huge development costs are accompanied by equally huge maintenance costs: maintenance of the on-board Space Shuttle software costs approximately $ 100 million a year [Leveson 95].

2- Driving forces

The situation portrayed in the previous section, together with technological trends, enable the following driving forces to be identified, which will be commented upon in this section:

- cost-effective highly dependable systems via re-use,
- evolution towards integration (vs. federation),
- fault evolution.

2.1- Cost-effective highly dependable systems via re-use

Although we don't know yet whether computerized critical systems satisfy their dependability requirements, the development and maintenance efforts which are devoted to them are clearly hardly sustainable for software which would be one or several orders of magnitude larger than the current ones, as are some critical systems currently under development.

A traditional engineering paradigm for reducing costs of new systems is re-use of components developed for previous systems. Re-use is indeed a largely employed approach in software as well. However, re-use for dependability needs special care, as the recent infamous failure of the Ariane launcher has demonstrated it [Lions 96].

2.2- Evolution towards integration (vs. federation)

Pre-computer realizations of safety-related systems were classically based on the *federation* of equipments implementing each one or several subfunctions of the system; one of the criteria for partitioning the system global function into subfunctions was that a failure of any equipment should be confined, and should not prevent the global function of the system to be performed, possibly in a degraded mode. When the transition was made to computer technology, the safety-related systems generally kept on the federation approach [Avizienis & Laprie 86]. Each subfunction was then implemented by a "complete" computer: hardware, executive and application software. Examples of this approach may be found in airplane flight control systems (e.g. the Boeing 757/767 [Spradlin 80]) or nuclear plant monitoring (e.g. the SPIN system [Remus 82]). A pre-requisite to confining computer failures is the auto-detection of the error(s) having led to failure. The auto-detection of errors due to design faults can be achieved through two main approaches:

- acceptance tests of the results under the form of executable assertions [Andrews et al. 84, Rabejac et al. 96], which constitute a generalization and a formalization of the likelihood checks which are classical in process control;
- diversified design leading to two software variants whose results are compared; real-life examples are the Airbus A-300 and A-310 [Traverse 88], or the Swedish railways' interlocking system [Hagelin 88].

The federation approach generally leads to a very large number of processing elements, larger than what would be necessary in terms of computing power; for instance, the Boeing 757/767 flight management control system is composed of 80 distinct functional microprocessors (300 when redundancy is accounted for). Although such large numbers of processing elements are affordable due to the low cost of today's hardware, this approach suffers severe limitations, as a consequence of the subfunctions isolation, which inhibits the cooperative use of total computer resources:

- limitations to evolving towards new functionalities, due to a restricted exploitation of the software possibilities;
- limitations to the improvement of reliability and safety, due to rigid redundancy management schemes, induced by the necessity of local redundancies for the equipments.

An additional step towards a better use of the possibilities offered by computers consists in having several subfunctions implemented by software, supported by the same hardware equipment, that is an *integration* approach.

2.3- Fault evolution

The usual perception of software, and thus design faults, as the current bottleneck of dependability is clearly supported by the statistics given in section 1. We have also stated in section 1 that design faults are the source of most interaction faults. Another major trend in fault characterization is the ever growing importance of temporary faults vs. permanent faults. Such a trend is not new as far as physical faults are concerned (see, e.g., [Siewiorek & Swarz 92]), but also apply to design faults.

The very notion of temporary design fault, especially for software [Elmendorf 72, Gray 86], may be felt as contradicting the usual perception of design fault. In fact, if it is not arguable that design faults are present as long as they are not fixed, it has to be recognized that most design faults manifesting in operation in large, complex, software are subtle enough in order that their activation conditions depend on equally subtle combinations of internal state and external solicitation, so that they can hardly be reproduced. If we call failure domain the set of inputs which activate faults, what precedes can be expressed in stating that the failure domain of temporary design faults can vary with the conditions of system solicitation, and be a null space under most operating conditions. This leads to an abstraction of the notion of fault, which is then defined by its failure domain; such an abstract view is however conform to reality, in the sense that a fault exists with respect to its capacity to cause a failure only if its failure domain is non-null.

3- Recommendations and directions

The above driving forces lead to the following recommendations and directions, in order to enable the development and deployment of computerized critical systems to be performed in a cohesive and coherent way:

- supplement off-line validation with on-line protection via fault tolerance,
- extend the applicability of dependability measures for dependability prediction,
- establish a theory of composability of dependability properties,
- build dependability-explicit development processes.

3.1- Off-line validation, and on-line protection via fault tolerance

As already stated, verification and validation costs have reached such amounts that the benefit in terms of reliability resulting from their increase is dubious, as we are likely in the diminishing returns region, besides that such an increase is hardly foreseeable from a strict economy viewpoint. Practical use of mathematically formal verification approaches may help improving the situation, but will certainly not provide for a breakthrough from a cost-effectiveness viewpoint, besides the fact that it is now largely recognized that mathematically formal verification of software does not mean absolute freedom from fault. Furthermore, reuse poses new challenges to verification and validation activities, be reuse based on traditional software, as unfortunately exemplified by the recent Ariane 5 failure, or based on object-oriented approaches, due to inheritance, dynamic binding and polymorphism (see, e.g., [Kung et al. 95]).

There is thus a growing need for providing software components with error-detection mechanisms which, in order to be effective, need to be designed along with the functional design. Exploitation of error signals in turns needs mechanisms for subsequent action, i.e. recovery. In short, there is a need for a shift from design-fault avoidance to design-fault tolerance [Avizienis 95], oriented towards software reuse.

The move towards integration mentioned in section 2.2 is an additional incentive for such a shift, in order to enable software of different criticalities to coexist on the same hardware platform, where the corresponding protection mechanisms would implement integrity policies similar to policies developed for security purposes [Vardanega et al. 95].

In order to avoid man-machine interaction faults becoming the next bottleneck in dependability, developing approaches for their tolerance is also needed [Maxion 1986, Rouse & Morris 1987, Mazet & Guillermain 96, Mo & Crouzet 96].

Finally, as interactions between safety and security can only develop, tolerance should extend to intentionally malicious faults as well [Joseph & Avizienis 1988, Rabin 1989, Deswarte et al. 1991]. Such interactions originate from, at least, two different trends: a) the incorporation of COTS (Commercial Of The Shelf) software components in critical systems, which may carry along malicious faults such as viruses, and b) the move towards open, networked, systems which will inescapably reach critical applications.

3.2- Dependability measures for dependability prediction

The imperfections of fault tolerance, i.e. the lack of *fault tolerance coverage,* constitute a severe limitation to the increase in dependability which can be obtained. Such imperfections of fault tolerance are due either a) to design faults affecting the fault tolerance mechanisms with respect to the fault assumptions stated during the design, the consequence of which is a lack of *error and fault handling coverage*, or b) to fault assumptions which differ from the faults really occurring in operation, resulting in a lack of *fault assumption coverage*, which can be in turn due to either i) failed component(s) not behaving as assumed, that is a lack of *failure mode coverage*, or ii) the occurrence of correlated failures, that is a lack of *failure independence coverage* (figure 10).

The influence of a lack of error and fault handling coverage [Bouricius et al. 69, Arnold 73] has been shown to be a drastic limit to dependability improvement. Similar effects can result from the lack of failure mode coverage: conservative fault assumptions (e.g., Byzantine faults) will result in a higher failure mode coverage, at the expense of necessitating an increase in the redundancy and more complex fault

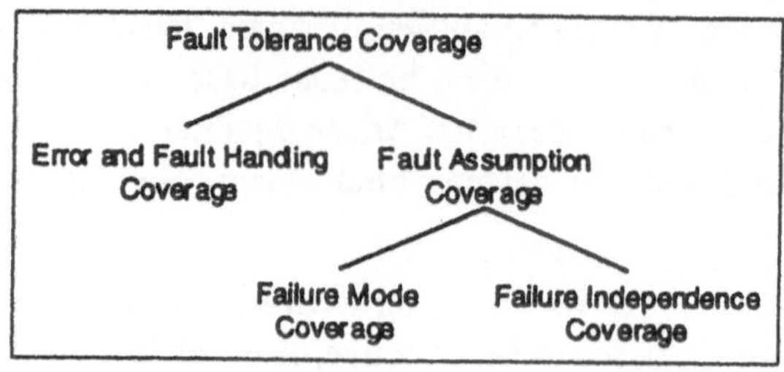

Figure 4 - Fault tolerance coverage

tolerance mechanisms, which can lead to an overall decrease in the system dependability [Powell 92].

The evaluation of the error and fault handling coverage has been recently devoted a large attention, be it via modeling [Dugan & Trivedi 89], or via fault-injection [Arlat et al. 90, Iyer 95]. However, those studies have largely focused on hardware fault tolerance, and additional work is needed for extending the results to software fault tolerance (in spite of the difficulties [Butler & Finelli 93]), or to security [Dacier et al. 96].

3.3- Composability of dependability properties

The traditional method for building computing systems is systematic composition of components. However, little is known about the composability of dependability properties [Nicolaidis et al. 89, Abadi & Lamport 90, McLean 94]. This is clearly an area of utmost importance, as being able to have insights into why certain properties are preserved or not preserved by certain forms of composition is a prerequisite to building dependable systems from dependable components.

3.4- Dependability-explicit development process

Design faults having their origin in the development process leads naturally to pay attention to development models. The situation varies largely depending on whether hardware or software is considered. Hardware development models traditionally incorporate reliability evaluation (see, e.g., [BSI 85]), but pay less attention to verification and to fault tolerance. On the other hand, traditional software development models (waterfall, V, spiral, etc.) incorporate verification and validation activities, but do not mention reliability evaluation nor fault tolerance.

It is our opinion that the means for dependability [Laprie 92], i.e. fault prevention, fault tolerance, fault removal and fault forecasting should be explicitly incorporated in a development model focused at the production of dependable computing systems. Such a model [Laprie et al. 95, chap. 3], which can be termed as *dependability-explicit development model,* is outlined in the annex.

4- Summary

The topics addressed in the paper and their relationship are summarized by figure 5.

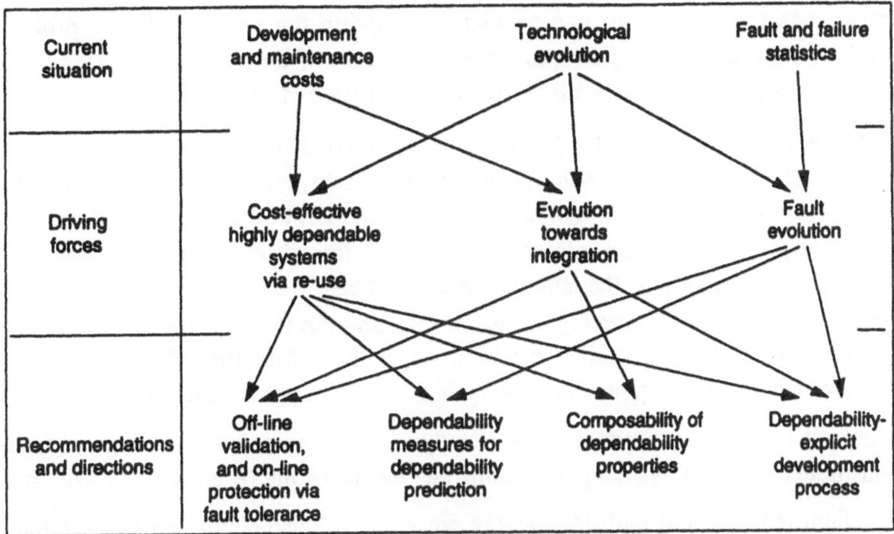

Figure 5 - Relationship between the current situation, the driving forces and the recommendations and directions

Annex: Dependability-explicit development model

The basic model (figure 6) identifies three classes of processes:

- the system creation process which builds on the classical development steps, i.e. requirements, design, realization, integration,
- the dependability processes: fault prevention, fault tolerance, fault removal and fault forecasting,
- other processes, such as quality assurance and certification.

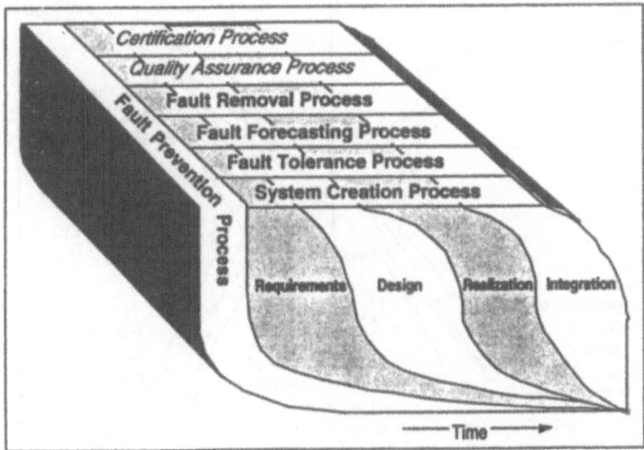

Figure 6 - Dependability-explicit development model

The system creation process sequences the other processes.

The fault prevention process plays a specific role, i.e. it defines and coordinates the activities of the system creation process together with the other processes. Three major classes of activities can be identified within the fault prevention process: choice of *formalisms and languages*, *organization* of the project, *planning* of the project and *evaluation of risks* incurred from the system development.

The fault tolerance process is composed of three main activities:

- the study of the *behavior in the presence of faults*, aimed at eliciting the faults against which the system will be protected via a fault tolerance approach,
- the *system partitioning*, aimed at structuring the system in error confinement area, and at identifying the fault independence areas,
- the *fault and error handling*, aimed at selecting the fault tolerance strategies, at determining the appropriate mechanisms, without forgetting the protection of those mechanisms against the faults which are likely to affect them.

The fault assumptions produced by the study of the behavior in the presence of faults constitute a basis for system partitioning, and inputs for the fault and error handling.

The fault removal and fault forecasting processes are also composed of three main classes of activities, respectively:

- *verification*, *diagnosis* and *correction* for the fault removal process,
- expression of dependability *objectives*, *allocation* among system components, *evaluation*, for the fault forecasting process.

Figure 7 gives a schematic view of those various activities and of their interactions. For the sake of simplicity, the interactions between the dependability processes are not represented, although they do exist via a natural recursion, which leads, for instance, to verify evaluation results, and to evaluate the progress of verification activities.

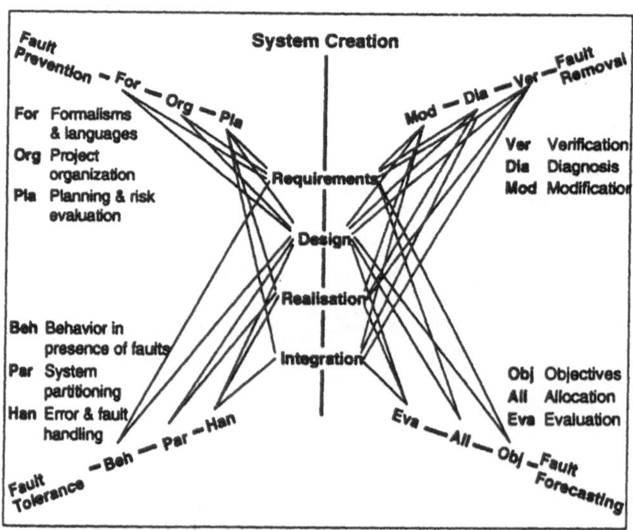

Figure 7 - Interactions between the processes

The model is not a classical life-cycle model, as figure 7 gives the logical sequencing according to which the activities are undertaken, but not their temporal sequencing. Especially, similarly to the model of DO-178B [RTCA 91], groups of activities can be instanciated several times in order to accommodate several approaches in the development of a given system, as illustrated by figure 8.

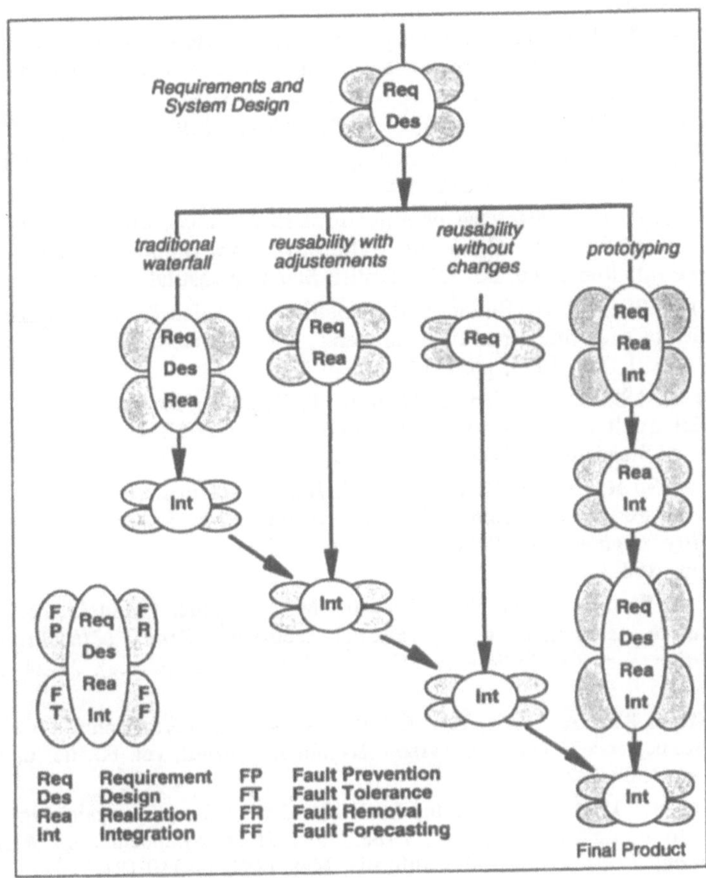

Figure 8 - Example of application of the dependability-explicit model

References

Abadi & Lamport 90 M. Abadi, L. Lamport, "Composing specifications", Tech. Report 66, Digital Equipment Corp. Systems Research Center, Palo Alto, California, Oct. 1990, 90p.

Andrews et al. 84 D. Andrews, A. Mahmood, E.J. McCluskey, "Executable assertions and flight software", in *Proc. AIAA/IEE 6th Digital Avionics Systems Conf.*, Dec. 1984, pp. 346-351.

Arlat et al. 90 J. Arlat, M. Aguera, L. Amat, Y. Crouzet, J.-C. Fabre, J.-C. Laprie, E. Martins and D. Powell, "Fault Injection for Dependability Validation — A Methodology and Some Applications", *IEEE Trans. on Software Engineering*, 16 (2), pp.166-182, February 1990.

Arnold 73 T.F. Arnold, "The Concept of Coverage and its Effect on the Reliability Model of Repairable Systems", *IEEE Trans. on Computers*, vol. C-22, no. 3, pp. 251-254, 1973.

Avizienis & Laprie 86 A. Avizienis, J.-C. Laprie, "Dependable computing: from concepts to design diversity", *Proceedings of the IEEE,* vol. 74, no. 5, May 1986, pp. 629-638.

Avizienis 95 A. Avizienis, "Building dependable systems: how to keep up with complexity", *Proc. 25th IEEE Int. Symp. on Fault Tolerant Computing (FTCS-25), Special Issue*, Pasadena, California, June 1995, pp. 4-14.

Bouricius et al. 69 W.G. Bouricius, W.C. Carter, P.R. Schneider, "Reliability Modeling Techniques for Self-Repairing Computer Systems", *Proc. 24th. National Conference*, pp. 295-309, ACM Press, 1969.

BSI 85 *Reliability of Constructed or Manufactured Products, Systems, Equipment and Components, Part 1. Guide to Reliability and Maintenability Programme Management,* Report no. BS 5760, British Standard Institution, 1985.

Butler & Finelli 93 R.W. Butler, G.B. Finelli, "The infeasibility of quantiying the reliability of life-critical real-time software", *IEEE Trans. on Software Engineering*, vol. 19, no. 1, Jan. 1993, pp. 3-12.

Craigen et al. 1993 D. Craigen, S. Gehrart, T. Ralston, "An international survey of industrial applications of formal methods", report NIST GCR 93/626, National Institute of Standards and Technology, March 1993.

Cramp et al. 92 R. Cramp, M.A. Vouk, W. Jones, "On operational availability of a large software-based telecommunications system", *Proc. 3rd Int. Symp. on Software Reliability Engineering (ISSRE'92)*, Research Triangle Park, North Carolina, Oct. 1992, pp. 358-366.

Dacier et al. 96 M. Dacier, Y. Deswarte, M. Kaaniche, "Models and tools for quantitative assessment of operational security", *Proc. 12th International Information Security Conference (IFIP SEC'96)*, Samos (Grèce), 21-24 Mai 1996, pp.177-186.

Davis & Giloth 81 E.A. Davis, P.K. Giloth, "No 4 ESS: performance objectives and service experience", *The Bell System Technical Journal*, vol. 60, no. 6, July-Aug. 1981, pp. 1203-1224.

Deswarte et al. 1991 Y. Deswarte, L. Blain, J.C. Fabre, "Intrusion tolerance in distributed computing systems", *Proc. 1991 IEEE Symposium on Research in Security and Privacy*, Oakland, California, May 1991, pp.110-121.

Dugan & Trivedi 89 J.B. Dugan, K.S. Trivedi, "Coverage modeling for dependability analysis of fault-tolerant systems", *IEEE Trans. on Computers*, vol. 38, no. 6, June 1989, pp. 775-787.

Elmendorf 72 W.R. Elmendorf, "Fault-tolerant progtamming", in *Proc. 2nd IEEE Int. Symp. on Fault Tolerant Computing (FTCS-2)*, Newton, Massachusetts, June 1972, pp. 79-83.

FIND/SVP 93 "The Impact of Online Computer Systems Downtime on American Businesses", FIND/SVP Survey, 1993.

Gray 86 J.N. Gray, "Why do computers stop and what can be done about it?", in *Proc. 5th Symp. on Reliability in Distributed Software and Database Systems,* Los Angeles, Jan. 1986, pp. 3-12.

Gray 90 J. Gray, "A census of Tandem system availability between 1985 and 1990", *IEEE Trans. on Reliability*, vol. 39, no. 4, Oct. 1990, pp. 409-418.

Hagelin 88 G. Hagelin, "ERICSSON safety system for railway control", ERICSSON Document ENR/TB 6078, Oct. 1986; also in *Application of design diversity in computerized control systems*, U. Voges, ed., vol. 2 of the Series on Dependable Computing and Fault Tolerance, Vienna: Springer Verlag, pp. 11-21.

Ichiyen & Joannou 91 N. Ichiyen, P. Joannou, "The CANDU appraoch to diftal safety systems", *Nuclear Engineering International*, Sep. 1991, pp. 35-37.

Iyer 95 R. K. Iyer, "Experimental evaluation", *Proc. 25th IEEE Int. Symp. on Fault Tolerant Computing (FTCS-25), Special Issue*, Pasadena, California, June 1995, pp. 115-132.

Joseph & Avizienis 1988 M.K. Joseph, A. Avizienis, "A fault tolerance approach to computer viruses", *Proc. 1988 IEEE Symp. on Security and Privacy*, Oakland, California, April 1988, pp. 52-58.

Kung et al. 95 D. Kung, J. Gao, P. Hsia, Y. Toyoshima, C. Chen, Y.S. Kim, Y.K. Song, "Developing an object-oriented software testing and maintenance environment", Communications of the ACM, October 1995, pp. 75-87.

Laprie 92a J.C. Laprie (Ed.), *Dependability: Basic Concepts and Terminology*, Springer-Verlag, Vienna, 1992.

Laprie et al. 95 J.C. Laprie, J. Arlat, J.P. Blanquart, A. Costes, Y. Crouzet, Y. Deswarte, J.C. Fabre, H. Guillermain, M. Kaâniche, K. Kanoun, C. Mazet, D. Powell, C. Rabéjac, P. Thévenod, *Guide de la Sûreté de Fonctionnement* (Dependability Handbook), Cépaduès Editions, 1995; in French

Laryd 94 A. Laryd, "Operating experioence of software in programmable equipment used in ABB atom nuclear I&C applications", IAEA TCM, Helsinki, Finland, June 1994.

Leveson 95 N. Leveson, *Safeware - System Safety and Computers*, Addison Wesley, 1995

Lions 96 J.L. Lions, "Ariane 5 flight 501 failure - Report by the inquiry board", ESA, July 1996.

Maxion 1986 R.A. Maxion, "Towards fault-tolerant user interfaces", *Proc. 5th IFAC Workshop on Safety of Computer Control Systems (SAFECOMP'86)*, Sarlat, France, Oct. 1986, pp. 117-122.

Mazet & Guillermain 96 C. Mazet, H. Guillermain, "Dependable systems: error tolerance and man-machine cooperation", *Proc. 3rd Conf. on Probabilistic Safety Assessment and Management (PSAM-III – ESREL'96)*, Crete, Greece, June 1996, pp. 406-411.

McLean 94 J. McLean, "A general theory of composition for trace sets closed under selective interleaving functions", Proc. 1994 IEEE Symp. on Research in Security and Privacy, Oakland, California, May 1994, pp. 79-93.

Mo & Crouzet 96 J. Mo, Y. Crouzet, "Human error tolerant design for air traffic control systems", *Proc. 3rd Conf. on Probabilistic Safety Assessment and Management (PSAM-III – ESREL'96)*, Crete, Greece, June 1996, pp. 400-405.

Nicolaidis et al. 89 M. Nicolaidis, S. Noraz, B. Courtois, "A generalized theory of fail-safe systems", *Proc. 19th IEEE Int. Symp. on Fault Tolerant Computing (FTCS-19)*, Chicago, USA, June 1989, pp. 398-406.

Norman 83 D.A. Norman, "Design rules based on analyses of human error", *Communications of the ACM*, vol. 26, no. 4, April 1983, pp. 254-258.

OFTA 94 French Observatory for Advanced Techniques, ARAGO 15, *Fault-Tolerant Computing*, Masson, Paris, 1994; in French: Observatoire Français des Techniques Avancées, *Informatique Tolérante aux Fautes*.

Powell 92 D. Powell, "Failure Mode Assumptions and Assumption Coverage", *Proc. 22nd IEEE Int. Symp. on Fault-Tolerant Computing (FTCS-22)*, Boston, July 1992, pp.386-395.

Rabejac et al. 96 C. Rabejac, J.P. Blanquart, J.P. Queille, "Executable assertions and timed traces for on-line software error detection", Proc. 26th IEEE Int. Symp. on Fault Tolerant Computing (FTCS-26), Sendai, Japan, June 1996, pp. 138-147.

Rabin 1989 M.O. Rabin, "Efficient dispersal of information for security, load balancing and fault tolerance", *Jounal of the ACM*, vol. 36, no. 2, April 1989, pp. 335-348.

Remus 82 L. Remus, "Methodology for software development of a digital integrated protection system", presented at the EWICS TC-7 meeting, Brussels, Jan. 1982, 19 p.

Rouse & Morris 87 W.B. Rouse, N.M. Morris, "Conceptual design of a human error tolerant interface for complex engineering systems, Automatica, vol. 23, no. 2, 1987, pp. 231-235.

RTCA 91 *Software Considerations in Airborne Systems and Equipment Certification*, RTCA paper no. 591-91/SC167-164, DO-178 B.5, 1991.

Ruegger 90 B. Ruegger, "Human error in the cockpit", Swiss Reinsurance Company, 1990.

Sheridan 1994 T.B. Sheridan, "Allocating functions among humans and machines", in "Improving function allocation for integrated systems design", Report from NATO Defence Research Group Panel-8/RSG.14, TNO Human Factors Research Institute, Soesteberg, The Netherlands, Nov. 1994, pp. 1-11.

Shooman 96 M.L. Shooman, "Avionics software problem occurrence rate", *Proc. 7th Int. Symp. on Software Reliability Engineering (ISSRE'96)*, White Plains, New York, Oct. 1996, to appear.

Siewiorek & Swarz 92 D.P. Siewiorek, R.S. Swarz, *Reliable Computer Systems, Design and Evaluation*, Digital Press, 1992.

Spradlin 80 R.E. Spradlin, "Boeing 757 and 767 flight management system", in *Proc. RTCA Technical Symp.*, Washington, Nov. 1980, pp. 107-118.

Toy 78 W.N. Toy, "Fault-tolerant design of local ESS processors", *Proceedings of the IEEE*, vol. 66, no. 19, Oct. 1978, pp. 1126-1145.

Traverse 88 P. Traverse, "AIRBUS and ATR System Architecture and Specification", *Dependability Computing and Fault-Tolerant Systems, Vol. 2, Software Diversity in Computerized Control Systems*, Springer Verlag, 1988, pp. 95- 104

Vardanega et al. 95 T. Vardanega, P. David, J.F. Chane, W. Mader, R. Messaros, J. Arlat, "On the development of fault-tolerant on-board control software and its evaluation by fault injection", *Proc. 25th IEE Int. Symp. on Fault-Tolerant Computing (FTCS-25)*, Pasadena (USA), 27-30 juin 1995, pp.510-515

Wood 94 A. Wood, "NonStop availability in a client/server environment", Tandem Technical Report 94.1, March 1994.

Session 3
Reliability and Safety Assessment

Systematic Development of Markov Models for the 1oo2D Programmable Electronic System Architecture - Analysis of Safety and Availability

William M. Goble
Moore Products Co., Sumneytown Pike, Spring House, PA, USA
Eindhoven University of Technology, Eindhoven, the Netherlands

Julia V. Bukowski
Department of Electrical and Computer Engineering
Villanova University, Villanova, PA, USA

Prof. Dr. Ir. A. C. Brombacher
Faculty of Mechanical Engineering
Eindhoven University of Technology, Eindhoven, the Netherlands

Abstract

Programmable Electronic Systems are being used in many industrial protection applications because they supply important benefits including calculation capability and automated documentation. There is a question though of whether these systems provide the necessary reliability and safety. Special circuit designs and special architectures are needed.

One architecture that has not been extensively analyzed is called the "1oo2D." It is constructed from two physical computers each with two logical channels, a logic solver channel and a diagnostic channel. In this paper three different Markov models will be developed for the 1oo2D architecture showing how the exact operation of the system will affect safety and availability.

Introduction

Programmable Electronic Systems (PES) are used in many industrial applications where protection systems can reduce the risk of a potentially hazardous industrial process. A PES offers several advantages for these safety protection applications. PES equipment offers graphical application design tools which reduce systematic errors. A PES is capable of extensive on-line diagnostics to detect electronic component failures. A PES has the calculation capability, the fast response time, and digital communications capability needed for many industrial processes.

These advantages are significant. However, the major disadvantage of PES equipment is the potential for dangerous failures. Electronic components have many failure modes some of which may prevent the PES from performing its protection function. In a majority of applications (de-energize-to-trip), failures which cause the output to stay energized are dangerous. Special circuit designs and architectures are needed to insure safety. As various PES design alternatives are considered, the safety and reliability of such designs must be evaluated.

Markov models offer a capable and appropriate tool to model the reliability and safety of a PES. These systems have multiple failure modes, self diagnostic capability which affects repair times, and fault tolerant architectures that maintain system operation when components fail. Common cause failures must be modeled. A Markov model can show all of this complexity in one drawing.

Systematic method

Markov modeling can be done more quickly and more accurately when a systematic method [1] is used to build the model. The following procedure is used:

1. Identify failure modes and failure rates of all system components.
2. Start the Markov model with a state where all components are successful.
3. For each system success state, build a checklist of all failure rates and failure modes of all operating components. Show each failure rate as an exit arc.
4. Repeat step 3 until no successful components remain.
5. For each state with a failed component, add appropriate repair rates.
6. Simplify model by merging states. Any states that have identical exit rates to the same states can be merged.

The use of these techniques greatly facilitates the construction of an accurate Markov model. The actual operation of the system under various component failure conditions becomes obvious.

Modeling Technique

The techniques developed by ISA's TR84.02 [Ref. 8] committee were used for the modeling. The notation and definitions are from Reference 8. For comparison purposes, we simplified the modeling by using a "single board" model. The system architecture is constructed from programmable electronics units consisting of input circuitry, a logic solver, and output circuitry (Figure 1). It was assumed that the failure of any component on the board causes the entire board to fail. The modeling technique can easily be extended to multi-board construction by adding additional failure rates. We also used the beta model [Ref.3] for common cause.

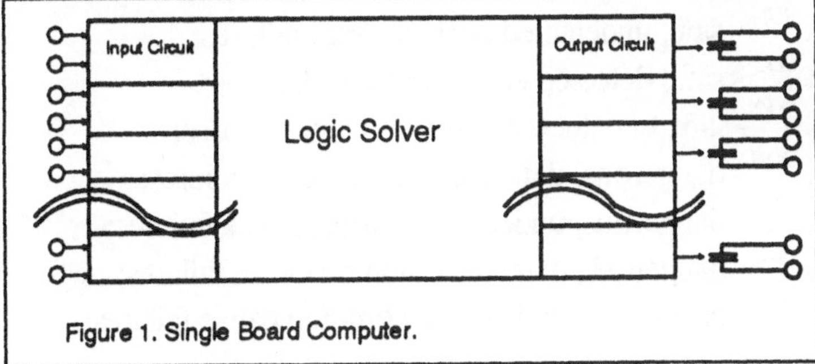

Figure 1. Single Board Computer.

Failure rates for each circuit section were estimated. These failure rates were first divided into two categories - safe, those failures that cause the unit output to fail de-energized and dangerous - those failures that cause the unit output to fail energized or frozen. Those two failure rates were partitioned again into ones detected by on-line diagnostics and ones not detected by on-line diagnostics. These categories are called detected and undetected. Thus far four failure rates (Figure 2) have been established:

λ^{SD} - safe, detected failures;
λ^{SU} - safe, undetected failures;
λ^{DD} - dangerous, detected failures; and
λ^{DU} - dangerous, undetected failures.

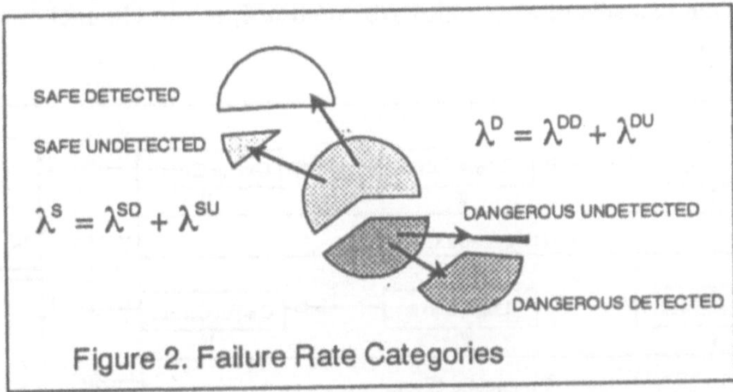

Figure 2. Failure Rate Categories

To properly account for common cause failures, each failure rate was partitioned into normal and common cause using the beta factor. For example:

$$\lambda^{SDN} = (1 - \beta)\,\lambda^{SD} \quad \text{, safe detected normal}$$

$$\lambda^{SDC} = \beta\,\lambda^{SD} \quad \text{, safe detected common cause}$$

This results in eight failure rates for each set of electronics in the PES:

λ^{SDN} - safe, detected normal stress failures;
λ^{SUN} - safe, undetected normal stress failures;
λ^{SDC} - safe, detected common cause failures;
λ^{SUC} - safe, undetected common cause failures;
λ^{DDN} - dangerous, detected normal stress failures;
λ^{DUN} - dangerous, undetected normal stress failures;
λ^{DDC} - dangerous, detected common cause failures;
λ^{DUC} - dangerous, undetected common cause failures;

The Markov models for three different versions of the 1oo2D architecture were developed using all eight failure rates for each successful unit in the system. The models are valid for an interval of time between major inspections. Thus, repairs are assumed only when on-line diagnostics detect a failure. All of the above assumptions are consistent with Reference 8.

1oo2D Architecture

The 1oo2D architecture was designed to provide high safety and high availability. It is typically implemented with two physical sets of electronics. Each set of electronics includes the input circuitry, a logic solver, and output circuitry. Each circuit has special diagnostic circuitry that combines to form another logical channel. When two sets of electronics are combined, a four channel architecture (Figure 3) is created.

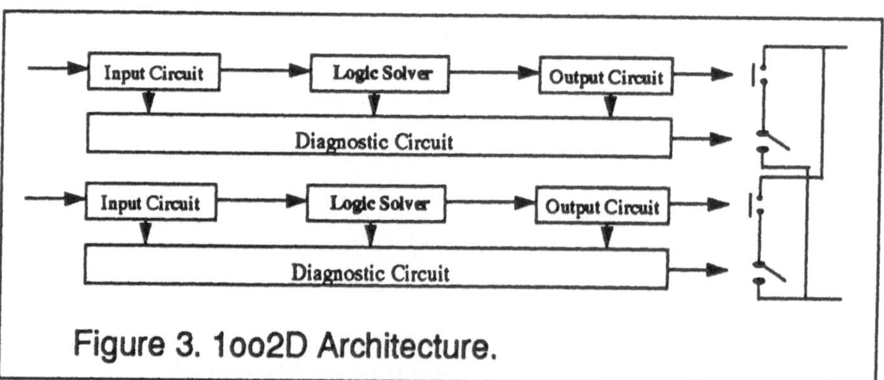

Figure 3. 1oo2D Architecture.

Conceptually, each of the two units in the PES reads inputs, calculates, and stores outputs. The diagnostic circuits monitor proper operation and will de-energize a second series output switch if a failure is detected. Any potentially dangerous failure is converted into a safe failure if detected by the diagnostics. If the diagnostics work perfectly, the system is fail safe. High availability is achieved through the parallel combination of the two sets of electronics. If one side fails safely, the other side maintains the load and the protection function.

The fundamental architecture requires good diagnostics. Diagnostic techniques have improved considerably however it is arguable if perfect diagnostics can be achieved. Therefore in order to assure high safety integrity, most actual implementations of the 1oo2D provide interprocessor communication between the logic solvers. A comparison of critical data between the two units provides protection independent of the diagnostics. When the comparison of either unit detects a mismatch, the system is de-energized (fail-safe).

Different operating modes

There are several possible variations in operation of the 1oo2D architecture. Three operational schemes were examined in this paper. The safety and availability of each variation were modeled for comparison purposes.

In one mode of operation called calculate / calculate mode, both units asynchronously process. No interprocessor comparison is performed. In normal operation all four switches are closed. A second mode of operation is called synchronous calculation mode. Both units synchronously read inputs, calculate and compare results, then generate outputs. Interprocessor communication allows synchronization and data comparison. All four switches are closed in normal operation.

A third mode of operation is called calculate / verify. One unit performs normal processing: read inputs, calculate and compare results, generate outputs. The other unit reads inputs, calculates and compares results, and then runs test and initialization routines. During normal operation the calculate unit keeps its switches closed and the verify unit keeps its switches open. Interprocessor communication allows data comparison. The calculate / verify role is constantly switched between units during normal operation. An instantaneous switch occurs when failure is detected.

The safety and availability of these different modes will vary. Each offers different failure degradation characteristics as shown in the Markov models.

Markov Model Development

Mode 1. Async. C/C - No IPC

The asynchronous calculate / calculate mode Markov model is published in reference 2. It is shown in Figure 4. In state 0, all components are operating properly. From this state the model must account for all failure rates in the checklist as shown in Figure 5. One set of common cause failure rates appears in the checklist since both successful units are failed by stress levels high enough for common cause [Ref. 3,4]. Two sets of normal failure rates represent the two successful operating units.

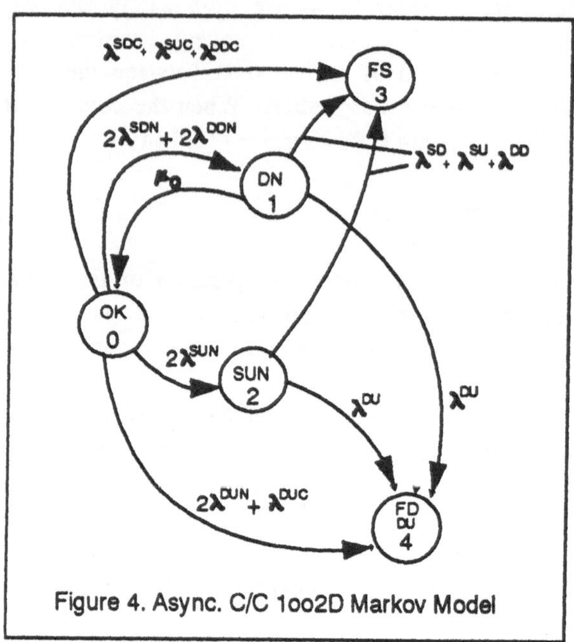

Figure 4. Async. C/C 1oo2D Markov Model

From state 0 any safe common cause failure causes the system to fail safely. A dangerous detected common cause failure will also fail safely since diagnostic circuits will de-energize outputs. A dangerous undetected common cause failure will prevent the system from protecting the process and is shown as a dangerous system failure (an arc from state 0 to state 4, fail-danger).

C 1	SDC	SUC	DDC	DUC	SDN	SUN	DDN	DUN
C 2					SDN	SUN	DDN	DUN

FIGURE 5. State 0 failure rate checklist.

Any detected failure from state 0 takes the system to a degraded state (state 1) where the system is still operating on one unit. An on-line repair rate for detected failures goes from state 1 to state 0. A safe undetected failure also degrades the system to single unit operation. Since the failure is not detected by diagnostics, no repair arc is provided. A dangerous undetected failure of one unit causes the entire system to fail dangerously. Since either unit can fail in this manner, a failure rate of $2\lambda^{DUN}$ is added to the arc from state 0 to state 4.

From either of the degraded states (state 1 and state 2) an additional safe or dangerous detected failure is safe. An additional dangerous undetected failure is dangerous.

Model 2. Synchronous C/C - IPC

The primary difference in the synchronous calculate / calculate mode is the addition of interprocessor communication and a comparison between data tables in each unit. If the comparison does not match, both logic solvers de-energize. A mismatch will occur if there are any undetected failures. The Markov model for this operating mode is shown in Figure 6.

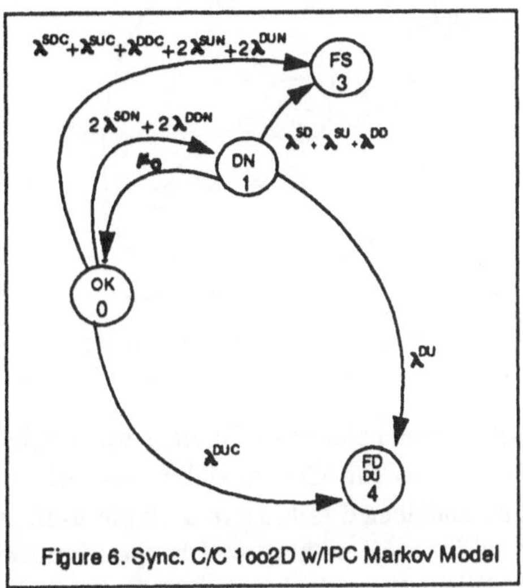

Figure 6. Sync. C/C 1oo2D w/IPC Markov Model

Comparing the Markov models, the differences include no degraded operating state when one unit fails safe undetected. Another significant difference is effect of dangerous undetected failures in one unit. With interprocessor communication and data table comparison, a mismatch due to a dangerous undetected failure in one unit will cause the system to fail safely not dangerously. Only a dangerous undetected common cause failure or a dangerous undetected failure from the degraded state can cause the system to fail dangerously

Mode 3. C/V - IPC

In the calculate / verify mode of operation, only the two switches from the calculate mode unit are energized. The verify mode unit has its output switches de-energized. When the two units switch modes, all switches are exercised. This constant dynamic situation makes diagnostics much simpler. This mode of operation also avoids certain software common cause failures characteristic of synchronous software systems. This operational mode does not automatically de-energize outputs if an undetected error causes a mismatch in compared data tables.

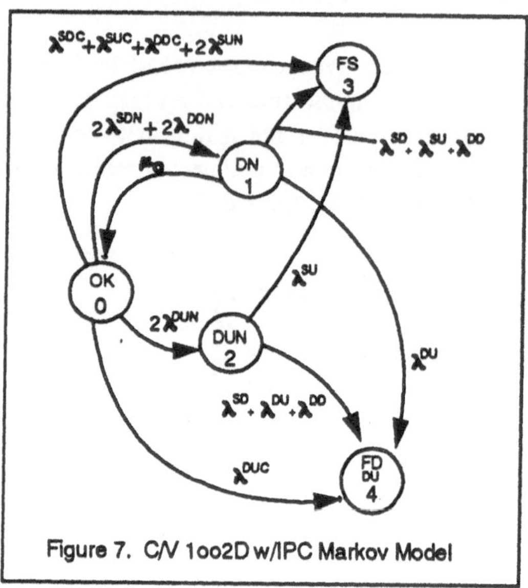

Figure 7. C/V 1oo2D w/IPC Markov Model

A version of the Markov model shown in Figure 7 was published in Reference 5. The significant difference in the C/V model is the effect of safe undetected failures and dangerous undetected failures in a single unit. If a safe undetected failure occurs in the verify unit, nothing will happen immediately. The system remains successful. When the calculation function switches to that unit, the system will fail safely. Assuming that the switching function is rapid, the safe undetected failure rate is placed in the arc from state 0 to state 3. When a dangerous undetected failure occurs in the calculate unit, the system is momentarily unable to respond to dangerous event. However, when the other unit switches into calculate mode the automatic protection function operates successfully. Again assuming that the switching function is faster than the amount of time required for the process to get into trouble, the system remains successful. This condition is represented in state 2.

Comparison

A qualitative examination of the models shows that the primary differences occur when an undetected failure exists. In order to quantify these differences the Markov models were solved using a sample set of input data as shown in Table 1. Failure rates are in units of FITS, failures per billion hours. An on-line MTTR of 8 hours was used. The models were solved using a spreadsheet to do matrix multiplication per well established methods [Ref. 7]. The spreadsheet model was solved for MTTF (Mean TIme To any Failure - a measure of availability), PFD

(Probability of Failure on Demand - a measure of safety), and HRF (Hazard Reduction Factor - inverse of PFD).

Results are shown in Table 2 for a one year time interval. Using the spreadsheet, comparisons were made over wide ranges of input data. With the exception of extremely poor diagnostic coverage and low common cause strength, the comparisons were always similar.

Table 1 - Input Data for comparison spreadsheet.

Variable	Symbol		Units
Input Circuit - safe failure rate	λ s ic	50	FITS
Logic Solver - safe failure rate	λ s mp	350	FITS
Output Circuit - safe failure rate	λ s oc	100	FITS
Input Circuit - dangerous failure rate	λ d ic	100	FITS
Logic Solver - dangerous failure rate	λ d mp	150	FITS
Output Circuit - dangerous failure rate	λ d oc	100	FITS
Software - safe failure rate	λ s sw	1000	FITS
Software - dangerous failure rate	λ d sw	0	FITS
Coverage - safe input failures	c s ic	0.9	Prob.
Coverage - dangerous input failures	c d ic	0.99	Prob.
Coverage - safe logic solver failures	c s mp	0.9	Prob.
Coverage - dang. logic solver failures	c d mp	0.99	Prob.
Coverage - safe output failures	c s oc	0.9	Prob.
Coverage - dang. output failures	c d oc	0.99	Prob.
Coverage - safe software failures	c s sw	0.9	Prob.
Coverage - dang. software failures	c d sw	0.99	Prob.
Quantity of input circuits	n	16	Quan.
Quantity of output circuits	m	8	Quan.
Prob. of common cause fail - hardware	β hw	0.05	Prob.
Prob. of common cause fail - software	β sw	0.1	Prob.

Table 2 - Comparison Results

	Async. C/C	Sync. C/C - IPC	C/V
MTTF	1,190,171	1,081,892	1,091,421
HRF	2251	46898	46898

Conclusions

Markov techniques are appropriate to analyze PES architectures like the 1oo2D. The multiple failure modes, effects of diagnostics, common cause failure and architectural details can be shown on a single drawing. These drawings can be qualitatively compared providing insight into operation. The models can easily be solved quantitatively over a wide range of possible inputs using ordinary personal computers. With these tools, design tradeoffs can be made.

In the case of the 1oo2D architecture, tradeoffs between safety and availability can be made. For safety critical applications however, the synchronous calculate / calculate mode or the calculate / verify mode offer higher levels of safety.

References

[1] Bukowski, J. V. and Goble, W. M., "The Reliability Analysis of PES Safety-Systems," *Proceedings of the Spring Symposium*, Instrument Society of America, Toronto, Canada, 1992.

[2] Goble, W. M., *Evaluating Control Systems Reliability - Techniques and Applications*, NC: Raleigh, ISA 1992.

[3] Stavrianidis, Paris, Common Cause Failure Models, Annex A, *Electrical (E) / Electronic (E) / Programmable Electronic Systems (PES) for Use in Safety Applications - Safety Integrity Evaluation Techniques*, Draft 4, ISA, June 1995.

[4] Brombacher, A. C., *Reliability by Design*, NY: New York, J. Wiley and Sons, 1992.

[5] Bukowski, J. V., and Goble, W. M. , "Using Markov Models for Safety Analysis of PES, "*Proceedings of the 50th Texas A & M Conference on Process Control*," ISA, 1995.

[6] Bukowski, J. V. and Goble, W. M., "Reliability Analysis of Controllers for Safety Shutdown Systems", *Proceedings - Ninth International Conference of the Israel Society for Quality Assurance (ISQA)*, Jerusalem, Israel, November, 1992.

[7] Maki, D.P. and Thompson, M., *Mathematical Models and Applications*, Prentice-Hall, Englewood Cliffs, NJ, 1973.

[8] *Electrical (E) / Electronic (E) / Programmable Electronic Systems (PES) for Use in Safety Applications - Safety Integrity Evaluation Techniques*, Draft 4, ISA, June 1995.

Failure Risk Estimation via Markov Software Usage Models

Walter J. Gutjahr
Department of Statistics, Operations Research and Computer Science
University of Vienna, Austria

Abstract

A software usage models describes the prospective use of a program in its intended environment and allows the generation of random test cases leading to unbiased estimates of the failure risk, i.e., the expected loss by program failure. We concentrate on usage models of Markov type and show that by suitable changes of the probabilities of state transitions during test, the precision of the risk estimate can be optimized. An algorithm for the computation of optimal transition probabilities is presented, and experimental results based on a C++ implementation of this algorithm are reported.

1 Introduction

Recently, software usage models of Markov type have found considerable interest (see [10, 11, 12, 9]). The purpose of a software usage model is to give a formal description of the expected operational use of a software system, i.e., its use in its intended application environment. Such a model is an essential prerequisite for statistical testing, a special variant of random testing allowing predictions on the operational behavior of the software ([8, 9]).

Markov software usage models aim at representing the (estimated) distribution of possible uses by means of a Markov chain. This probabilistic concept combines the ability of picturing relative complex dynamic use structures with the advantage of still being mathematically tractable. A Markov usage model is based on a directed graph $G = (V, A)$, where

- V is a set of *nodes*, representing usage states (e.g., program invocation, program termination, input/output screens), and
- A is a set of *arcs*, representing state transitions which always correspond to specific operations of the program. An arc from state i to state j can be denoted by the ordered pair (i, j).

Furthermore, each arc (i, j) is labelled by a *transition probability* $p(i, j)$. This value indicates the relative frequency of a transition to state j, given that the current state is state i, during the operational use of the program.

Fig. 1 shows a (very simple) example of a Markov usage model: A program has a main menu, from which the user may select two special functions. The probabilities for the selection of function 1 resp. function 2 are estimated to be 0.6 resp. 0.3. After the execution of the selected function, the program returns to the main menu. In about 10 percent of all cases, the user decides not to make a (new) function call, but to terminate the program.

A Markov usage model of more realistic, but still moderate complexity is shown in Fig. 2. It is models a (small) part of a train schedule program of the Austrian Federal Railways.

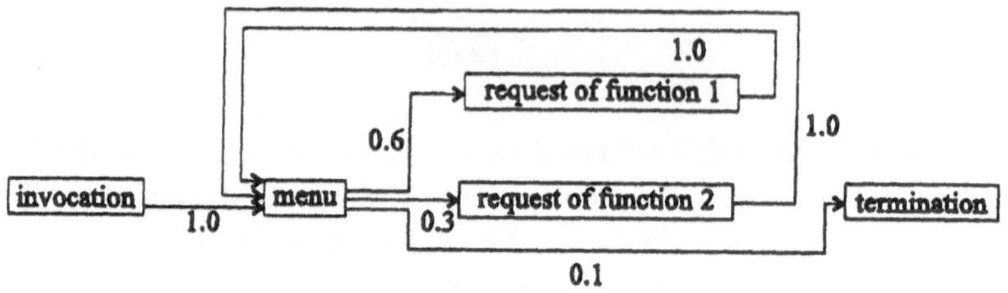

Fig. 1. A simple Markov usage model. The values assigned to the arcs are transition probabilities; the arcs themselves are operations of the program.

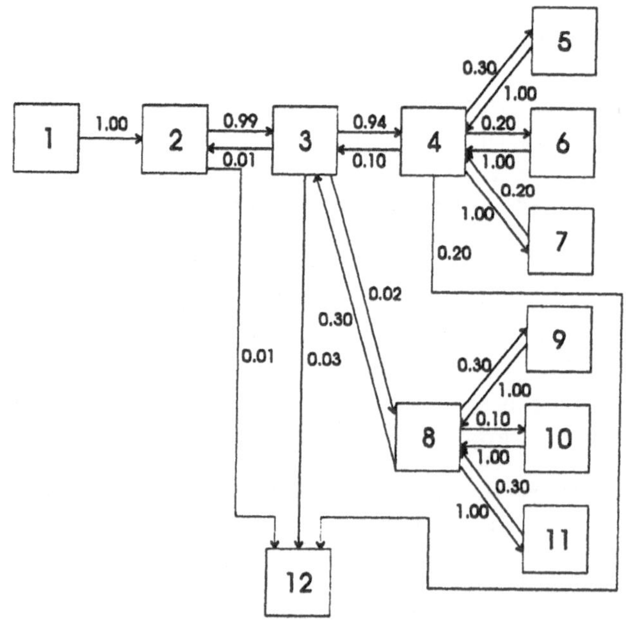

Fig. 2. A Markov usage model with $n = 12$ nodes.

We always assume that the states (nodes) of G are numbered by $1, \ldots, n$. State 1 is the *initial state* corresponding to program invocation, state n is the *final state*, corresponding to program termination. In order to obtain a complete Markov chain, we add an arc with assigned probability one from node n to node n; by this self-loop, n gets an *absorbing state* which cannot be left anymore.

Since program failures cannot be completely excluded, each program has a *failure risk*, defined as the expected loss by program failure during a single program execution in the operational environment (see [1, 6]). A straightforward method for getting an unbiased estimate of the risk consists in statistical testing with an input distribution corresponding to the operational use, and to take the average loss occured during test as a risk estimate. On the base of a Markov usage model, a statistical test case is obtained as a state sequence or *path* $(X_0, \ldots, X_q)$ of the usage chain, beginning at the initial state and ending at the final state, where each transition (X_{k-1}, X_k) is selected according to the operational transition probabilities $p(i, j)$.

Walton, Poore and Trammell [9] have outlined a serious drawback of this method: There can be critical operations (connected with high failure probability and/or high loss in case of failure) that are nevertheless invocated infrequently during the operational use of the program. Then, in a Markov chain model, the corresponding arcs get only small probabilities of being activated, which may exclude these operations from test. Walton et al. suggest an adjustment of the usage distribution to ensure that such operations are tested more frequently, but they admit that this approach limits statistical inferences on the program behavior in its operational environment.

The intention of this paper is to develop a technique of changing the transition probabilities *and* compensating the effect of this change in such a way that

- critical operations are tested with sufficiently high frequency, and
- the risk estimate derived from the test remains unbiased.

To optimize the benefit of our technique, we try to find *that* probability change that allows *the most precise risk estimation*. (A similar approach for the more special case of a program with n alternative program functions has been developed in [2].)

Let us emphasize that the drawback of conventional Markov usage models mentioned in [9] is especially prohibitive in the case of *safety-critical software*, where there are always operations with low use frequency and high criticality. So, without a technique overcoming this drawback, the applicability of Markov usage models to this type of software would be very limited. The technique outlined here provides a satisfactory solution to this problem.

2 Change of transition probabilities

Our method relies on the *importance sampling* technique of Rare Event Simulation, which prescribes that a probility change in a simulation has to be compensated by adjusting suitable weights to the outcomes of the simulation run. For a Markov usage model, one obtains the following compensation for a shift of the transition probabilities from $p(i,j)$ to $t(i,j)$: Let loss(X) denote the loss occuring in a single test execution along path $X = (X_0, \ldots, X_q)$, resp. loss$(X) := 0$ if no failure occurs along X. Then (cf., e.g., [4]),

$$S := \prod_{k \geq 1} \frac{p(X_{k-1}, X_k)}{t(X_{k-1}, X_k)} \cdot \text{loss}(X) \tag{1}$$

is an unbiased estimate of the risk. This can be seen as follows. Let P resp. T stand for the operational distribution, resp. the test distribution obtained by the probability change. By E_P and E_T we denote the mathematical expectation under distribution P and T, respectively. Obviously,

$$p(X) := \prod_k p(X_{k-1}, X_k) \quad \text{resp.} \quad t(X) := \prod_k t(X_{k-1}, X_k)$$

is the probability that path X occurs, given that the states are selected according to the probabilities $p(i,j)$ resp. $t(i,j)$. So for the expected value of S under distribution T,

$$E_T(S) = E_T\left(\frac{p(X)}{t(X)} \cdot \text{loss}(X)\right) = \sum_x t(x) \cdot \frac{p(x)}{t(x)} \text{loss}(x)$$

$$= \sum_x p(x) \,\text{loss}(x) = E_P(\text{loss}(X))$$

holds. The rightmost expression, however, is just the risk. Let us mention that in importance sampling, the weight $p(X)/t(X)$ is usually called the *likelihood ratio*.

Taking the average value of S for all random test cases (selected according to the changed transition probabilities) yields the overall risk estimate.

For example, the transition probabilities in Fig. 1 could be changed to, say,

$$t(\text{menu, req\_function\_1}) := 0.2, \quad t(\text{menu, req\_function\_2}) := 0.8$$

(other probabilities as before). Assume that, selecting states according to these new probabilities, we obtain the path

$$X = (\text{invoc, menu, req\_function\_2, menu, req\_function\_1, menu, term}),$$

and that an execution of the program with inputs driving it along X reveals a program failure of a severity estimated by 10 cost units. Then this loss has to be weighted by

$$\frac{0.3}{0.8} \cdot \frac{0.6}{0.2} = 1.125,$$

such that we derive a risk estimate of $S = 1.125 \cdot 10 = 11.25$ from this test case. Of course, the risk estimate corresponding to a test case is zero, if the test case does not reveal a failure. If N test cases are selected, executed, and evaluated in the way described above, yielding risk estimates $S(1), \ldots, S(N)$, then the overall risk estimate is given by the arithmetic mean $\bar{S}$ of the values $S(1), \ldots, S(N)$.

3 Optimizing the precision of the risk estimate

Now let us turn to the question how to choose the transition probabilities in order to get a risk estimate with maximum precision. As a measure for the unprecision of an (unbiased) estimator we take its *variance*, as customary in software reliability (cf. [7]). The variance $var_T(S)$ of S for test cases selected according to test distribution T would satisfy our requirements on a criterion; however, $var_T(S)$ cannot be determined in advance, since S depends on loss(X), and loss(X) is unknown before the test.

Therefore, a Bayesian approach is applied: For each fixed x, loss(x) is conceived as a random variable, whose distribution reflects the *prior information* of the tester on failure probabilities and amount of loss in case of failure. To be more explicit: It is assumed that the operation corresponding to arc (i, j) has a prior probability $f(i, j)$ of failing, and causes a loss of $l(i, j)$ in case of the failure event. Failures of different arcs (i, j) occur independently from each other. The values $f(i, j)$ and $l(i, j)$ have to be estimated for each arc. Although this model is only an approximation to reality, it already allows the test designer to take account of critical operations, which is not possible in the conventional Markov usage model framework.

Denoting the expectation with respect to the (estimated) prior loss distribution by E_Q, the aim is to minimize $E_Q(var_T(S))$ by an appropriate choice of the test distribution T. (Notice that E_Q acts, for *fixed* path x, on the loss, while E_P and E_T act on the path.)

In the sequel, we assume that the loss occuring at a specific program execution mainly depends on the path x triggered by the input data. Then, for fixed x, loss(x) can be considered (in a first approximation) as independent of the actual input causing the execution of x. Of course, this is a simplification: For example, it may happen that input_1 and input_2 lead to the same path x, but input_1 is processed correctly, while input_2 is not. Nevertheless, a Markov usage model with a sufficiently high granularity (cf. [9]) has the property that inputs triggering the same path are processed in a very similar way, such that they can be considered as "near–to–homogenous" in their failure behavior. (For the notion of homogenous sets of inputs, see [3]). We emphasize that this assumption is not required for the unbiasedness of the estimator (1).

By $A(x)$, the set of transitions (i, j) contained in path x is denoted. According to the model described above, we obtain now

$$\text{loss}\,(x) = \sum_{(i,j)\in A(x)} \text{fail}\,(i, j)\, l(i, j), \tag{2}$$

where

$$\text{fail}\,(i, j) = \begin{cases} 1, & \text{if transition } (i, j) \text{ fails,} \\ 0, & \text{else,} \end{cases}$$

such that $E_Q(\text{fail}\,(i, j)) = f(i, j)$. Since

$$var_T(S) = E_T(S^2) - [E_T(S)]^2 = E_T(S^2) - (\text{risk})^2,$$

minimization of $E_Q(var_T(S))$ with respect to T is equivalent to minimization of $E_Q(E_T(S^2))$. By a short calculation, one finds

$$E_Q(E_T(S^2)) = E_P\left(\frac{p(X)}{t(X)} \cdot E_Q([\text{loss}(X)]^2)\right). \qquad (3)$$

Furthermore, because of (2), $E_Q([\text{loss}(X)]^2)$ is equal to

$$E_Q\left(\sum_{(i,j)} \text{fail}(i,j)\,[l(i,j)]^2 + \sum_{(i,j)\neq(k,m)} \text{fail}(i,j)\,\text{fail}(k,m)\,l(i,j)\,l(k,m)\right)$$

$$= \sum_{(i,j)} f(i,j)\,[l(i,j)]^2 + \sum_{(i,j)\neq(k,m)} f(i,j)\,f(k,m)\,l(i,j)\,l(k,m), \qquad (4)$$

all sums being only over arcs contained in $A(X)$. On the usual assumption that prior failure probabilities are small quantities (only programs that are already relatively stable are worthwile to be exposed to tests for risk estimation!), the second sum on the rightmost side of (4) is negligible, compared to the first sum. Omission of this second sum, insertion into (3) and re-insertion of the expressions for $p(X)$ and $t(X)$ leads to the following stochastic optimization problem:

$$\text{Minimize } G(T) := E_P\left(\prod_{k\geq 1} \frac{p(X_{k-1}, X_k)}{t(X_{k-1}, X_k)} \cdot \sum_{(i,j)\in A(X)} f(i,j)\,(l(i,j))^2\right). \qquad (5)$$

Therein, the variables $t(i,j)$ $(1 \leq i,j \leq n)$ to be optimized have to satisfy the constraint of being transition probabilities of a Markov chain with absorbing state n. Additionally, the constraint $t(i,j) = 0$ if and only if $p(i,j) = 0$ must be satisfied, otherwise the likelihood ratio could become infinite.

It can be shown that the function $G(T)$ in (5) is *convex* in the variables $t(i,j)$. This fact allows an efficient solution of (5) by well-known stochastic optimization techniques.

4 Numerical computation of the optimal transition probabilities

For the numerical solution of (5), we use, in our implementation, an optimization algorithm of Frank–Wolfe–type (see [5], section 10.10.3). Basically, we proceed as follows: A sample of R independent paths according to distribution P is drawn, and the minimization of the expected value of the product in (5) is replaced by the minimization of the average value of the product over the paths in the sample. This is done iteratively. We start with the $[n \times n]$-matrix T of the variables $t(i,j) := p(i,j)$ as the initial solution. The solution is improved successively: For each fixed line i of the current matrix T, consider the partial derivatives $D(i,j)$ of $G(T)$ to the variables $t(i,j)$ $(j = 1,\ldots,n)$. Take that $j = j^*$ for which $D(i,j)$ is minimal, and modify the current solution T by augmenting the value of $t(i,j^*)$ by a certain stepsize. The partial derivatives $D(i,j)$ can be computed explicitly. We use the result of this computation in the following, more detailed description of the algorithm:

procedure optimize_test (P, F, L)
/* $P = (p(i,j))$, $F = (f(i,j))$, $L = (l(i,j))$ */
{ **for** $r := 1$ to R
  { draw path X according to P;
    compute $Y(r) := \prod_{k\geq 1} p(X_{k-1}, X_k) \sum_{(i,j)\in A(X)} f(i,j)\,(l(i,j))^2$;
    for each (i,j)
      compute $M(r,i,j)$:= number of transitions $(X_{k-1}, X_k) = (i,j)$; }
  set $T := P$;
  for $s := 1$ to S
    for $i := 1$ to n
    { **for** each j with $p(i,j) > 0$
        compute $D(i,j)$:= arithmetic mean over $r = 1,\ldots,R$ of
        $-Y(r) \prod_{k,m} (t(k,m))^{-M(r,k,m)}\, M(r,i,j) \,/\, t(i,j)$;
      determine that $j = j^*$ for which $D(i,j)$ is minimal;
      for each j with $p(i,j) > 0$
      { $t(i,j) := (1 - 1/(2s))\, t(i,j)$;
        if $(j = j^*)$ $t(i,j) := t(i,j) + 1/(2s)$; } }
  return T; }

By some minor modifications, both the runtime efficency and the solution quality of this algorithm can still be improved.

5 Experimental results

For judging the solution quality of the approximation algorithm optimize_test described in the previous section, one may start with the small example of a Markov usage model in [9], p. 102, since it is still possible to calculate the solutions of (5) *exactly*, for this case. The Markov chain in the example has four states (invocation, main_menu, display, and termination). The operational transition probabilities, the failure probabilities and the failure severities (losses in case of failure) are estimated as follows:

$$P = \begin{bmatrix} 0 & 1 & 0 & 0 \\ 0 & 0 & 0.5 & 0.5 \\ 0 & 1 & 0 & 0 \\ 0 & 0 & 0 & 1 \end{bmatrix}, \quad F = \begin{bmatrix} 0 & 0.1 & 0 & 0 \\ 0 & 0 & 0.4 & 0.1 \\ 0 & 0.1 & 0 & 0 \\ 0 & 0 & 0 & 0 \end{bmatrix}, \quad L = \begin{bmatrix} 0 & 1 & 0 & 0 \\ 0 & 0 & 5 & 1 \\ 0 & 1 & 0 & 0 \\ 0 & 0 & 0 & 0 \end{bmatrix}.$$

Because of the condition $t(i,j) = 0 \Leftrightarrow p(i,j) = 0$, the test transition probabilities are identical to the values $p(i,j)$, except $t(2,3) =: t$ and $t(2,4) = 1 - t$. Explicit solution of (5) yields $t^* = 0.620$ as the optimal value for t.

Our C++ implementation of optimize_test finds for $R = 100$ and $S = 100$ an approximation to the exact value of t^* that is is sufficiently good for

R	S	K	t^*	Runtime (in sec)
100	100	1	0.635	6
1000	100	1	0.617	59
100	100	10	0.621	56

Table 1. Application of optimize_test to the 4–state Markov usage model: results and runtimes on a PC with Intel 80486 DX processor.

practical purposes (see line 1 of Table 1). Of course, increasing the sample size R improves the solution quality (see line 2 of Table 1). Note, however, that storing the entries $M(r,i,j)$ requires space of order $O(R \cdot n^2)$, which can be prohibitive for large Markov usage models. So we have investigated an alternative: Instead of generating a very large sample of paths and storing the quantities $Y(r)$ and $M(r,i,j)$ for all these paths, we generate only a more limited sample, perform the approximation steps of optimize_test, and *repeat* this process K times, such that newly drawn samples can influence the current approximate solution. In order to obtain convergence of the algorithm, the stepsize for the modification of the $t(i,j)$ has to be decreased in the successive iterations $k = 1, \ldots, K$. We have chosen a stepsize of $1/(2sk)$ in the kth iteration. The effect of this procedure is here even more convincing than that of working with a large sample as a whole (see line 3 of Table 1). However, the last observation cannot be generalized: for other values R and S, we have also found cases where augmenting K by a certain factor produced slightly worse results than augmenting R by the same factor.

Another situation where exact solution values can be determined is the case of Markov chains consisting of edge–disjoint paths from node 1 to node n. Here, Theorem 2 in [2] yields an explicit formula for the optimal test probabilities. We have tested some examples of this type and found a good agreement between theoretical and experimental values.

Finally, we present the outcomes of optimize_test for the Markov usage model of Fig. 2. The values $f(i,j)$ and $l(i,j)$ have been estimated as follows:

$F =$

	1	2	3	4	5	6	7	8	9	10	11	12
1	0	l	0	0	0	0	0	0	0	0	0	0
2	0	0	m	0	0	0	0	0	0	0	0	l
3	0	l	0	m	0	0	0	l	0	0	0	l
4	0	0	l	0	m	m	l	0	0	0	0	l
5	0	0	0	l	0	0	0	0	0	0	0	0
6	0	0	0	l	0	0	0	0	0	0	0	0
7	0	0	0	l	0	0	0	0	0	0	0	0
8	0	0	l	0	0	0	0	0	l	h	l	0
9	0	0	0	0	0	0	0	l	0	0	0	0
10	0	0	0	0	0	0	0	l	0	0	0	0
11	0	0	0	0	0	0	0	l	0	0	0	0
12	0	0	0	0	0	0	0	0	0	0	0	0

$L =$

	1	2	3	4	5	6	7	8	9	10	11	12
1	0	m	0	0	0	0	0	0	0	0	0	0
2	0	0	m	0	0	0	0	0	0	0	0	l
3	0	m	0	m	0	0	0	m	0	0	0	l
4	0	0	m	0	m	m	m	0	0	0	0	l
5	0	0	0	m	0	0	0	0	0	0	0	0
6	0	0	0	m	0	0	0	0	0	0	0	0
7	0	0	0	m	0	0	0	0	0	0	0	0
8	0	0	m	0	0	0	0	0	l	h	m	0
9	0	0	0	0	0	0	0	m	0	0	0	0
10	0	0	0	0	0	0	0	m	0	0	0	0
11	0	0	0	0	0	0	0	m	0	0	0	0
12	0	0	0	0	0	0	0	0	0	0	0	0

Therein, the constants l (low), m (medium) and h (high) have the following values: In F, $l = 0.001$, $m = 0.01$, and $h = 0.1$. In L, $l = 1$, $m = 10$, and $h = 100$. We have assumed that arc (8,10) corresponds to a newly implemented (i.e., error–prone) function with high criticality. Here, explicit solutions of (5) are not available anymore, so we cannot compare the output of optimize_test with the true solution values. For $R = 100$, $S = 100$ and $K = 10$, we obtained, after 196 seconds of computation time, the following output matrix T:

$T =$

	1	2	3	4	5	6	7	8	9	10	11	12
1	.00	1.0	.00	.00	.00	.00	.00	.00	.00	.00	.00	.00
2	.00	.00	.99	.00	.00	.00	.00	.00	.00	.00	.00	.01
3	.00	.01	.00	.83	.00	.00	.00	.11	.00	.00	.00	.05
4	.00	.00	.11	.00	.31	.20	.20	.00	.00	.00	.00	.18
5	.00	.00	.00	1.0	.00	.00	.00	.00	.00	.00	.00	.00
6	.00	.00	.00	1.0	.00	.00	.00	.00	.00	.00	.00	.00
7	.00	.00	.00	1.0	.00	.00	.00	.00	.00	.00	.00	.00
8	.00	.00	.29	.00	.00	.00	.00	.00	.15	.28	.28	.00
9	.00	.00	.00	.00	.00	.00	.00	1.0	.00	.00	.00	.00
10	.00	.00	.00	.00	.00	.00	.00	1.0	.00	.00	.00	.00
11	.00	.00	.00	.00	.00	.00	.00	1.0	.00	.00	.00	.00
12	.00	.00	.00	.00	.00	.00	.00	.00	.00	.00	.00	1.0

Comparing the values $t(i,j)$ with the values $p(i,j)$ unveils that the intended effect of favoring the test of critical operations has been achieved: While for, say, 100 test cases in total, the generation of test sequences according to P gives the critical function (8,10) only a small probability of being tested (note that already (3,8) has a small probability!), transition (8,10) obtains a fair chance in a test sequence generation according to T.

Since in a careful implementation, the runtime of optimize_test grows only linearly in the size (number of entries) of the input matrices, even large Markov usage models can be treated in reasonable time. The bottleneck is the storage requirement of order $O(Rn^2)$: For large n, R has to be chosen relatively small, which reduces the solution quality. This can partially be compensated by increasing the value K. In any case, the number $R \cdot K$ of generated test sequences should be *essentially* larger than the number of test cases to be executed. Otherwise, it may happen that infrequently used, but error–prone paths do not occur in the sample and get therefore no opportunity to shift the numbers $t(i,j)$ to more favorable values.

An approach for the solution of the last-mentioned problem is to reformulate (5) in such a way that instead of E_P, the expected value $E_{\tilde{P}}$ with respect to another distribution $\tilde{P}$ giving *all* possible state transitions fair chances is used. For the sake of brevity, we omit the details of this more sophisticated technique.

References

[1] Ehrenberger, W. D., "Combining probabilistic and deterministic verification efforts", *Proc. SAFECOMP '92* (ed. H. Frey), 299 - 304, Oxford: Pergamon Press (1992).

[2] Gutjahr, W. J., "Optimal test distributions for software failure cost estimation", *IEEE Trans. Software Eng.*, vol. 21 (3), 219 - 228 (1995).

[3] Hamlet, R., Taylor, R., "Partition testing does not inspire confidence", in: *Proc. 2nd Workshop an Software Testing, Verification, and Analysis*, July 1988, 206 - 215 (1988).

[4] Heidelberger, P., "Fast simulation of rare events in queuing and reliability models", *ACM Trans. on Modeling and Computer Simulation*, vol 5 (1), 43 - 85 (1995).

[5] Murty, K. G., *Linear Complementary, Linear and Nonlinear Programming*, Heldermann (1988).

[6] Sherer, S. A., *Software Failure Risk*, New York, London: Plenum Press (1992).

[7] Thayer, Th. A., Lipow, M., and Nelson, E. C., *Software Reliability*, Amsterdam: North-Holland (1978).

[8] Thevenod-Fosse, P., "From random testing of hardware to statistical testing of software", *Proc. IEEE COMPEURO '91*, 200 - 207 (1991).

[9] Walton, G. H., Poore, J. H., Trammell, J., "Statistical testing of software based on a usage model", *Software - Practice and Experience*, vol. 25 (1), 97 - 108 (1995).

[10] Whittaker, J. A., "Markov chain techniques for software testing and reliability analysis", Ph. D. dissertation, University of Tennessee (1992).

[11] Whittaker, J. A., Poore, J. H., "Markov analysis of software specifications", *ACM Trans. Software Eng. and Method.*, vol. 2 (1), 93 - 106 (1993).

[12] Woit, D. M., "Operational profile specification, test case generation, and reliability estimation for modules", Thesis, Queen's University, Kingston, Ontario, Canada (1994).

Design Methodologies and Systems Safety

Daniel E. Sniezek
Lockheed Martin Federal Services
Owego, NY USA

1 Introduction

System Safety in the computer control industry is becoming an important issue. Computers are moving from desks to factory floors as process controllers. Computer functionality and reliability are becoming very important on the factory floor, especially when used in safety systems. It is no longer acceptable for a computer to stop processing data. On the factory floor, stopping processing is not only a problem of availability of equipment, but of safety.

Microprocessors are being used in basic process control systems and have been very successful. Due to this success, microprocessors are being used as safety systems. These safety systems are referred to as Process Electronic Systems (PES). When designing a PES, it is important to understand what the requirements of the design are early in the project. International standards are making design metrologies of software critical to the certification of PES designs used in safety.

There is much the software world can learn from the world of hardware. In the spirit of Systems Science, one discipline can learn from another. This paper will look at Design For Manufacturability and see how these concepts can be used in software development.

Using another Systems Science approach, this paper will look at the whole area of design in general (Generic Design). Understanding where PES design relates to a search for the solution space will help understand why some methodologies work better than others.

Classical Design will be briefly presented, then a look at new methodologies in PES design will be explored. Special consideration will be given to requirement and specification definition. According to the Health and Safety Executive of the United Kingdom, 44% of all accidents are due to specifications not having the proper function or integrity[1].

2 Generic Design

To create something is a goal of many. Design is a process of solving a problem. Problems relate to people because only people perceive problems. The design process relates to peoples' thoughts and actions. Gause and Weinberg define a problem

---

[1] Gruhn, P. ISA S84..."I have to do what?", ISA Safety Standard May 1996: 42-44

GENERIC DESIGN CONTINUUM[2] (Figure 1)

Design Factors	Indigenous	Satisficing	Optimizing
Means of search	Natural selection via experimentation	Synthesis guided by intuition	Creation and analysis of abstract models
Convergent characteristics	Divergent or chaotic	Converges to a region	Converges to a global or local optimum
Goal of solution	Comfortable	Satisfactory	Optimal
Complexity	Highly complex with many interactions	Manageably complex with few interactions	Simple: with few variables and simple interactions
Performance measure	Not formal	Semi-formal but not exactly defined	Formal and exactly defined
Design decisions	Democratic with many participants	Oligarchic with few participants	Autocratic with very few participants
Design awareness	Unselfconscious	Self-conscious	Unselfconscious
Design methodology	Tradition	Rules of thumb	Formal rules
Design space	Global context space	Eco-space	Solution space
Design frozen	Never	Early in the life of the form	Very early before the form is released to its environment

as "... a perceived difference between things as they are and things as they should be". A good design is a design that perception is reality.

2.1 Search of Solution-Space

The process of designing can be viewed as a search for an apple on a tree starting at the trunk. The solutions to the problem are the apples on the tree. Remembering that different problems have the possibility of multi-solutions, one solution or even none. The search for the apple is an exciting process of understanding performance functions in the requirements phase, then finding solutions.

To define a generic design problem, one must define the constraints, the response to inputs and the method of determining preferences of the solutions within all possible solutions. The designer starts by attempting to remove the uncertainty from the problem statement. This is referred to as the requirements phase. Understanding the true problems is critical to the success of the design. The design phase is where the designer determines the form of the solution.

---

2 Gause, Donald C., Minch, Eric, "Design Processes: A Space Perspective", Design and Systems, Vol. 12, 1990, p 33-50

2.2 Design Strategies

Generic design is a continuum that starts with natural selection via experimentation that leads to an indigenous design through satisficing to optimizing. Different design strategies are more affective depended on design factors. As shown in Figure 1, design factors for indigenous design take on a high degree of complexity using natural selection taking a lengthy time looking for a comfortable solution.

The optimizing design strategy has a very small set of variables with little interactions that leads to a single local optimum solution. Satisficing design takes on a number of possible solutions using "rules of thumb" as a design methodology. This paper will hope to show Programmable Electronic Systems (PES) take on the design factors of a satisficing design.

2.3 People Relationships

The relationship the designer has with the perceivers in the design process will affect the design either positively or negatively. The users are people who react to the system, but do not change the system. Users are people who are affected in any way by the system. The designers adapt the system and change its identity or characteristics.

2.3.1 Client and Designers

The client is the group of users who can hire or fire the designers from the design project. The clients perceive a need and have the power to commission designers to solve the problem.

The client is usually the first person the designer has contact with in the design process. The preliminary information about the project is usually given by the client. The client, in most cases, is not the final user of the system and the designer must take this into consideration.

2.3.2 Maintainers and Designers

The maintainers are the users that keep the system operating after implementation. When maintainers are overlooked, the system may function well at implementation, but will surely fail due to lack of adjustment or repair.

An example of overlooking the maintainers is explained by Donald E. Petersen, former CEO of Ford Motor Company, when he admits that Ford had problems with their computer based control system when it was first implemented. Petersen says that Ford "... hadn't found the time to teach the repairmen how to solve some extremely sophisticated computer-related problems in our cars." What made matters worse, "... more and more owners had to come back several times for the same problem, and, predictably, they started to get mad".[3] The designer must consider what happened after

[3] Petersen, D. E. and Hillkirk, J., A Better Idea: Redefining the Way Americans Work, Houghton Mifflin Company, Boston, Mass, 1991:151

implementation. The maintainers must be trained about the new system or the system must be designed to take advantage of the knowledge base of the maintainers.

3 Classical Design of PES

Programmable Electronic Systems (PES) Design is a small part of Generic Design with design characteristics as shown by the highlighted areas of the design continuum in the previous table.

Design of PES takes on the attributes of a satisficing design with some attributes in the indigenous and optimizing design. PES design takes on the means of the search as a synthesis guided by past experience and in many cases intuition. With the advent of computerized models such as Markov and Monte Carlo, analysis of Probability of Failure on Demand (PFD) can be determined. These models are not used to optimize the design but to validate the choice of PES configurations of sensors, process controllers and actuators.

PES design converges on a region of solutions. When designing to a Safety Integrity Level (SIL) of 1, there are actually a number of satisfactory solutions. There is not only one. The designer is looking to get in a region not looking for the optimum design. If they were looking for the local optimum they may never find it for the field is moving so fast. Designers must understand the idea of a regional search, for if they are looking for the optimum, they may never find it. This is when the force fit comes in. If the design is better than what is currently available and meets the SIL targets, then the project should continue. Don't wait for the optimum solution! The real goal of a PES design is to achieve a satisfactory solution.

PES design is becoming more complex. Complexity is reduced through the use of standards. Standards simplify the design into more modular forms that can be reused over again with some past history. We attempt to reduce the interactions but they are never really reduced. In an optimizing design, the system must have few variables and simple interactions. Even though the designer attempts to simplify the PES design, especially in safety systems, many variables and interactions still exist.

The measure of a PES design process are semi-formal and are not defined in a repeatable mathematical sense. The measure often takes on a rule of thumb methodology. The PES designer looks at other companies that do similar designs and use "Good Engineering Practices" with a rule of thumb. The formal rules of science are sometimes used but on only small parts of the overall PES system such as determining the maximum current for an output module. To determine all possible software errors is beyond the formal rules so a methodology for verification and validation is used during design . In the Generic Design Continuum, the PES design takes on the design factors of a satisficing design.

3.1 Programmable Electronic Systems Design

The Classical PES design takes on the traits of a satisficing design as described earlier. Designer/engineers take on a project assigned to them by their manager and attempt to optimize the system within the scope of the project. In reality, the designer/engineer attempts to optimize but will settle on any design that will satisfy the client's request.

The client is usually a production representative from management who wishes to change the current process to improve some characteristic of the process. The client has little experience in the design processes or knowledge of "what is possible". The client simply has a problem and wishes it solved. The client's primary concern is productivity of the current process for the purpose of manufacturing more widgets. The client is usually not concerned with the quality of those widgets.

The operators are the user group who will manipulate the system when it is implemented. In the "typical" design process the operator is never consulted until the design is implemented.

The designer of the PES is part of the Engineering Department and is usually known as an engineer. In some cases, the designer/engineer is actually not educated in the engineering discipline but may have considerable experience and is called an engineer.

The designer/engineer uses his/her past experience and educational background to search for a solution to the project. The designer/engineer uses his/her contacts through vendors to determine what products have already been developed to solve the project. There is normally not one product that can accomplish the task assigned, so the designer/engineer will combine products and/or develop new products that can be combined with existing products to solve the client's problem. The designer/engineer is judged by speedy and low cost design implementation. As a result, the operator's input is often overlooked, only the client provides the "real" problem.

If the client is not knowledgeable enough to clearly explain the "real" problem, the design may be filed in the designer's drawer forever. The client may settle for a lesser problem that can be explained.

The nature of the design task is moderately complex with few interactions. The designer/engineer attempts to optimize the design but seldom looks at the interactions. The designer/engineer's concern is local to the plant facility. They are seldom concerned about the global affects of the design change. The designer/engineer usually has knowledge of the local laws dealing with the design but the local laws do not take into account the global affect of changes in design. An example is a local law which allows manufactures to use fluorocarbons in spray containers.

The designer/engineer wishes to freeze the design as quickly as possible so that any change in the design is not their responsibility. As stated before, little effort is put into understanding the system requirements from all users' perspectives. The clients view is simply implemented.

The designer/engineer is measured on speed of implementation and keeping costs within budget. This limits the designer/engineer who wishes to have the system function at implementation. The maintainers of the system are seldom consulted, if ever. This makes the system less likely to function throughout its expected life.

4 Improving PES Design

Why change? The process above worked very well with PES designs for many years. The PES systems worked very well and continue to do so. Why change now? The major reason for change is the realization that PES designs are getting to be very complex. The systems of the past had fewer than 100 I/O points. Today systems are running into the thousands.

PES designs are now getting into areas where only hardwire systems were used. PES designs are now being used for safety integrated systems (SIS). These systems in the past used relays or electrical/mechanical switches. The PES design now has to not only be reliable but safe. The world is now developing standards to regulate the use of PES designs in safety applications such as IEC 65A - 1508.

The new world wide standards are setting targets for safety integrity levels that relate to systems having the ability to function reliably in safety situations. These same standards reflect the inadequacy of reliability analysis to show the ability of software to function without flaw. This is why standards are insisting on a design structure that consistently produce safe functioning PESs.

4.1 Slow down the requirement development

If a designer/engineer in PES Design is to accomplish breakthroughs in the design process he/she must slow down the requirements phase of the project.

The initial requirements were derived from the designer/engineer through the client. It is important that the designer/engineer feels free to slow down the process at this point to get other users involved in the development of requirements.

Microsoft actually has a method for slowing the requirements stage by allowing the requirements to be not complete before moving to the next stage. This sounds strange when dealing with PES systems that are so critical to peoples' lives. The key is that Microsoft prioritizes the design requirements. The top priority improvements are always done prior to the release date. The less important, not critical requirements can be implemented in the next iteration.

4.2 Human Systems

The designer/engineer will have to work with the existing organizational structure. This may mean meeting with the operators and/or maintainer's supervisor explaining why they need to be at the design meetings.

Having final users at the design meeting may be difficult for the users because they have been told for years to keep quiet and not rock the boat. It may be the first time they were ever asked to attend a meeting at the company for any reason. The final users' (operators') input is very important to the success of any project, especially at implementation time. The people who know the processes in the manufacturing facility best are the operators and maintainers. The designer/engineers must listen carefully to their concerns. This is known by Management Science as the Behavioral Science

Approach in which employees are provided the responsibility and satisfaction of participating in their work environment.[4]

4.2.1 Understanding Users through Brainstorming

After the designer/engineer has formed the design team, the first process should be to brainstorm with the users of the new process. A user is anyone that will be affected by the new process.

During the brainstorming meeting the designer/engineer will simply ask the team who will be affected by this new design. The leader should have a few individuals in mind before the meeting so as to keep the meeting moving.

The list could be quite long, but the designer should keep writing until the group feels comfortable that most of the users are listed.

Using the list the team generated, choose all users that the new process should treat favorably, unfavorably and ignored. There may be a number of users the team wishes to simply ignore and that is fine. The users that are to be treated unfavorably are those that the system is designed to have discouraging features and functions for.

4.2.2 Team Efforts

A PES Project should have most user parties represented: operators, who are the final users of the process; maintainers, who will keep the process adjusted; other designer/engineering functions such as mechanical or electrical engineers and; in larger projects, an accounting representative and/or purchasing representative maybe helpful.

The team should not be greater than eight people but all users that are to be treated as friendly and unfriendly users should be represented during the meetings. This problem can be solved by assigning each member specific user constituency.

In the beginning of the project, it is important to determine who has what responsibility. In some cases, it is known that the production operator is representing all operators in the facility who will use the system. In other cases, the purchasing agent may have two functions, that of contacting vendors for the best pricing as well as representing the janitorial maintenance user group.

4.3 Design For Manufacturability and PES

Using Systems Science one looks at methodologies from seemingly unrelated fields that can be utilized into the field of interest. When designing a PES concept for hardware development of Design for Manufacturability (DFM) is readily understood and proven to be effective. This paper will show how concepts for DFM can and are used in software development.

---

[4] Daft, Richard, Management, The Dryden Press, New York, New York, 1988:51

4.3.1 Reduce the Number of Parts

In hardware this helps to reduce inventory. It further helps in the design process by having standard parts to design from. With hardware, the invention of IC chips helped standardize design by making transistors come in a set number of patterns thus reducing the number of configurations needed. In software, designers must to the same.

The history of software of having experts who developed code specific for each application was very valued. Today this is not the case. The code needs to be universal, i.e., standardized. Having standardized code helps eliminate the need to verify or validate the function every time it is used. In the software world this is known as Coding Standards. The soon to be international standard for PESs used in safety IEC-1508 requires the use of Coding Standards, Library of trusted/verified modules and components, Certified Tools, Modular Approach, and structure programming in Section 3 Annex A.

The reduction of software parts reduces the complexity of the design. Software designers should take advice from their hardware counterparts and not use another part if not required. To software this means not using unneeded instructions if not needed. Especially if the functions already exists. This is why the RISC was originally developed. It is known as Reduced Instructions Set methodology.

4.3.2 Parts fit/Snap Together

Snap fit part help reduce the number of fasteners needed in the assembly operations. This is why chips have a limited number of packaging configurations. In software this is related to the method of assembly of code modules. The code modules should snap together without the need for interface modules. This requires code models to have standard interface capability. A good measure for this activity is to count the number of modules needed to interface modules. The goal should be to reduce this to zero.

4.3.3 No Fasteners

The goal in DFM for hardware is to have no fasteners. In software this relates to no module should be used only once. If it is the modules it interfaces with should be redesigned.

4.3.4 No Assembly Tools Required

In DFM, the use of people to assemble and do repairs makes the need for the rule "No assembly tools required". With software, this relates to the need for people to generate code through the use of modules. Also, the maintainability of software is critical here.

In software, no tools required relates to the use of compilers and assemblers. This relates the use of CASE tools. Ideally CASE tools will take drawings and generate the code without the use of compilers and assemblers. Well transparent t the user. The CASE tool idea is not yet available, but similar software generating technology will some day be available.

Systems Science and Systems Thinking should be used to understand how seemingly unrelated disciplines have similarities. In complex systems it is important to learn from dissimilar disciplines, but similar in design continuum.

4.4 Creative Thinking in a Structured Environment

Design walkthroughs was very well documented by Weinberg. With the advent of IEC-1508 and other standards being imposed on designers of PES systems used in safety it seems creativity is being lost. This need not be the case. With the use of Change Management or Configuration Management, a design can be creative as well as structured. This at first glance seems contradictory, but it need not be.

The method is to allow the designer to have total control of the design as long as it does not affect the other systems. When using the standard V diagram, a system is defined then subsystems are allocated from the system level design. Subsystems then have a set of requirements that include interface specifications that shows how the subsystem relates to the other subsystems and overall system. The designer has the freedom to design creatively within those constraints. This design technique has been used in the medical industry for years. They call it Form, Fit or Functions. As long as the design does not change in any of these parameters the designer is free to change the design thus giving the designer the ability to create in a structured environment.[5]

4.5 Management Responsibilities

PES design cannot overlook the responsibility of management. Providing funding and manpower is a normal part of management's job.

Leadership in design consists of developing an environment that allows people on the design team to be empowered. Empowered people are free to present ideas, free to steal ideas from others, free to move about to develop ideas, free to ask questions, free to comment, and free to make choices. The design/engineer is usually the leader who has the responsibility to "free" people.

Leadership provides the designer/engineer with authority over team participants. His/her skills to get individuals in the group to see the importance of the project is highly valued. This type of leadership is referred to by Weinberg as technical leadership.[6] Technical leaders have traits of being coaches, being respected, and having an investment in the design. Technical leaders take the responsibility for failure but always give the team the credit for a success.

---

[5] Weinberg, G. M and Freedman, D. P., Handbook of Walkthroughs, Inspections, and Technical Reviews, Winthrop, New York, New York, 1983

[6] Weinberg, G. M., Becoming a Technical Leader, Dorset House Publishing, 1986.

The biggest problem a designer/engineer has is the attitude of a company toward change. The climate is usually felt at implementation time when the operators and their supervisors must change the way they have always done things. The designer/engineer can change this trend by getting the operators and supervisors to be involved in the project at requirement development time. Management must allow for hourly operators to be part of the team.

Management must allow the teams to meet. Meetings may occur off-site. Sending hourly operators on business trips to vendor locations is new and unknown to classical management. The technical review may require the operators to travel to review the system. Management should encourage these types of activities and budget for them.

5.0 Conclusion

Systems Science and Systems Thinking helps one use knowledge from studying the whole as in Generic Design and PES design being a subset called satisficing. Systems Science helps transfer knowledge in one discipline to knowledge in another such Design For Manufacturability and Software design. There is much more to learn from Systems Thinking and this paper has only touched upon a very small part.

Future research may relate to using visualization techniques used in developing specifications and fuzzy logic in safety systems. Yes, this is a very scary topic for some, but as safety systems get more complex, there may not be a solid, crisp answer.

The world of safety, thanks to international standards development has realized the need for Design Methologies especially in software development of Programmable Electronic Systems. The use of Design Methodologies will make the world a safer place and reduce cost and time to implement designs. An added benefit will be the reuse of software in future systems.

Reference List

1. Gruhn, P. ISA S84..."I have to do what?". ISA Safety Standard May 1996: 42-44
2. Gause, Donald C., Minch, Eric, "Design Processes: A Space Perspective", Design and Systems, Vol. 12, 1990, p 33-50
3. Petersen, D. and Hillkirk, J. A Better Idea: Redefining the Way Americans Work. Houghton Mifflin Company, Boston, MA, 1991:151
4. Daft, R. Management. The Dryden Press, New York, NY, 1988:51
5. Weinberg, G. Becoming a Technical Leader. Dorset House Publishing, 1986
6. Weinberg, G. and Freedman, D. Handbook of Walkthroughs, Inspections, and Technical Reviews. Winthrop, New York, NY 1983

Session 4
Industrial Applications and Experience

Reliability and Vulnerability Assessment as Decision Support during Purchase and Design of Complex, Technical Systems

A practical case for the Railway Signalling System at Gardermobanen

Remi Eriksen[*] *Terje Andersen*[*] *Claus Feyling*[**] *Hans-Joachim Petersen*[***]

[*] *Det Norske Veritas*
1322 Høvik, Norway

[**] *NSB Gardermobanen*
0107 Oslo, Norway

[***] *Siemens Transportation Systems*
38023 Braunschweig, Germany

Abstract

This paper describes the work carried out in connection with a reliability and vulnerability study of the Signalling System at "Gardermobanen" - a highspeed rail link between the centre of Oslo and the Main Airport in Oslo to be built and operated by NSB Gardermobanen - in order to ensure that the specified success criteria and quality goals are met. Focus in this paper is put on events and failures causing degraded operational modes or traffic restrictions such as reduced speed, single track operation, full stop in traffic, etc., and not on safety related aspects. Failure statistics from the Norwegian State Railways (NSB) show that the signalling system is one of the largest contributors to train delays (30-40%). The main results of a reliability and vulnerability analysis of the Signalling System at Gardermobanen carried out at an early stage of the development project to assist the purchaser (NSB Gardermobanen) as well as the supplier of the system (Siemens) are given together with an outline of the proposed reliability management programme for the further work. Our experience with respect to this type of work is presented together with a discussion of how similar reliability management programmes could be improved.

1 Introduction

In 1854 the first railway line in Norway opened between Kristiania (Oslo) and Eidsvoll at the southern end of the lake Mjøsa. Today Norway's first high speed rail connection is being built between the same two places as part of the development of a new Oslo Airport at Gardermoen.

The Gardermoen line from Oslo to Eidsvoll is 65.5 km long and double tracked apart from the last 3.5 km south of Eidsvoll which is single tracked. The line is dimensioned for maximum speeds in excess of 200 km/h with train following intervals as low as 2 minutes. Airport trains will be run at 10 minute intervals with various other trains scheduled inbetween. This combination of high train speeds and short train following intervals requires a sophisticated and powerful signalling system in order to handle the traffic safely with high operational availability. The signalling system will be delivered by Siemens subcontracting Ansaldo on outdoor ATC equipment.

The reliability assessment, cost control and optimisation process for the signalling system was started early in the life cycle of the Gardermobanen line. Quantitative reliability and availability measures were worked out for most of the sub-systems and functions and included as a part of the contract. Quantification was important to facilitate:

- Specification of contracts, verification and validation of deliveries
- Optimisation of system configurations regarding operational reliability and safety
- Choice of strategy regarding operation and maintenance
- Planning of the supply of spare parts and other resource management.

2 Success Criteria and Quality Goals

An extract of NSB Gardermobanen's traffic regularity requirements for the airport rail link is given below:

> *"Airport trains running directly between Oslo S and Gardermoen (48 km) shall not use more than 19 minutes. Airport trains stopping at Lillestrøm shall not use more than 21 minutes. The rail link is dimensioned for a minimum train-following time of 2 minutes between Oslo S and Gardermoen, and for a minimum train-following time of 4 minutes north of Gardermoen. The time tables and the scheduling of rolling stock shall be such that delay of one departure does not influence other departures in the system.*
>
> *The components used in the remote control and interlocking system shall be considerably more robust against failures delaying the train operation than the present system.*
>
> *Airport trains to/from "Oslo Lufthavn Gardermoen" shall have a punctuality equal to 95% or better within 1 minute, and 99.5% or better within 3 minutes delay. The availability shall be 99.5% as a minimum (i.e. only 5 of 1000 departures will be cancelled, i.e. one return departure per week)."*

The airport rail link system consist of many sub-systems, and consequently the above success and quality goals were mapped into a set of sub-system specific reliability requirements.

3 Present Signalling System Experience

Failure of signalling system has proved to be the most frequent cause of train delays (30–40%) in Norway. The root cause of the failure may be outside the signalling system itself, but the result is that proper clear aspect ("go-signal") for the train can not be set. The existing signalling systems at NSB (Norwegian State Railways) is in general of single station relay based interlocking system being remote controlled from an operation centre. The track between the stations are protected by track circuits.

Altogether, there is about 1.5 failures annually per km line (mainly single track) caused by the signalling system at NSB and approximately 25% of these failures affect train regularity.

Figure 1 shows a percentage distribution of the main category of causes of signalling failures at NSB (definition of categories are given below) [1].

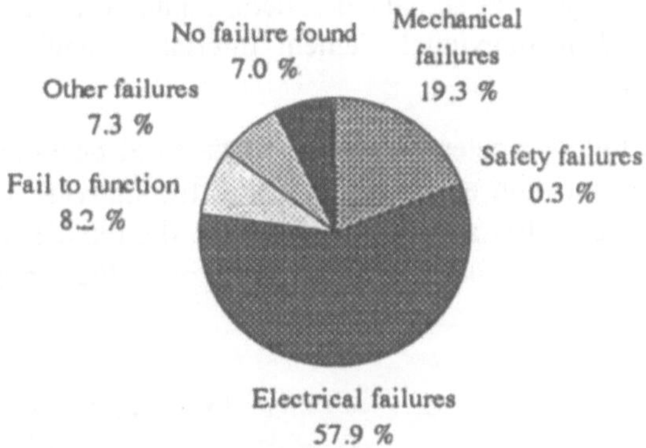

Figure 1: *Percentage distribution of the main causes to signalling system failures at NSB.*

Mechanical failures: Breakdown, deformation, ice blocking, etc.

Electrical failures: Contact failures, low ballast resistance, conductor break, etc.

Fail to function: Signalling system not manoeuvrable, point not in control, etc.

Safety failures: Track circuit fails to detect rolling stock on track, erroneous go signal, etc.

In order to obtain high regularity in the train traffic and fulfil the success criteria of Gardermobanen it is therefore of utmost importance that the signalling system is made as reliable as possible, with reduced vulnerability to external failures. This was a main decision factor in selecting the signalling system. Additionally, NSB Gardermobanen initiated a quantitative reliability and vulnerability study of the signalling system, including the power supply, cooling and ventilation of computer houses [2].

4 System Description

The term signalling system is a generic term for systems located on railway stations and on railway lines for supervision and control of the rail traffic. The main objective of the signalling system is to ensure safe rail traffic combined with high speed, punctuality and good capacity utilisation of the rail infrastructure. The system divides into the following parts [3]:

Interlocking System:

Systems on the railway stations and along the track to ensure that the lines and tracks are free of other trains before setting a green signal and to display those signals to the trains in traffic. A sketch of the system configuration for Gardermobanen is given in Figure 2. The signalling system is normally divided into internal and external parts according to the following principles:

External parts: External objects as signals, derailers, points, track circuits, balises and cables as well as their individual element interface modules towards the interlocking system (computer).

Internal parts: Interlocking computers along the track, local operation equipment and internal data communication equipment between the individual interlocking computers and towards the central signalling computers, the remote control system (CTC - Centralised Traffic Control) and the computers in the Automatic Train Control (ATC) system.

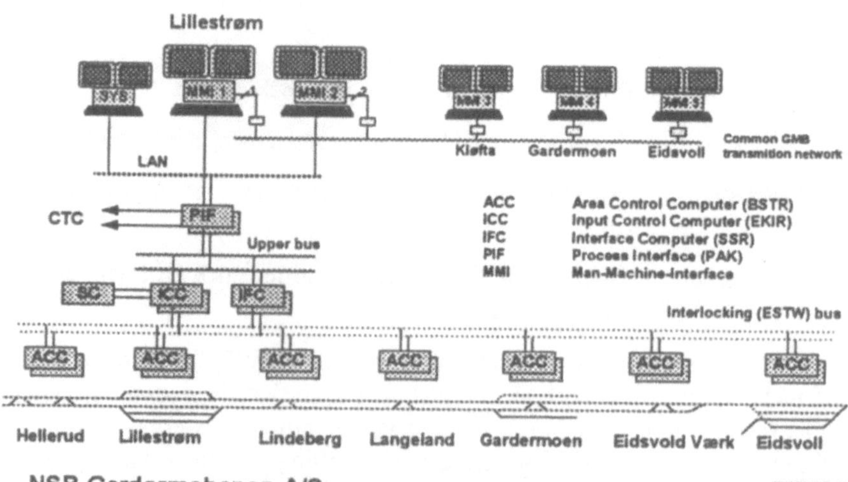

Figure 2: *The interlocking system for Gardermobanen including local operation facilities. Track topology is only indicative.*

ATC (Automatic Train Control):

This is a system for control of the train traffic, where the appropriate safety information for the train is transmitted between the track and the train and presented to the driver at a display in the cabin. If the speed of the train is not adjusted to the ATC-message, the brakes will be automatically applied. A sketch of the system configuration for Gardermobanen is given in Figure 3. There will be one ATC Interface (ATCI) computer per sub-interlocking, each with an optical sub-net (PTS- Profibus Transmission System) for distribution of ATC messages.

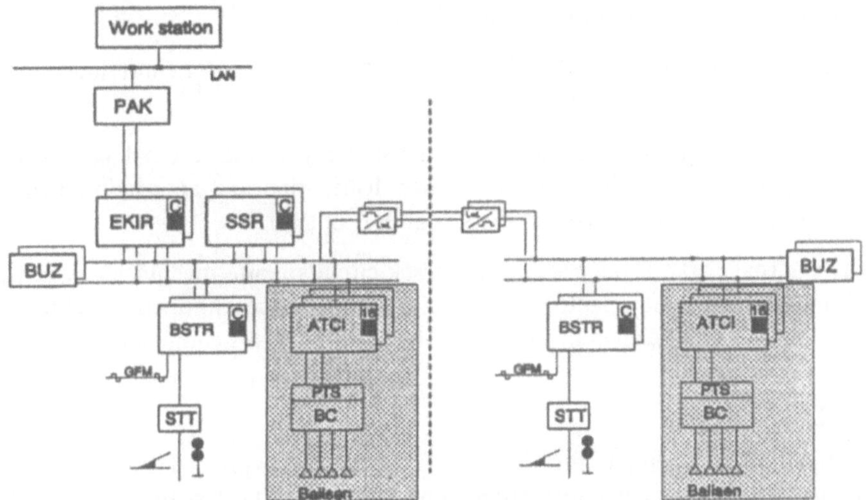

Figure 3: *An outline of the Interlocking and ATC system (shaded) at Gardermobanen. BUZ = Sub-interlocking bus, STT = EIM = Element Interface Module, BC = Balise Controller. See also Figure 2 for abreviations.*

CTC (Centralised Traffic Control):

The CTC system enables remote control and monitoring of the train traffic. The system provides the ability to remotely set the signals and the train routes on the tracks in question for control of the train traffic and for making priorities between trains. The system also includes interfaces towards information systems, time table databases, etc. The most important elements in the CTC system are:

- MMI: Man-Machine Interfaces.
- COM: Computer enabling linking of train numbers, automatic train management and other support functions for the traffic controllers.
- TTC: Timetable databases, tools for updating the databases.
- ADM: Computer to overview data and the status of all units in the system. Includes tools for administration and updating of the overall system.
- PIF: Process interface towards the interlocking system, ATC system and other external systems.
- X-term: A modem for presentation of overall status of the train traffic in the whole control area on LCD large-screens.
- LAN: Communication bus for the internal network at the operation central.

5 Reliability Requirements

5.1 Interlocking System and CTC system

Quantitative reliability and availability requirements were set to the internal parts of the systems [3]. The overall availability shall not be below 99.995%, i.e. in average 30 minutes or less per annum. Additionally, minimum requirements were set regarding MTBF values:

- MTBF for the internal part shall be minimum 52000 hours for long downtimes, with an MTTR of approximately 2 hours.
- MTBF for the internal part shall be minimum 8000 hours for short downtimes, with an MTTR less than 5 minutes.

The requirements apply to the total contract delivery of the internal part of the interlocking system including internal power distribution and air conditioning of the buildings where this is a part of the signalling contract.

For the external parts of the systems, i.e. track circuits, point machines and visual signalling equipment no quantitative reliability requirements were specified in the contract, but the supplier had to submit reliability data for this equipment.

5.2 ATC System

The reliability requirement for the ATC system was expressed in terms of a minimum MTBF value per information point, with an MTTR less than 2 hours [3]. The values consider all systems needed to operate an information point with minimum 2 controlled and 2 fixed balises. The ATC system requirement was chosen to match the reliability requirement for the interlocking system.

6 Reliability and Vulnerability Analysis

The analysis was carried out during start-up of the development project, i.e. during conceptual design. The study comprised the interlocking system, the automatic train control (ATC) system and the remote control system (CTC) as well as the internal power distribution and the climate control system.

6.1 Method

A technical system review followed by a failure mode, effects and criticality analysis (FMECA) were executed. These two analyses formed the basis on which the availability of the signalling system was estimated. Both the Reliability Block Diagram (RBD) method and the Markov Model method were used as modelling tools [4,5,6]. An example of an RBD for the CTC system is given in Figure 4. Where relevant, the predicted reliability performance was compared against contractual requirements. However, due to confidentiality aspects, detailed results are not presented in this paper.

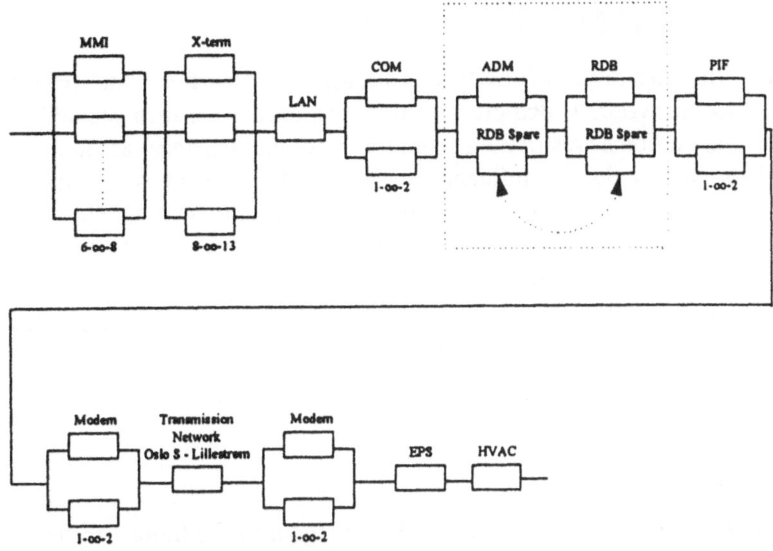

Figure 4: Example: *Simplified reliability block diagram for the CTC system.*

The results of the study were given in terms of average fraction of time the system spends in various degraded operational modes (see Section 6.3), the average visit frequency, and the average duration of a visit to an operational mode.

The various failure modes of the signalling system and unwanted external events may cause a range of degraded operational modes. How the train operation at degraded modes will be carried out in practice, depends on where the failure has occurred, and the traffic load on the line at the time of the failure. The possibility to use alternative lines, time slack in the time table of the trains as well as the possibility to cancel trains, will influence the choice.

6.2 Failure Classification

Failures of the railway signalling system were divided in two categories depending on whether the failures could be blamed to the signalling system itself (A) or they were due to external conditions that could not be related to the quality of the signalling system (B):

A. (1) Internal technical failures traceable to component or module failures in the signalling system.

   (2) Systematic weaknesses due to lack of non-fulfilment of contractual specifications, i.e. failures not directly caused by the signalling system equipment, but which may be traced to the equipment's' lack of suitability for the operating conditions or robustness against external influences, and where specifications for the delivery have not been fulfilled.

B. External influence and mis-operation. Failures caused by external conditions and operator failures that can not be blamed on the supplier.

6.3 Operational Modes

During normal operation of the system no unintentional functional restrictions of the interlocking system which prevents traffic at Gardermobanen should be present. However, operational restrictions connected to maintenance as well as minor restrictions, which do not influence the traffic (i.e. delays < 2 min.), may be present. Some types of degraded signal aspects are estimated as non-critical with minor effect on traffic and therefore included in the normal operation mode.

Some of the typical degraded operational modes of the railway line resulting from failures of the various parts of the signalling system are listed below.

(I) Interlocking system

1. *Failure at a single block section:* The go signal to the trains can not be set or shown to trains for one block distance of one track. The traffic may proceed in one of the following ways:

   a) single line operation between two stations with reduced capacity;

   b) train operation with reduced speed (max. speed 40 km/h) based on orders over the train radio or telephone.

2. *Point failure or track circuit failure at point*: Operation of a point or position control of the point is failing, or the track circuit connected to a point is not released. NSB Gardermobanen will have a secondary interlocking logic at point loops between the main tracks ensuring the traffic to proceed on the adjacent track provided certain conditions are fulfilled. The traffic may proceed in one of the following ways:

   a) single track operation with highly reduced capacity;

   b) if the failure is caused by a failed track circuit, or the point in question is to a side-track, train operations will proceed with reduced speed (max. 40 km/h) based on orders over the train radio.

3. *Loss of a functional ACC:* The consequences will depend upon what objects are controlled from the ACC. Track selectivity of ACCs is sought. However, a failure of an ACC will in many situations result in stop signals given to trains on both tracks, and traffic on the track controlled by another ACC has to, if possible, proceed by radio/telephone orders on the track not controlled of the failed ACC.

4. *Loss of a ACC-cluster:* The loss of an ACC-cluster caused by interruption in power supply to the ACC-computers or the Element Interface Modules (EIM) will cause loss of control of the track covered by relevant ACC-cluster with a high risk of loss of both interlocking buses. Both tracks on the relevant section of the line are shut down until the failure has been repaired.

5. *Loss of one or more ACC-clusters on the section from Oslo S to Gardermoen:* This failure mode will cause a total failure of the ACC house in question and all other ACC-houses communicating with the interlocking centre at Lillestrøm through this ACC-house until repair of one of the relevant buses is carried out. The situation will in most situations be equivalent to full stop in the traffic between Oslo S and Gardermoen. Loss of upper bus or both ICCs, IFCs or PIF-units at Lillestrøm or power supply to computers and/or interlocking bus communication equipment at Lillestrøm causes total loss of the interlocking system.

(II) ATC-system

1. *Failures of one information point.* Failure of one information point gives emergency break of first train, subsequent trains proceed with reduced speed (max. speed 40 km/h) based on orders over the train radio or telephone on the applicable track and direction of traffic until a new operational information point has been passed (1.2-1.7 km).
2. *Failure of one balise controller.* Failure of one balise controller gives emergency break of first train, subsequent trains proceed with reduced speed (max. speed 40 km/h) based on orders over the train radio or telephone on the applicable track and direction of traffic for minimum two information points (1.2-1.7 km), i.e. main/distant signal and repeater balise group for main signal in opposite direction.
3. *Failure of Profibus per PTS.* Failure of a Profibus transmission link or PTS-group (e.g. failure of power supply) gives emergency break of first train, subsequent trains proceed with reduced speed (max. speed 40 km/h) based on orders over the train radio or telephone. This applies to both tracks within an area covered by one PTS, i.e. up to a distance of 6 km.
4. *Failure of ATCI.* Failure of an ATCI-computer which gives "stop" aspect (max. speed 40km/h) to all controlled balises on both tracks in both directions within the area covered by an ACC-cluster, i.e. up to a distance of 13 km.

Additionally, repair of failures of balises or cable connections in the track requires track occupation during the repair period.

(III) Remote Control System (CTC)

Principally, there are two operational modes for the CTC system with regard to control of the Gardermoen line; operational or non-operational. There are a number of degraded operational situations with non-operational support functions, database for timetables etc. or reduced capacity in operating consoles. Present indications are that those situations will not be very critical for operation of the track.

If the Gardermoen line can not be controlled from the operations centre at Oslo-S, the traffic at the Gardermoen line can normally be controlled from the local operation centre at Lillestrøm. Under such conditions, it is important that the interlocking system at the Oslo-S station is operational.

6.4 Reliability Data Sources

The reliability calculations carried out in this study were based upon data from various sources. The sources utilised are briefly described below:

- Reliability data from the supplier regarding MTBF values of the signalling and interlocking equipment including computers,
- Evaluated and modified experience data from NSB regarding failures due to external impact,
- Operational experience of similar equipment in Finland (VR), Sweden (SL, Stockholm Tunnelbana and Roslagsbanan), Switzerland (SBB) and Holland (NS, Rotterdam Area),
- EIREDA [7] and OREDA [8] regarding internal power distribution equipment such as supply boards, fuses, transformers, rectifiers and circuit breakers,
- Existing availability assessment of the external power supply,
- Expert judgements and own estimates based on experience and various external sources regarding air conditioning, cabling and external accidents as fire, flooding, etc.

It is difficult to assess the relevance and validity of the data that were used. The majority of the data were quoted by the suppliers, Siemens and Ansaldo, which are expected to have the best available knowledge regarding the reliability and availability of the signalling equipment and the various computers which are part of the delivery.

Failure data based on NSB's operating experience were to a large extent derived from equipment based on an older generation of technology, which partly are used under other operating conditions. This was taken into account in our failure rate assessments.

Availability calculations for the cooling and air conditioning of the ACC equipment were based on our own assumptions with regard to whether recovery can be made before the temperature reaches critical limits (i.e. causing failure of interlocking equipment) combined with external or generic data sources. The same applies to the data used to estimate the probability of damage of the signalling system due to fire, flooding and other types of external disturbances.

6.5 Main Results

Overall, the predicted values showed approximately 1.0 failure per km track and year for the signalling system (interlocking, ATC and CTC) including category A as well as category B failures. This is lower than the failure rate per km track at NSB today. The distribution of failures with respect to various consequence classes is given in Figure 5.

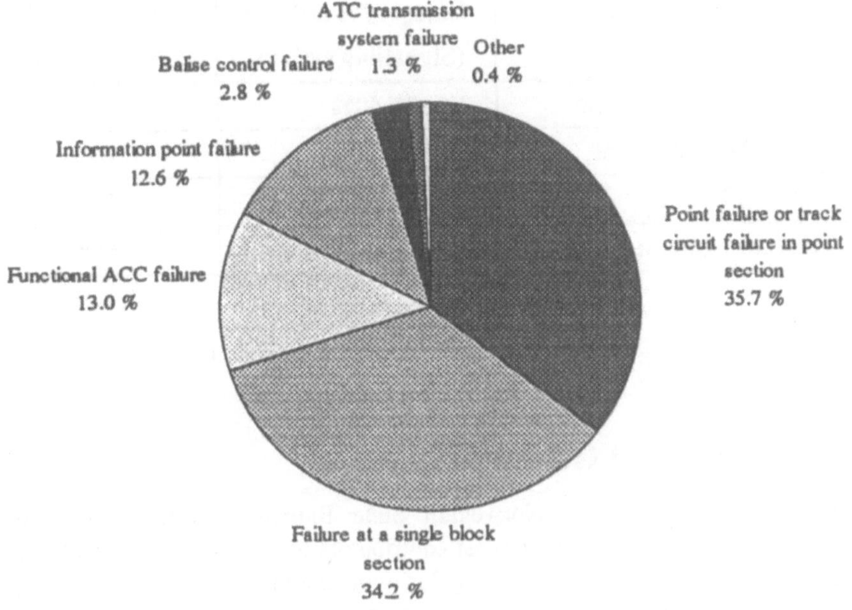

Figure 5: *Contributors. Signalling system failures at Gardermobanen.*

For the internal system including the central and area control computers (IFC, ICC and ACCs), the study revealed the following elements as the main contributors to system unavailability:

- Loss of cooling system. A loss of cooling will cause the temperature to increase. If cooling recovery is not obtained in time, the temperature raise will cause functional failure of the interlocking system. Availability increasing measures could be justified by the results of this study and has partly been implemented.
- Failure of internal power distribution. Such failures were found to have large consequences on the train traffic. Redundant circuits were incorporated into the design as a results of this study.
- Failure of one area control computer (ACC). All signals connected to the ACC in question are changing to stop. Switching to redundant control computer will cause downtime for at least 2 minutes in order to regain status and full control of the various track objects (signals, points, track circuits). Train routes must be set manually by the operator after start-up. Various compensating measures were evaluated.

For the external part of the system the main contributors to failures were track circuits and point equipment.

The contribution from the two failure categories (A and B) to degraded operation is given in Table 1.

Type of system	Category A failures (Signalling system)	Category B failures (External impacts)
Interlocking system	40%	60%
ATC equipment	33%	67%
CTC system	25%	75%
Total for signalling system	**40%**	**60%**

Table 1: *Contributors to signalling system unavailability (time in degraded operation).*

7 Summary and Conclusions

Failure statistics from the Norwegian State Railways (NSB) show that the signalling system is one of the largest contributors to train delays. Considering the strict requirements to train traffic regularity on the Gardermoen line, serving the new Oslo Airport, it is of utmost importance that the reliability and the survivability of the signalling system at Gardermobanen are given special attention and assured through a well designed system supported by proper evaluations and analyses.

The reliability assessments carried out during the purchase and design of the Signalling System for Gardermobanen have been a *trustworthy and efficient support* for the purchaser - NSB Gardermobanen. In this respect, we would like to highlight the following:

- NSB Gardermobanen has gained a relatively high degree of confidence with respect to the *overall* reliability performance of the signalling system.
- NSB Gardermobanen has obtained knowledge about unwanted critical failure situations and their impact on train traffic regularity as well as the effect of alternative design solutions and concepts.
- The proposed degree of redundancy and diversity has been reviewed and verified.
- Increased redundancy in the internal power distribution concept could be justified.
- Increased balise controller redundancy could *not* be justified.
- Support during recast of contract requirements for the ATC system.

Furthermore, the reliability and vulnerability assessment has been used to optimise the maintenance and spare part activities, the verification and validation process as well as the mechanical completion and installation program up to the opening of the line in 1998. The reliability and vulnerability assessment has later (1996) been incorporated in a larger analysis of all disciplines (signalling, power supply, telematics, track and trains) also taking into account external lines (Asker - Oslo S) where the airport rail link will run.

The success of the reliability work carried out for Gardermobanen is to a large extent due to the following:

- Early specification of quantitative reliability requirements which were incorporated into the contract.
- Reliability and vulnerability analysis carried out at an early stage of the development project. Decisions made at any instant of time have an impact on product reliability and cost at that time and in subsequent phases of the life cycle. However, it is during the concept, definition and design phase, the impact on the product and its life cycle cost is greatest.
- A total system approach was chosen, i.e. the signalling system was seen in context with the overall goal - an airport rail link with high regularity and punctuality.

As there will be diverging interests in such projects, we would like to stress the need for a common understanding of the reliability requirements put into contractual terms as well as the procedures, methods and decisions rules to be used during the verification and validation process of the delivery.

For the further work the following Reliability Programme for Gardermobanen was proposed:

1. Follow-up of the reliability and vulnerability study of the signalling system during detail design.
2. Organisation of the verification work.
3. Classification of failures.
4. Identification of test objects.
5. Definition of test parameters and decision criteria:
   (a) Assumptions regarding the life time model.
   (b) Specification of test criteria and consequences of possible outcomes.
6. Collection of reliability data:
   (a) Establish a failure reporting system.
   (b) Failure registration and reporting.
   (c) Handling of failure reports.
7. Analysis and documentation of test results.
8. System for remedial measures, decisions and follow-up.

A verification and validation process should encourage the supplier to improve the equipment rather than introducing punitive fees. Punitive fees will seldom compensate for loss in profit due to an unreliable system - in this case due to rail traffic disturbances and bad market reputation.

The main challenge during the remaining work is to establish a failure reporting system that is quick, precise and manageable, and to development of a failure classification system in order to avoid blaming the contractor of conditions outside his control.

8 References

/1/ NSB, Banedivisjonen. Signalfeil, Årsaker, Statistikk og Tiltak. Utgave 1 31.01.95. NSB internal document.

/2/ NSB Gardermobanen AS. Reliability and Vulnerability Analysis of the Signalling System at Gardermobanen. Report No.: 95-3464 prepared by Det Norske Veritas AS. Confidential project document.

/3/ NSB Gardermobanen AS. Contract Documents - governing requirements for safety and availability of the signalling system. Confidential project document.

/4/ Ascher and Feingold. Repairable Systems Reliability - Modelling, Inference, Misconceptions and their Causes. Marchel Dekker Inc, USA 1984

/5/ Martz and Waller. Bayesian Reliability Analysis. John Wiley and Sons, USA 1982.

/6/ Høyland and Rausand. System Reliability Theory - Models and Statistical Methods. John Wiley and Sons, USA 1994

/7/ OREDA Participants. OREDA-92 - Offshore Reliability Data Handbook, 2nd Edition. Det Norske Veritas AS, Norway 1992.

/8/ EIReDA - European Industry Reliability Data Bank, Volume 2, 2nd edition, Société Francaise d'Etudes et Réalisations, PARIS 1995.

Safety Analysis and Evaluation of an Air Traffic Control Computing System

Nicolae Fota[◊*], Mohamed Kaâniche[◊], Karama Kanoun[◊] and Alain Peytavin[◊◊]

[◊]LAAS-CNRS
7, Av. du Colonel Roche
31077 Toulouse—France
fota@laas.fr

[◊◊]CENA
7 Av. Edouard Belin, BP. 4005
31055 Toulouse—France
{kaaniche;kanoun}@laas.fr

[*]SOFREAVIA
3 Carrefour de Weiden
92441 Issy les Moulineaux—France
peytavin@cena.dgac.fr

1 Introduction

The French Air Traffic Control is based on an automated system referred to as CAUTRA (Coordinateur AUtomatisé du Trafic Aérien). The CAUTRA is implemented on a distributed fault-tolerant computing system installed on five en-route traffic control centers and one centralized operating center, that are connected through an aeronautical telecommunication network. The CAUTRA mission is to provide computerized means for the safe and efficient movement of aircrafts. The main services provided are flight plans processing, radar data processing and air traffic flow management.

However, the computing system failures could temporarily prevent the system from performing some or all of its required functions. The impacts of failures on the traffic safety depend on the criticality of the affected functions and the duration of the service interruption. In order to analyse and evaluate these impacts, we have defined a global approach that can be decomposed into two parts. The first part is aimed at a preliminary Failure Modes Effects and Criticality Analysis of the global CAUTRA: this study led us to identify the main subsystems that have a significant impact on the traffic safety. The second part of the approach focuses on the dependability modeling and evaluation of each subsystem and the combination of the dependability measures evaluated for each subsystem to obtain global measures characterizing the impact of the CAUTRA failures on the traffic safety. This approach is presented in [1]. In this paper, we focus on one subsystem of the CAUTRA centralized operating center, referred to as "STIP", which performs the centralized acquisition, processing and distribution of the flight plan information to the en-route traffic control centers.

This paper is decomposed into seven sections. Section 2 presents the STIP architecture. Section 3 outlines the failure and repair assumptions considered. Section 4 discusses the classification of the STIP failures according to their impact on the traffic safety. Section 5 summarizes the modeling approach used to describe the STIP behaviour and evaluate dependability measures. Section 6 comments some quantitative results. Finally, Section 7 concludes and outlines some directions for the future work.

2 The STIP Architecture

The STIP is implemented on a duplex architecture (Figure 1) composed of two redundant computers (Co the operational one and Cs the spare one), two redundant disks (Do and Ds where each one is composed of a couple of mirrored disks), two replicas of the application software (So and Ss), and an I/O board (B) for the connection of the communication lines and peripherals to the STIP operational configuration. So periodically sends a copy of its current state to Ss through a local data link. Do and Ds can be accessed by either Co or Cs. Two configuration modes are possible: "local disk association" when Co is connected to Do, and "remote disk association" when Co is connected to Ds. The computers — disks interconnections are implemented with coupler devices. The couplers and the dedicated data link are supposed to be failure-free since their failure rates are significantly small compared to the failure rates of the rest of components.

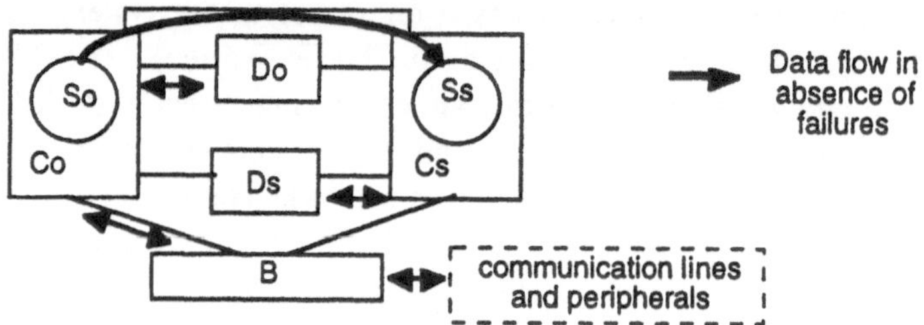

Figure 1 - Overview of the STIP Architecture

3 Failure-Repair Assumptions

So & Ss. Two parameters are relevant to identify software failures and their impact on the air traffic control: the amount of dynamic data lost and the duration of service interruption. The dynamic data correspond to information about the flight plans processed by the system that are vital for the air traffic control services.

Most of the So failures can be recovered without a significant loss of data either by restarting the software, or by rebooting Co. However, some So failures may lead to the partial or total loss of the dynamic data. In this case, several recovery actions are attempted by the operators: 1) the reconfiguration of Ss from the spare mode to the operational mode and the switching of I/O links; 2) the reboot of Co and its association to the remote disk followed by the restart of So (this sequence of recovery actions is denoted as *"asso"*); 3) the restart of So after *partial* elimination of some dynamic data (denoted as *"rpl"*) ; 4) the restart of So after *total* elimination of the dynamic data (denoted as *"rtl"*). It is noteworthy that the data lost has to be manually recreated by the controllers after the service restoration.

Three failure modes of the So replica have been identified:

- *"So and Ss common mode failures"* which are due to some erroneous dynamic data processed by both replicas; they are immediately diagnosed by the operators as they lead to the simultaneous failure of both replicas; these failures may be recovered by an "rpl", followed by an "rpt" if the first recovery action fails;

- *"local failures - immediately diagnosed"*; these failures affect the dynamic data processed by So without affecting those processed by Ss and they are immediately diagnosed by the operators. The reconfiguration of Ss (from the spare to the operational mode) and the switching of the I/O links are first attempted followed, in case of failure, by an *"asso"* recovery action;
- *"local failures - not immediately diagnosed"*: other software failures whose origin is not immediately identified by the operators. The four recovery procedures mentioned above are consecutively applied until the service restoration.

With regards to the Ss replica, it is assumed that any failure of Ss leads to the loss of the dynamic data processed by this replica. Moreover, Ss can be restarted only if So is operational.

Co and Cs. A computer failure leads to the stop of the software replica hosted. Moreover, when Co fails, Cs becomes the operational computer after the reconfiguration of the system.

Do and Ds. The failure of Do or Ds results from the unavailability of the associated couple of mirrored disks. The failure of a disk causes the associated software replica to stop and leads to the total loss of the dynamic data processed by that replica. When Do fails, the reconfiguration of Ss into So and the switching of the I/O links are first attempted followed, in case of failure, by an *"asso"* recovery action.

B. Two failure modes may affect the I/O board: 1) "a switching failure" when a system reconfiguration is attempted; this failure mode precludes the use of Cs, however Ds may still be accessed via a remote disk association; 2) "the interruption of I/O links" that may occur at any time and leads to the global system unavailability .

Repair resources. We assume the availability of two repairmen for hardware failures: one for the I/O board and another one for the disks and the computers.

4 Impact of the STIP Failures on the Traffic Safety

The STIP failures impact on the traffic safety is measured through the assessment of the degradation of the controllers capability in performing safe control. During the global CAUTRA analysis, we have identified five classes of service degradation that are summarized in Table 1. The level of degradation is dependent on the amount of data lost, the duration of service interruption and the availability of alternative facilities to continue the air traffic control service in the degraded operation mode (control services provided by systems outside the air traffic control computing system). Levels IS1 and IS5 correspond respectively to the most and the least critical failures.

Table 2 gives a classification of the STIP failures and their correspondence with the safety levels defined in Table 1. This classification, which has been validated by experts involved in the system design and operation, takes into account the criticality of the services provided by the STIP. It is noteworthy that the "swl" failures are not critical when the rest of the CAUTRA subsystems are not failed. However, their impact becomes more significant when some of these subsystems are in a degraded mode (See [2] for more details).

Safety levels	Definition
IS1	Failure conditions which lead to a long interruption of the service and which prevent continued safe control
IS2	Failure conditions which lead to a highly degraded service until the application of alternative recovery means and the service continues in a full degraded mode with these means
IS3	Failure conditions which lead to a degraded service; the service can be continued in a degraded mode with the alternative means
IS4	Failure conditions which have minor effects on the service
IS5	No impact

Table 1 - Failures classification according to their impact on the air traffic safety

Notation	Definition	IS
swl	short interruption (≤15 min), without loss of data	5
lwl	long interruption (> 15 min), without loss of data	4
pl	interruption (independent of the duration), with partial loss of the data	3
tl	interruption (independent of the duration), with total loss of data	2

Table 2 - Classification of the STIP failures and corresponding IS levels

Table 2 leads to the partitioning of the STIP states into five state classes. Several paths between these classes may be observed. The transition from one class to another one corresponding to a more degraded service are due to the failure of recovery actions or the occurrence of additional hardware or software failures. To evaluate quantitative measures characterizing the impact of the STIP failures on the traffic safety, we have further considered the following grouping of states: E2 = {tl} ; E3 = {tl, pl} ; E4 = {tl, pl, lwl}. Therefore, when the STIP occupies Ei, the minimum level of degradation is equal to ISi. Two measures can be evaluated:

- MTFFi (Mean Time to First Failure): mean time to the first visit to Ei (i = 2, 3, 4);
- UAi (Unavailability): Ei sojourn probability in steady-state (i = 2, 3, 4).

To evaluate these measures, we model the system behaviour resulting from the failure-repair assumptions presented in Section 3.

5 Modeling Approach

5.1 Principles

The STIP and the other CAUTRA sub-systems have complex hardware and software fault-tolerant architectures, characterised by a large number of components with complex behaviour and multiple interactions. Over the years, several model types such as combinatorial models (reliability block diagrams, fault trees, etc.), and state-space models (homogeneous Markov chains and their extensions) have been used to model systems and to evaluate various dependability measures [3]. Because they are able to capture various functional and stochastic dependencies among components (such as shared repair facilities), state-space models, in particular homogeneous Markov chains, are commonly used to model the dependability of fault-tolerant systems [4]. To alleviate the problem of specification and generation of complex models, higher level model types such as stochastic Petri nets [5] are generally used since they a) allow the definition of a more compact representation of a model (closer to the real system behaviour), b) provide some model structural verification facilities, and c) can be automatically converted to Markov models. In our study we use the SURF-2 dependability modeling software tool, developed at LAAS-CNRS [6] which allows the construction and processing of Generalized Stochastic Petri Nets (GSPN). Nevertheless, the specification and the construction of complex GSPN models is a fastidious and an error-prone task. To master the models complexity, we have defined a modular and an incremental modeling approach based on:

- a *specification formalism* allowing the structured high-level description of each component behaviour, accounting for its interactions with the other components;
- a *set of transformation rules* allowing the specification formalism to be directly converted to a GSPN model;

The model is specified, built and validated in an incremental manner. At the initial step, the behaviour of the system is described taking into account the failures of only one selected component, assuming that the rest of the components are in an operational nominal state. The failures of the other components are integrated progressively. At each integration step, the GSPN model is derived and validated. The validation is carried out at the GSPN level (structural verifications) and also at the Markov level in order to check the different scenario represented by the model. When the Markov chain size increases, the exhaustive analysis of the Markov chain is impractical. In this case, sensitivity analyses are used to check the validity of the model assumptions.

To our knowledge, most of the studies that addressed the direct generation of dependability models from the translation of a specification formalism focus on the construction of Markov models without considering the intermediate step of GSPN generation. This latter step offers some model verification facilities allowing the user to gain confidence in the models on which judgements about the system dependability will be based (see for instance [7-10]).

Because of the lack of space, this paper is restricted to a brief presentation of the specification formalism through an example taken from the STIP; only the elements necessary for understanding the example are outlined. The transformation rules for the conversion of the specification formalism to a GSPN model are not presented.

5.2 Presentation of the Specification Formalism

The aim of the specification formalism is to provide a structured description of the system behaviour allowing an optimal GSPN model to be directly derived from the specification by the automatic translation of the formalism elements. The optimality criteria we have considered are: 1) conciseness and ease of model specification to enhance readability, 2) flexibility allowing, for instance, an easy modification of the model when new assumptions are to be considered, and 3) reusability of some parts of the formalism to facilitate the validation of the model and the specification and modeling of alternative fault-tolerant architectures based on similar assumptions.

5.2.1 Main Concepts

Our formalism, called an *evolution diagram*, provides a high-level description of each system component behaviour. It is composed of a graphical representation with textual definitions of some elements of the graph. An evolution diagram describing the behaviour of a given component is a directed graph made up of two types of vertices: *phases* (represented by circles) and *evolution functions* (represented by rectangles), which are connected by edges. An unique evolution function is associated to each phase. A phase designates a state or a class of states of the component having some common features, while the associated evolution function describes the possible state changes from that phase. The evolution of a given component state may be due to the occurrence of local events (generated by the component itself) or external events (generated by other components). It is noteworthy that some local events may be common to several components (e.g. a common-mode failure affecting two software replicas). An evolution function defines on one hand, the local events that may occur during the sojourn of the component in one of the states of the associated phase (activation of faults, local repair actions, etc.), the enabling conditions associated to these events and the stochastic parameters characterizing each event. On the other hand, it describes, in a decision diagram-like form, each event occurrence consequences on the component itself and on its environment (various consequences are possible depending on the state of the component environment).

Different states from which the same kinds of events may occur can be identified in a component behaviour. Because the graphical representation of these different states increases the number of vertices of the evolution diagram which, therefore, becomes more difficult to understand, we chose to represent such a state class using one phase and defined Boolean *variables* to distinguish a particular state in the class when needed. These Boolean variables are not graphically represented but they are defined in the textual description of the evolution functions.

Moreover, to enhance the readability of the evolution diagrams associated to each component, only phase evolutions due to the occurrence of local events are represented graphically. Component state changes resulting from interactions between components are described in the textual definition of the evolution functions. Bold characters are used to identify these interactions.

Figure 2 gives an example of an evolution diagram. The example describes the behaviour of the STIP So software replica that results from the failure-repair assumptions presented in Section 3. Only the evolution function associated to phase "So_ok" is presented and commented in the following.

5.2.2 Evolution Function Structuring

An evolution function attached to a phase P is denoted by OUT_P. Four levels can be distinguished in the specification of an evolution function.

- **level 1**, identified by the symbol "⇒", defines the local events that may occur from the states of P and the associated enabling conditions (if any). If an event is common to OUT_P and OUT_Q (Q being a phase of another component), it is identified by the symbol: "⇒(OUT_Q)".

Example: three events may occur from the phase S_ok corresponding to the three failure modes identified in Section 3. The first event is shared by phases So_ok and Ss_ok.

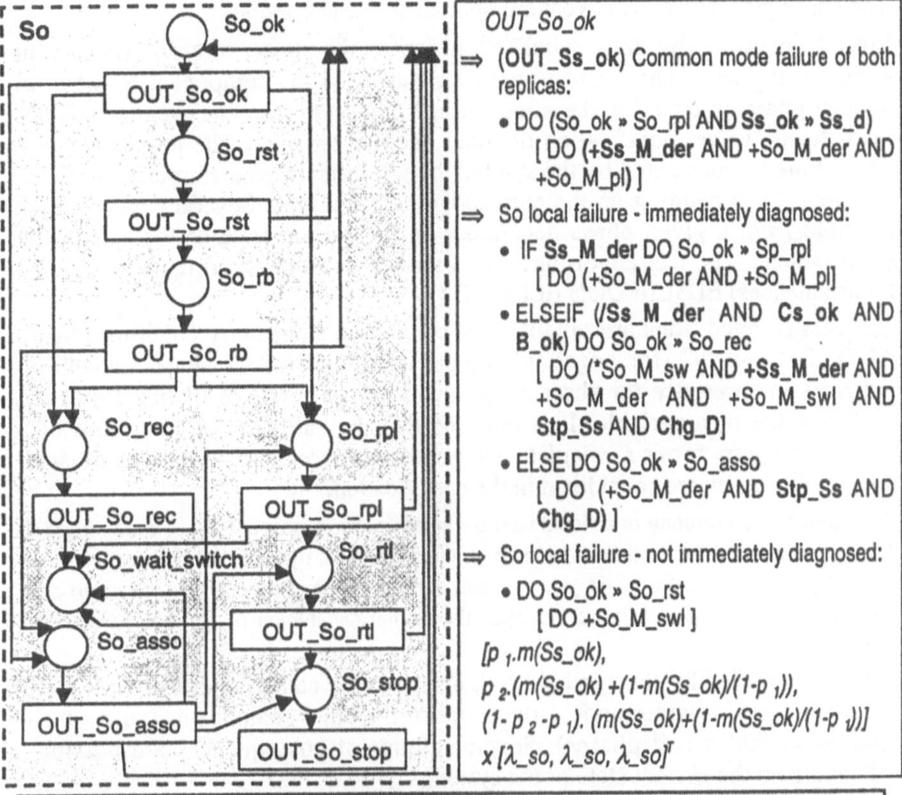

Notation	Phases and variables semantics
So_ok	So is in the "Up" state
So_rst	So is being restarted
So_rb	So while rebooting
So_rec	So while reconfiguration of Ss from spare to operational mode
So_wait_switch	So waiting for I/O switching
So_asso	So involved in a remote disk association
So_rpl	So is being restarted after partial dynamic data elimination
So_rtl	So is being restarted after total dynamic data elimination
So_stop	So is stopped (waiting for repair)
Ss_ok	Ss in the "Up" state
Ss_d	Ss is down (either stopped or under restart)
Cs_ok	Cs is in the "Up" state
B_ok	B is in the "Up" state
So_M_der	So variable characterizing the state of the data processed by So
So_M_sw	So variable used to memorise the need for an I/O switching
So_M_swl	So variable identifying an "swl" type service degradation
So_M_pl	So variable identifying a "pl" type service degradation
Ss_M_der	Ss variable characterizing the state of the data processed by Ss

Figure 2 - So evolution diagram and one of its evolution functions

The consequences of each event are defined at levels 2 and 3;

- **level 2**, identified by the symbol "•", describes the impact of each event on the component phase. The event occurrence may lead the component to move to another phase or keep it in the same phase. In the latter case, the event occurrence leads to the value modification of some internal variables. The phase change is either timed (represented by the symbol "»") or instantaneous (represented by the symbol "›"), depending on the corresponding event type. Several evolutions are possible from a given phase depending on the current state of the component environment. The test conditions are described as follows: IF (conditions) DO (consequences) ELSEIF DO() ELSE DO ().

  The sets of conditions are obtained by combining elementary conditions using AND, OR and "/" (for negation) logical operators. An elementary condition indicates the presence (or absence) of a component in one of its phases, or the logical value of a variable. It is required that the entire set of conditions of a decision diagram forms a *complete and exclusive system* in order to ensure that all the possible cases are considered in the specification.

  Example: the occurrence of a "local failure - immediately diagnosed" failure mode leads to three possible phase evolution depending on the state of the dynamic data processed by Ss (given by the value of the Ss_M_der variable) and the states of Cs and B. If the spare data are damaged, then So is restarted after the partial elimination of some dynamic data (So_ok » So_rpl);

- **level 3**, identified by the symbol "[ ]", describes the consequences of each event on the internal variables and on the states of the other components (i.e. on their internal variables and phases). Several groups of consequences are possible, depending on the current state of the system;

  Example: the failure mode "So local failure - immediately diagnosed" occurrence affects the dynamic data processed by So (+So_M_der) and leads the STIP to a "pl" degradation class (+So_M_pl). The "+" operator is used to set the value of a logical variable to "True";

- **level 4**, identified by the symbol "$[\ ] \times [\ ]^{T}$", specifies in a matrix format the stochastic parameters associated to each event. The first array defines the probabilities associated to each event (1 is the default value) while the second one gives the occurrence rates for the timed events only (the symbol "∞" is used for the instantaneous events).

  Example: the parameters associated to the three So failure modes are specified as follows:
  $[p_1.m(Ss\_ok), p_2.(m(Ss\_ok)+(1-m(Ss\_ok)/(1-p_1)), (1-p_1-p_2).(m(Ss\_ok)+(1-m(Ss\_ok)/(1-p_1))]$
  $\times [\lambda\_so, \lambda\_so, \lambda\_so]$

  λ_so is the failure rate of So. p_1 and p_2 are the probabilities of occurrence of a common mode failure and of a " local failure - immediately diagnosed" respectively when Ss is in the "ok" state. If Ss is in the "down" state, then the probabilities associated to the second and the third failure modes have to be updated. Function "m(Ss_ok)" returns the value 1 if Ss_ok is true and 0 if not. This function allows state dependent parameters to be specified. Therefore stochastic dependencies between component behaviours can also be specified in the evolution functions definition.

In addition to the concepts introduced above, the user can define *procedures* which can be reused by several evolution functions allowing the optimisation of the evolution functions description. The procedures are specified with a decision diagram-like format. The only condition imposed is that the associated set of conditions should be complete and exclusive. For instance, Stp_Ss and Chg_D are two procedures used in the STIP specification. Stp_Ss is invoked when the occurrence of an event (for instance, the failure of Cs or Ds) causes the Ss replica to be stopped. Chg_D describes the switching of the disks roles (switch Do to Ds and vice-versa). Stp_Ss is defined as follows:

Stp_Ss: IF Ss_ok DO Ss_ok › Ss_d ELSE DO NIL

The NIL operator is introduced to satisfy the completeness property. It denotes that for the remaining set of conditions no action is performed.

6 Application to STIP and Evaluation Results

We have applied the incremental modeling approach presented in Section 5 in order to build a GSPN model describing the behavior of the STIP and evaluate quantitative measures characterizing the impact of the STIP failures on the traffic safety. Due to the complexity of the STIP, it is not possible to present the corresponding GSPN models. Table 3 outlines the different steps considered and the size of the corresponding Markov chains. Each step led to the validation of the model and the evaluation of MTTF and UA measures (see Section 4). At each step, we check that the assumptions considered at the previous step are also satisfied. Model construction and processing have been carried out with the tool SURF2.

Modeling steps	# Markov chain states
So	5
So - Ss	12
So - Ss - B	38
So - Ss - B - Co - Cs	104
So - Ss - B - Co - Cs - Do - Ds	212
So - Ss - B - Co - Cs - Do - Ds - Rep	256

Tableau 3 - The STIP modeling steps and size of the corresponding Markov chains

Two kinds of quantitative analyses have been carried out. Firstly we conducted sensitivity studies in order to identify the model parameters that have the most impact on the quantitative measures. These studies revealed the major impact of the software failure rates, compared to the rest of the parameters. Secondly, we analysed several operating configurations of the STIP in order to evaluate their impact on the air traffic safety. These configurations are obtained from the "reference" model corresponding to the assumptions presented in Section 3 by modifying some recovery scenarios or some model assumptions. The configurations studied are listed hereafter:

- "cold" spare instead of a "warm" spare;
- "hot" spare instead of a "warm" spare;
- try a "remote disk association" before the reconfiguration of the software replicas, whenever this is possible ("Rec->Asso");
- try a reconfiguration of the software replicas or a "remote disk association" instead of a reboot, each time the spare dynamic data are not damaged ("Shunt_Reboot") ;
- consider three repairmen instead of two: one for the I/O board and two for the computers and the disks.

Thanks to the specification and model construction method, the modification of the reference model to account for these alternative configurations was relatively easy. The MTFFi and UAi values corresponding to each configuration are listed in Table 4.

	MTFF4	MTFF3	MTFF2	UA4		UA3		UA2	
				Pr. (E^{-6})	min/an	Pr. (E^{-6})	min/an	Pr. (E^{-6})	min/an
Reference	1324.9	3985.2	38422.1	165.9	87	41.5	21	4.0	2
Cold spare	743.0	1235.2	1885.1	304.4	159	159.5	83	112.9	59
Hot spare	1324.8	3985.1	38422.1	165.8	87	41.5	21	4.0	2
Rec->Asso	641.7	3986.7	38482.6	392.3	206	41.5	21	4.0	2
Shunt_Reboot	1324.9	3984.6	38332.6	108.1	56	41.6	21	4.0	2
3 Repairmen	1324.9	3985.2	38421.9	165.9	87	41.5	21	4.0	2

Table 4 - MTFF and UA for different STIP configurations

It can be noticed that the use of a warm spare instead of a cold one improves the safety measures: MTFF4, MTFF3 and MTFF2 increase by a factor of respectively 2, 3 and 20, while UA4, UA3 and UA2 decrease by a factor of respectively 2, 4 and 30. The improvement is more significant for class 2 which includes the most critical failures with respect to the traffic safety. It is noteworthy that the difference between these configurations decreases with the improvement of the disks reliability. Moreover, similar results are obtained for the "hot spare" and the "cold spare" configurations and also for the three and two repairmen cases. The configuration "Rec->Asso" leads to a small degradation of class 4 safety measures which concern the least critical failures. Finally, the configuration Shunt_Reboot leads to a significant improvement of UA4 (35%) against a weak decrease of MTFF2 (0.2%).

To conclude, the results given in Table 4 show the benefit of using a "warm spare" instead of a "cold spare". The other alternatives do not have a significant impact on safety. It is noteworthy that the STIP configuration that is currently operational is based on a cold spare. It is expected that the warm spare configuration will be introduced in the next release.

7 Conclusion and Future Work

In this paper, we analysed the failure impact of one subsystem of the French air traffic control computing system on the traffic safety. To master the complexity of this system, we presented a modeling approach that is based on a specification formalism allowing the structured description of the system behaviour and the automatic generation of a GSPN model from the specification. The aim of the formalism is to assist the modellers in the construction of dependability models. Moreover, we analysed several system operating configurations and evaluated the impact of each of them on the traffic safety. The results show the benefit of using a "warm spare" configuration instead of the "cold spare" configuration that is currently used. These results will be used to support the definition of the future architecture of the system. In the future, we will focus on the modeling and analysis of the CAUTRA subsystems implemented in the five en-route traffic control centers and the combination of the results obtained with those presented in this paper.

References

1. N. Fota, M. Kaâniche, K. Kanoun, and A. Peytavin, "The Air Traffic En route Control Computing System Dependability: Analysis and Modeling," Proc.10ème Colloque national de Fiabilité & Maintenabilité, Saint-Malo, France, 1996 (*in French*).

2. N. Fota, M. Kaâniche, K. Kanoun, and A. Peytavin, "Analysis of the Global Air Traffic en-route Control Computing System for Dependability Evaluation," LAAS-CNRS, LAAS Report n° 95280, june 1995 (*in French*).

3. A. L. Reibman and M. Veeraraghavan, "Reliability Modeling: An Overview for System Designers," *IEEE Computer*, vol. 24, pp. 49-57, 1991.

4. M. Malhotra and K. S. Trivedi, "Power-Hierarchy of Dependability-Model Types," *IEEE Transactions on Reliability*, vol. 43, pp. 493-502, 1994.

5. M. A. Marsan, G. Balbo, G. Franceschinis, and S. Donatelli, *Modelling with Generalized Stochastic Petri Nets*: John Wiley & Sons, 1995.

6. C. Béounes, M. Aguéra, J. Arlat et al. "SURF-2: A Program for Dependability Evaluation of Complex Hardware and Software Systems," Proc. 23rd Int. Symp. on Fault-Tolerant Computing (FTCS-23), Toulouse, France, 1993.

7. S. Berson, E. de Souza e Silva, and R. R. Muntz, "A Methodology for the Specification and Generation of Markov Models," in *Numerical Solution of Markov Chains*, W. .Stewart, Ed.: Marcel Dekker, 1991, pp. 11-36.

8. M. Bouissou, H. Bouhadana, M. Bannelier, and N. Villatte, "Knowledge Modelling and Reliability Processing: Presentation of the FIGARO Language and Associated Tools," Proc. 10th Int. Conf. on Computer Safety, Reliability and Security (SAFECOMP'91), Trondheim, Norway, 1991, pp. 69-75.

9. A. Goyal and S. S. Lavenberg, "Modeling and Analysis of Computer System Availability," *IBM Journal of Research and Development*, vol. 31, pp. 651-664, 1987.

10. J. A. Carrasco and J. Figueras, "METFAC: Design and Implementation of a Software Tool for Modeling and Evaluation of Complex Fault-Tolerant Computing Systems," Proc. 19th Int Symp. Fault-Tolerant Computing (FTCS-19), Vienna, Austria, 1986, pp. 424-429.

Creating Markov Models for Applications in the Process Industry

Ir. M.J.M. Houtermans, W.M. Goble MSc., Prof.Dr.Ir. A.C. Brombacher
Eindhoven University of Technology
Eindhoven, The Netherlands

Abstract

Safety Instrumented Systems or Emergency Shut-down Systems play an important role in the oil, gas & process industry. New standards, like the ISA SP84 or the IEC65, are under development to come to an unambiguous method to evaluate these systems in terms of safety. Markov modelling is a technique that plays an important role in these new standards. This paper represents a program that makes it possible to automatically generate Markov models for Safety Instrumented Systems and it explains the requirements and assumptions that have to be made. A practical case study is performed to show the results of the program and to explain the voting aspect which many systems use.

1 Introduction

Each plant in the industry is among other things equipped with protection systems, usually referred to as Safety Instrumented Systems. The function of these systems is to protect people, environment and investments. These Safety Instrumented Systems are today the subject of discussion, both from the peoples as from the industries point of view. The people do not accept risks for life and environment from industrial activities. They know there will always be a certain risk, but it has to be as low as possible. This results in strong demands from the public opinion to analyse the probability or likelihood of industrial accidents more accurately [1]. The owners of plants have to maintain a balance between investments and pay-off of these investments in terms of safety. Safety Instrumented Systems are expensive, both initial investments as well as maintenance. They do not want to make unnecessary investments in protection systems.

Until now there is no unambiguous method to evaluate Safety Instrumented Systems on reliability and safety aspects. Different companies and institutes use different techniques as a safety assessment tool. Qualitative as well as quantitative techniques are common practice to evaluate the same kind of systems. Rouvroye [2] made a comparison of these techniques and came to the conclusion that Markov modelling enhanced with techniques to analyse parameter variation, is the most comprehensive one. The Markov technique allows the modelling of failure and repair, on-line and periodic testing, functional and common cause failures and analysis as function of time. There are computer programs available that can perform calculations if the Markov model of the emergency shut-down system is available but creating these models is still being done by hand. This is not only very time consuming but can also become more complex as the emergency shut-down system becomes more complex. The influence of human factors, during the creation of these Markov models, increases the likelihood of making mistakes. This is one of the main reasons why users of the current methodologies (FMEA, FTA, etc.) do not yet believe in the real advantages of Markov modelling as a better technique to perform safety assessment. Creating the Markov models automatically will be a good step to overcome this problem.

2 Markov Modelling in General

The basic principle of Markov modelling is that a system can reside in different states. Each state is defined by (a combination of) one or more internal failures. These combinations of internal failures can be combined to a level called *systems states*. Creating a Markov model starts with identifying all these different system states. A system state is a representation of a combination of working and failed modules[1] and can thus represent a

- Fully operational system
- Partially failed system, but still fulfilling its function
- Totally failed system.

A Markov model consists of *Markov states* and the transitions between these states. A Markov state can be a single system state or a combination of system states. The driving force for the transition from one Markov state to an other is the failure or repair probability of the modules. There are two reasons why a transition from one state to an other can occur:

- Firstly, an operating module in a state has failed
- Secondly, a failed module in a state has been repaired.

3 Markov Modelling of Safety Instrumented Systems

In general, the modules of a system have only one failure mode. This means that the module is fully operational or fully failed. In many cases this is a valid assumption. Looking at Safety Instrumented Systems then different failure modes can have different effects on system functionality. These different failure modes should be taken into account during the creation of the Markov model.

Each module of a Safety Instrumented System can fail in three different ways:

- A failure of a module that causes the Safety Instrumented System to shut-down the process without a process demand is categorised as *safe* (s).
- A failure of a module that causes the Safety Instrumented System not to respond to a process demand and that is detected with on-line diagnostics is categorised as *dangerous detected* (dd).
- A failure of a module that causes the Safety Instrumented System not to respond to a process demand and which is not detected with on-line diagnostics is categorised as *dangerous undetected* (du). These failures can only be found during a thorough periodic test.

---

[1] A system is built from modules and modules are built from components. Every where in the text it is possible to replace the word module by component. Module has been chosen above component because usually a safety system is evaluated on module level.

The above results in the following system states:

- Fully operational state
- Partially failed state, but still fulfilling its function
- Failed safe state
- Failed dangerous state.

The maximum number of different system states a Safety Instrumented System can reside in equals 4^N, where N is the number of modules and 4 represents the number of possible modes a module can be in. The way the modules are configured and the combination of failure modes determines eventually the state the system will be in. Creating a Markov model with all its states and all its transitions between these states means determining all possible combinations of failed and working modules that lead to different system states.

4 Creating Markov Models

To create a Markov model one has to identify all possible system states a system can reside in. Starting point for identifying the system states of a Safety Instrumented System is usually a Process & Instrumentation Diagram, see Figure 1. In this diagram the engineering department gives a detailed overview of the way the different pieces of equipment physically are connected. The identification of these system states can be performed by a lot of different techniques. These techniques are for example FMEA or expert opinion, all with their own advantages and disadvantages. The only minus most of these techniques have is that it is almost impossible to automate this process of identifying all possible system states. The authors of this paper chose as starting point a reliability block diagram of the Safety Instrumented System. A reliability block diagram makes it possible to automatically translate it into a Markov model.

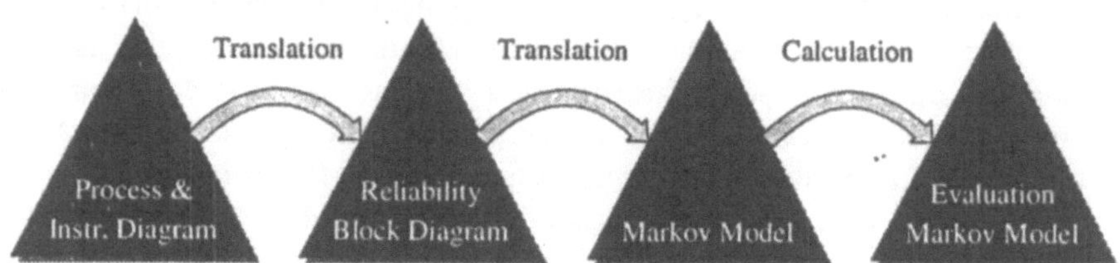

Figure 1 Different phases in the evaluation process [3]

4.1 Reliability Block Diagram

A reliability block diagram is a graphical representation of a system. Reliability block diagrams originally come from electrical engineering where they are used to represent electronic circuits. They are very useful to represent and clarify the redundancy of a system. Each component, or very often to simplify the complexity of the system, a combination of components (modules), is in the diagram represented by a block. The connections between these blocks represent the way the system works but they do not necessarily represent the real physical connection. To determine whether the system is still working a signal enters at the left side hand of the diagram and passes through the system so it can exit on the right. Redundancy usually enables the signal to follow different paths to the right side of the diagram. If enough modules in the system failed, so the signal has no paths left to exit the diagram on the right the system is said to be failed. Starting with a Reliability Block Diagram brings about that the block diagram should meet certain requirements to be a good representation of the Safety Instrumented System. Demands concerning the reliability block diagram are [4]:

- Each block should represent the maximum number of components in order to simplify the diagram
- The function of each block should be easily identified
- Blocks should be mutually independent (A failure in one should not effect the probability of failure in another)
- Blocks should not contain any significant redundancy, otherwise the addition of failure rates would not be valid
- The reliability block diagram must be able to handle voting [5].

This last demand will be pointed out further because voting has an important influence on the way the Safety Instrumented System will be translated into the reliability block diagram. Usually Safety Instrumented Systems are built from a safety point of view, which means that when a safe failure occurs the process is shut-down by the safety system. Only in case of voting it is possible that there is more then one safe failure needed to cause a shut-down. The following will explain the voting phenomena.

For the Safety Instrumented System in Figure 2A counts that one safe failure of a module is needed to shut the process down. For safe failures, the sensor, the two legs[2] and the final element are connected in series. For Safety Instrumented System B of Figure 2 counts that at least two modules must fail safe, one in each leg, to cause the system to shut-down the process. Only one safe failure is needed for a shut-down if the sensor or the final element fails safe. For safe failures the two legs are connected parallel.

The above also brings about that the reliability block diagram used can only represent the system from a safety point of view or from a dangerous point of view. In practice this would mean that two reliability block diagrams are needed to represent one emergency shut-down system. For the program there is only one reliability block diagram needed. This reliability block diagram is built from a dangerous point of view. By giving each module in the block diagram the right settings the program can handle safe, dangerous failures and repair the right way.

This brings about that besides the way the modules are connected there somehow has to be a way to inform the program, in case of voting, how many safe failures are needed for a shut down. There must be a way to handle voting.

---

[2] A leg is in this case an input module, a main processor and an output module

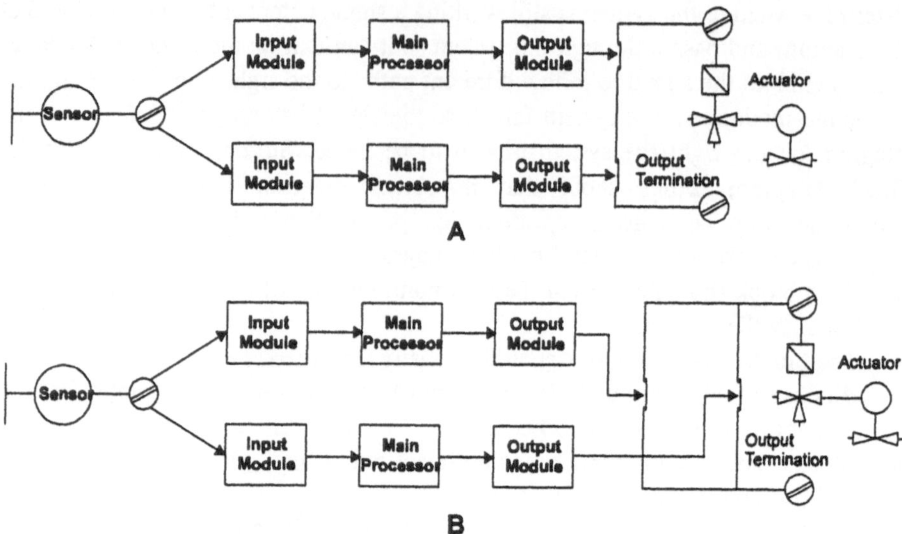

Figure 2 Two Safety Instrumented Systems [4]

4.2 Handling Voting

System A of Figure 2 is in voting terms a so called 1-out-of-2 system and system B a 2-out-of-2 system. For the program both systems from Figure 2 are translated into the reliability block diagram of Figure 3.

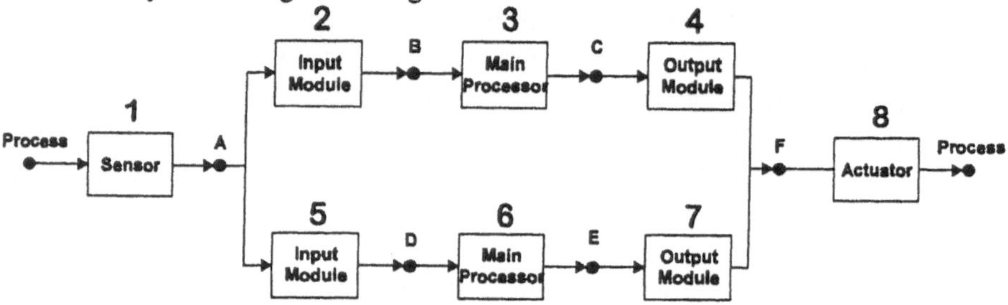

Figure 3 Reliability Block Diagram

The reliability block can be stored in the program. To handle safe failures and if necessary voting the program also needs information about this voting. The program can find this information in a so called *Voting Table* which is given to the program, see Table 1. From this voting table the program gets the information whether the Safety Instrumented System shuts the process down (puts the process into a safe state) because of a safe failure.

There are three fields in the *Voting Table*. The first field *ID* is for identification purposes only. The second field *Elements* contains all the elements that belong to this ID. These elements can not only be modules from the reliability block diagram but can also be previous ID's. The elements that belong to an ID are taken together because there is a certain relationship between these elements. This relationship is given in the third field *Down if*. This field contains the number of safe failed elements in this ID that are at least needed to explain the program that this ID is down. With other words each row is a little voter itself and the outcome can be down or not down. The real purpose of a voting table can best be explained by example, see paragraph 6. A table like this has the following advantages:

- It is possible to add as much rows as needed
- It is possible to make any combination of elements, it does not matter how the elements are connected (if they are) in the reliability block diagram
- It is always possible to add a last row which covers the total Safety Instrumented System

Table 1 General Voting Table [6]

ID	Elements			Down if
:	:	:	:	:
ID1	Element1, ..., ElementN			X
:	:	:	:	:

5 Program Assumptions

To make the program intelligent each module has certain specific properties so the program knows how to react on safe and dangerous failures and repair. For more details about handling failures and repair see [3]. The following assumptions are made to automate the process of Markov synthesis:

- All transition rates (failure and repair) are constant
- The different Markov or system states the system can be in are as described in paragraph 4
- The modules can fail in three different ways as described in paragraph 3. This does not mean that each module will or can always fail in these three different ways. For example most Safety Instrumented Systems are built de-energise-to-trip. In this case it is not unreasonable to make the assumption that a power supply can only fail safe
- Once a module has failed in either of the three possible ways it can not fail again in one of the remaining two ways, it resides in this original failure mode unless it gets repaired or replaced.
- A failed module can be repaired in three different ways:
  - On-line repair. For example in case of a detected failure of a redundant module it is sometimes possible to repair on-line
  - Periodic repair (testing). Failures not detected by on-line test can only be repaired during a thorough periodic test of the complete system
  - Safe repair. When the process has been shut-down after a safe failure, the repair is called safe repair.

  How a failed module will be repaired depends on the system state the system will reside in after the module failure and the repair policy of the company.
- Once the system has entered the fail safe or dangerous state the only transition out is to the operational state.

- The system is said to be failed if the reliability block diagram is said to be failed (no signal path left). The state the system will transition to depends on the last failed module.
- Starting point for the creating the Markov model is one reliability block diagram as described in paragraph 4.1
- The outcome of the program should be a Markov model with all the different states the system can reside in and all possible transitions between these states to come to the most accurate Markov model possible

6 Practical Example

As a practical example a Markov model is created for each Safety Instrumented System of Figure 2 with the corresponding reliability block diagram of Figure 3. The way each system reacts on different module failures is modelled like in the ISA SP84 standard [4]. The models represented are so called 2-order Markov models. This means that the impact on system level is shown of a maximum of two module failures in sequence. Common cause and functional failures are not taken into account. The following symbols and typology is used:

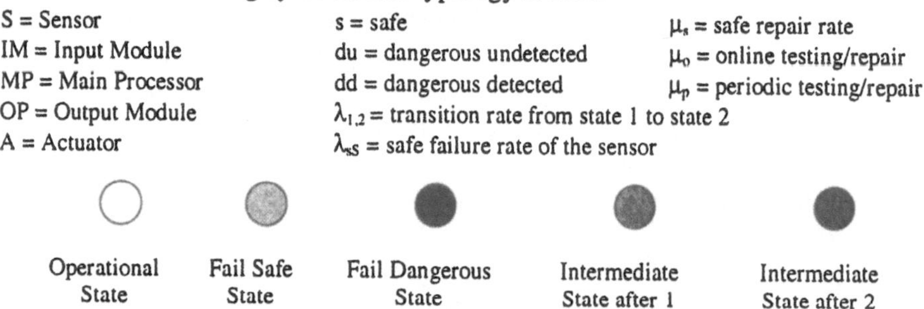

Starting with the 2-out-of-2 configuration, dangerous undetected failures of any of the modules leads immediately to a fail dangerous situation. A dangerous detected failure of the sensors or the actuators will lead to a safe state given the assumption that the operator will press the emergency stop. A dangerous detected failure of any module in a leg will lead to an intermediate state which can be repaired by an online repair. For safe failures the system will react as shown in the voting table, see Table 2. In the second row the elements 'Input Module (2)', 'Main Processor (3)' and 'Output Module (4)' are taken together and stored with

Table 2 Voting Table 2-out-of-2 system

ID	Elements	Down if
Leg1	Input Module (2), Main Processor (3), Output Module (4)	1
Leg2	Input Module (5), Main Processor (6), Output Module (7)	1
Leg3	Leg 1, Leg 2	2
Leg4	Sensor (1), Leg 3, Actuator (8)	1

ID 'Leg1'. The '1' in the *Down if* Column means that at least one element, of these three elements, must fail safe so Leg1 is said to be down (has failed safe, in voting terms it is a 1-out-of-3). This is logical because the modules are connected in series. In the third row the same counts for the elements 'Input Module (5)', 'Main Processor (6)' and 'Output Module (7)'. 'Leg2' is said to be down if at least one of these three elements has failed safe. Since this is a 2-out-of-2 configuration these two legs ('Leg1' and 'Leg2') are taken together with ID 'Leg3' and is said to be down if both 'Leg1' and 'Leg2' are down. To detect whether the Safety Instrumented System has failed a last row has been added to the table which covers it all. 'Leg4' consists of the elements 'Sensor (1)', 'Leg3' and 'Actuator (8)'. As soon as one of these three elements has failed safe the Safety Instrumented System resides in a safe state. Because it is always possible to make a last row like this the program only needs to check whether this last row is down to find out the status of the Safety Instrumented System after a safe failure. The Markov model belonging to this system is given in Figure 4.

The Voting table of the 1-out-of-2 system is much easier compared to the voting table of the 2-out-of-2 system. This can be explained because every safe failure of any module will lead to a fail safe state of the system. The voting table is only one leg with all the modules as elements and in the *Down if* column a 1, see Table 3. A dangerous detected failure of any module will lead to the fail safe state. Dangerous undetected failures of any of the modules Input Module, Main Processor or output module will not like in the 2-out-of-2 configuration lead to a fail dangerous situation but to an intermediate state. These on first sight, small, different assumptions will lead to a totally different Markov model, see Figure 5.

Table 3 Voting Table 1-out-of-2 system

ID	Elements	Down if
Leg1	Sensor (1), Input Module (2), Main Processor (3), Output Module (4), Input Module (5), Main Processor (6), Output Module (7), Actuator (8)	1

7 Conclusions

The program, as it is presented in this paper, is a first attempt to automate Markov synthesis. As starting point for the development of the Markov model one Reliability Block Diagram from a dangerous point of view of the system is being used. With a so called voting table it is possible to handle any voting between modules and safe and dangerous failures the right way. The creation of the reliability block diagram is done by hand and asks special attention but it is still easier than creating the Markov model by hand. The program can handle different ways of failure and repair for each module.

In the future the program needs to be enhanced with handling functional and common cause failures. Also a good user interface and a connection with a calculation engine are a must for the program to be useful in the future.

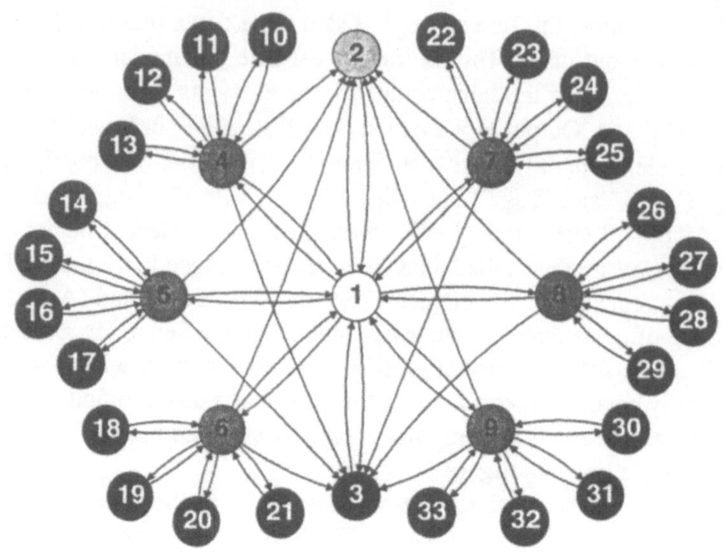

$\lambda_{1,2}=\lambda_{sS}+\lambda_{ddS}+\lambda_{sA}+\lambda_{ddA}$ $\quad\lambda_{1,3}=\lambda_{duS}+2\lambda_{duIM}+2\lambda_{duMP}+2\lambda_{duOM}+\lambda_{duA}$

$\lambda_{1,4}=2\lambda_{sIM}$ $\quad\lambda_{1,5}=2\lambda_{ddIM}$ $\quad\lambda_{1,6}=2\lambda_{sMP}$

$\lambda_{1,7}=2\lambda_{ddMP}$ $\quad\lambda_{1,8}=2\lambda_{sOM}$ $\quad\lambda_{1,9}=2\lambda_{ddOM}$

$\lambda_{2,1}=\mu_s$ $\quad\lambda_{3,1}=\mu_p$ $\quad\lambda_{4,1}=\mu_s$ $\quad\lambda_{5,1}=\mu_o$ $\quad\lambda_{6,1}=\mu_s$ $\quad\lambda_{7,1}=\mu_o$ $\quad\lambda_{8,1}=\mu_s$

$\lambda_{9,1}=\mu_o$ $\quad\lambda_{10,4}=\mu_s$ $\quad\lambda_{11,4}=\mu_o$ $\quad\lambda_{12,4}=\mu_s$ $\quad\lambda_{13,4}=\mu_o$ $\quad\lambda_{14,5}=\mu_s$ $\quad\lambda_{15,5}=\mu_o$

$\lambda_{16,5}=\mu_s$ $\quad\lambda_{17,5}=\mu_o$ $\quad\lambda_{18,6}=\mu_s$ $\quad\lambda_{19,6}=\mu_o$ $\quad\lambda_{20,6}=\mu_s$ $\quad\lambda_{21,6}=\mu_o$ $\quad\lambda_{22,7}=\mu_s$

$\lambda_{23,7}=\mu_o$ $\quad\lambda_{24,7}=\mu_s$ $\quad\lambda_{25,7}=\mu_o$ $\quad\lambda_{26,8}=\mu_s$ $\quad\lambda_{27,8}=\mu_o$ $\quad\lambda_{28,8}=\mu_s$ $\quad\lambda_{29,8}=\mu_o$

$\lambda_{30,9}=\mu_s$ $\quad\lambda_{31,9}=\mu_o$ $\quad\lambda_{32,9}=\mu_s$ $\quad\lambda_{33,9}=\mu_o$

$\lambda_{4,2}=\lambda_{sS}+\lambda_{ddS}+\lambda_{sIM}+\lambda_{ddIM}+\lambda_{sMP}+\lambda_{ddMP}+\lambda_{sOM}+\lambda_{ddOM}+\lambda_{sA}+\lambda_{ddA}$

$\lambda_{4,3}=\lambda_{duS}+2\lambda_{duMP}+2\lambda_{duOM}+\lambda_{duIM}+\lambda_{duA}$

$\lambda_{5,2}=\lambda_{sS}+\lambda_{ddS}+\lambda_{sIM}+\lambda_{ddIM}+\lambda_{sMP}+\lambda_{ddMP}+\lambda_{sOM}+\lambda_{ddOM}+\lambda_{sA}+\lambda_{ddA}$

$\lambda_{5,3}=\lambda_{duS}+2\lambda_{duMP}+2\lambda_{duOM}+\lambda_{duIM}+\lambda_{duA}$

$\lambda_{6,2}=\lambda_{sS}+\lambda_{ddS}+\lambda_{sIM}+\lambda_{ddIM}+\lambda_{sMP}+\lambda_{ddMP}+\lambda_{sOM}+\lambda_{ddOM}+\lambda_{sA}+\lambda_{ddA}$

$\lambda_{6,3}=\lambda_{duS}+2\lambda_{duIM}+2\lambda_{duOM}+\lambda_{duMP}+\lambda_{duA}$

$\lambda_{7,2}=\lambda_{sS}+\lambda_{ddS}+\lambda_{sIM}+\lambda_{ddIM}+\lambda_{sMP}+\lambda_{ddMP}+\lambda_{sOM}+\lambda_{ddOM}+\lambda_{sA}+\lambda_{ddA}$

$\lambda_{7,3}=\lambda_{duS}+2\lambda_{duIM}+2\lambda_{duOM}+\lambda_{duMP}+\lambda_{duA}$

$\lambda_{8,2}=\lambda_{sS}+\lambda_{ddS}+\lambda_{sIM}+\lambda_{ddIM}+\lambda_{sMP}+\lambda_{ddMP}+\lambda_{sOM}+\lambda_{ddOM}+\lambda_{sA}+\lambda_{ddA}$

$\lambda_{8,3}=\lambda_{duS}+2\lambda_{duIM}+2\lambda_{duMP}+\lambda_{duOM}+\lambda_{duA}$

$\lambda_{9,2}=\lambda_{sS}+\lambda_{ddS}+\lambda_{sIM}+\lambda_{ddIM}+\lambda_{sMP}+\lambda_{ddMP}+\lambda_{sOM}+\lambda_{ddOM}+\lambda_{sA}+\lambda_{ddA}$

$\lambda_{9,3}=\lambda_{duS}+2\lambda_{duIM}+2\lambda_{duMP}+\lambda_{duOM}+\lambda_{duA}$

$\lambda_{4,10}=\lambda_{sMP}$ $\quad\lambda_{4,11}=\lambda_{ddMP}$ $\quad\lambda_{4,12}=\lambda_{sOM}$ $\quad\lambda_{4,13}=\lambda_{ddOM}$

$\lambda_{5,14}=\lambda_{sMP}$ $\quad\lambda_{5,15}=\lambda_{ddMP}$ $\quad\lambda_{5,16}=\lambda_{sOM}$ $\quad\lambda_{5,17}=\lambda_{ddOM}$

$\lambda_{6,18}=\lambda_{sIM}$ $\quad\lambda_{6,19}=\lambda_{ddIM}$ $\quad\lambda_{6,20}=\lambda_{sOM}$ $\quad\lambda_{6,21}=\lambda_{ddOM}$

$\lambda_{7,22}=\lambda_{sIM}$ $\quad\lambda_{7,23}=\lambda_{ddIM}$ $\quad\lambda_{7,24}=\lambda_{sOM}$ $\quad\lambda_{7,25}=\lambda_{ddOM}$

$\lambda_{8,26}=\lambda_{sIM}$ $\quad\lambda_{8,27}=\lambda_{ddIM}$ $\quad\lambda_{8,28}=\lambda_{sMP}$ $\quad\lambda_{8,29}=\lambda_{ddMP}$

$\lambda_{9,30}=\lambda_{sIM}$ $\quad\lambda_{9,31}=\lambda_{ddIM}$ $\quad\lambda_{9,32}=\lambda_{sMP}$ $\quad\lambda_{9,33}=\lambda_{ddMP}$

Figure 4 Markov model of the 2-out-two-2 system

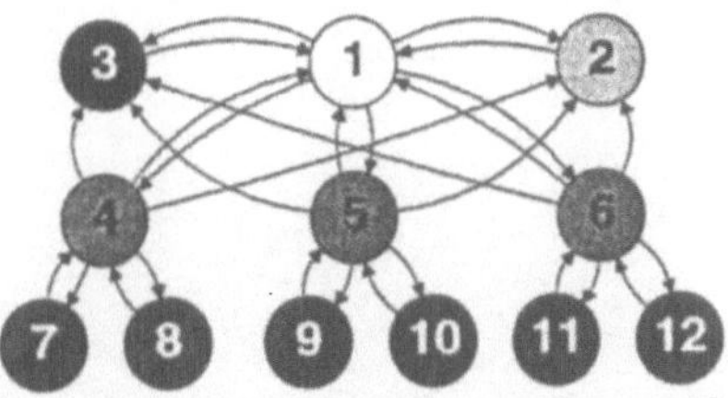

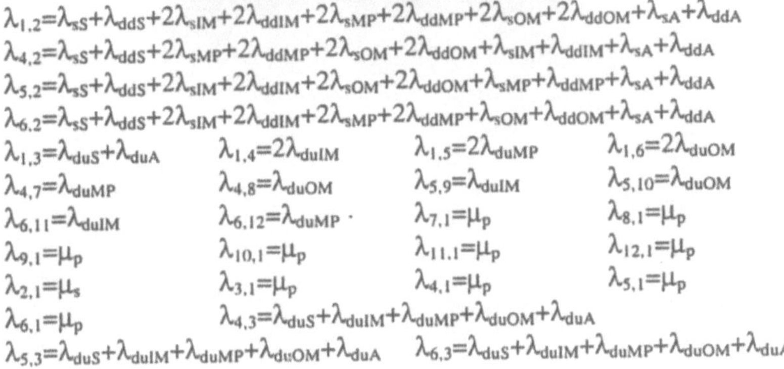

$\lambda_{1,2}=\lambda_{sS}+\lambda_{ddS}+2\lambda_{sIM}+2\lambda_{ddIM}+2\lambda_{sMP}+2\lambda_{ddMP}+2\lambda_{sOM}+2\lambda_{ddOM}+\lambda_{sA}+\lambda_{ddA}$

$\lambda_{4,2}=\lambda_{sS}+\lambda_{ddS}+2\lambda_{sMP}+2\lambda_{ddMP}+2\lambda_{sOM}+2\lambda_{ddOM}+\lambda_{sIM}+\lambda_{ddIM}+\lambda_{sA}+\lambda_{ddA}$

$\lambda_{5,2}=\lambda_{sS}+\lambda_{ddS}+2\lambda_{sIM}+2\lambda_{ddIM}+2\lambda_{sOM}+2\lambda_{ddOM}+\lambda_{sMP}+\lambda_{ddMP}+\lambda_{sA}+\lambda_{ddA}$

$\lambda_{6,2}=\lambda_{sS}+\lambda_{ddS}+2\lambda_{sIM}+2\lambda_{ddIM}+2\lambda_{sMP}+2\lambda_{ddMP}+\lambda_{sOM}+\lambda_{ddOM}+\lambda_{sA}+\lambda_{ddA}$

$\lambda_{1,3}=\lambda_{duS}+\lambda_{duA}$ $\lambda_{1,4}=2\lambda_{duIM}$ $\lambda_{1,5}=2\lambda_{duMP}$ $\lambda_{1,6}=2\lambda_{duOM}$

$\lambda_{4,7}=\lambda_{duMP}$ $\lambda_{4,8}=\lambda_{duOM}$ $\lambda_{5,9}=\lambda_{duIM}$ $\lambda_{5,10}=\lambda_{duOM}$

$\lambda_{6,11}=\lambda_{duIM}$ $\lambda_{6,12}=\lambda_{duMP}$ $\lambda_{7,1}=\mu_p$ $\lambda_{8,1}=\mu_p$

$\lambda_{9,1}=\mu_p$ $\lambda_{10,1}=\mu_p$ $\lambda_{11,1}=\mu_p$ $\lambda_{12,1}=\mu_p$

$\lambda_{2,1}=\mu_s$ $\lambda_{3,1}=\mu_p$ $\lambda_{4,1}=\mu_p$ $\lambda_{5,1}=\mu_p$

$\lambda_{6,1}=\mu_p$ $\lambda_{4,3}=\lambda_{duS}+\lambda_{duIM}+\lambda_{duMP}+\lambda_{duOM}+\lambda_{duA}$

$\lambda_{5,3}=\lambda_{duS}+\lambda_{duIM}+\lambda_{duMP}+\lambda_{duOM}+\lambda_{duA}$ $\lambda_{6,3}=\lambda_{duS}+\lambda_{duIM}+\lambda_{duMP}+\lambda_{duOM}+\lambda_{duA}$

Figure 5 Markov model of the 1-out-of-2 model

References

1. Brombacher A.C., Rouvroye J.L., Spiker R. Th. E., How safe is "safe", uncertainty in Safety, Proceedings NIRIA symposium on process safety, Utrecht, the Netherlands,1994 1

2. Rouvroye, J.L., Brombacher A.C., Spiker R. Th. E., et al, Uncertainty in safety, New techniques for the assessment and optimisation of safety in process industry. SERA-Vol. 4, Safety Engineering and Risk Analsysis, ASME, San Francisco, 1995 1

3. Houtermans, M.J.M., Nieuwenhuizen, J.K., Brombacher, A.C., Automatic Synthesis of Markov Models, Paper submitted to the PSAM III Conference, Crete, Greece, 1996 3, 6

4. ISA Technical Report, Electrical (E) / Electronic (E) / Programmable Electronic Systems (PES) for Use in Safety Applications - Safety Intergrity Evalutation Techniques, Draft 4, ISA, 1995 4, 5

5. Houtermans, M.J.M., Development of a program for more accurate safety assessment of emergency shut-down systems. Master Thesis, Eindhoven University of Technology, Eindhoven, The Netherlands, 1995 4

6. Houtermans, M.J.M., MUST, Internal Report, Eindhoven University of Technology, Eindhoven, The Netherlands, 1996 6

Session 5

Railway Applications and Experience

Specifying Railway Interlocking Requirements for Practical Use

Lars-Henrik Eriksson

Logikkonsult NP AB
Swedenborgsgatan 2
S-118 48 STOCKHOLM
SWEDEN

E-mail: lhe@lk.se

Abstract

An essentially complete formal specification of safety requirements for railway interlockings has been developed. The work is part of as project with the Swedish National Rail Administration investigating the feasibility of using formal methods for the analysis of interlockings in a production setting. An overview of the specification is given and two ongoing case studies on verifying interlockings using the specification are described. Verification is done using the very fast Stålmarck theorem prover for propositional logic. The current limits of the technology is discussed.

1. Introduction

The Swedish National Rail Administration (Banverket) has for some years been using formal methods as one of the techniques to verify new software releases for the Ebilock 850 and 950 interlockings made by ABB-Daimler Benz Transportation (Adtranz) Signal AB in Sweden. Formal methods are also used by the manufacturer itself during software development.

In both cases, the source code of the interlocking logic software is analysed using a software tool – CVT [1] – specifically developed by Logikkonsult NP AB for this purpose. The tool accepts programs written in Adtranz Signal's proprietary programming language STERNO²L and can carry out various kind of analyses.

So far, the use of formal methods has been limited to specific low-level properties of the software - in particular that a certain kind of execution error can never occur. Banverket are also using formal methods to compare different versions of software modules to pinpoint functional changes of the software.

Banverket is currently investigating the possibility of full-scale use of formal methods to communicate safety requirements of interlockings and to verify the correct realisation of these requirements. To this end, Logikkonsult NP AB has developed a set of formal specifications for safety properties of interlockings under a contract with Banverket. In this paper we will briefly describe that specification and its use. A complete presentation of the specification is given in references [2,3] (in Swedish). The contract also calls for the formal verification of an actual interlocking using the specification. As of this writing, that work is still in progress but some preliminary results can be presented.

2. The specification

In contrast with previous work on specification and/or verification of interlockings [4,5,6,7], this work has from the beginning been intended to describe a set of requirements which is at the same time general and complete. The scope of the specification is essentially limited only by the fact that Banverket's present interest is concentrated on small rail yards[1]. Features relevant mainly to larger yards have been excluded from the specification. To facilitate the acceptance and practical use of the specification, it has been written using concepts traditionally used in Swedish signalling practise.

The specification describes *functional* safety requirements only. Safety requirements relating to the construction of the interlocking, such that certain failure modes must not lead to dangerous situations, are not included. Also, safety requirements about the internal workings of track side objects are not included. The track side objects are regarded as black boxes which the interlocking requires to be in certain states, and which report back to the interlocking whether those states have properly been assumed or not.

The specification has been written using an extension to first-order predicate logic known as "Delphi logic". Formal specification using this logic is supported by the Delphi tool[2] [8,9], a prototype specification tool developed jointly by Logikkonsult NP AB and Ellemtel Telecommunication System Laboratories in Sweden. In the case of the present specification, the important extension involves a way to describe state changes, so that the behaviour of a system under sequences of external events can be described, even though the logic is not temporal.

A Delphi specification consists of two parts: the *conceptual model* and the *rule set*. The conceptual model describes the concepts that are used in the specification, what properties they have and how they are related. The rule set describes (the requirements on) how the system should react to events in the environment.

The conceptual model of the present specification describes the different physical objects and abstract concepts used by interlockings, such as rail yards, signals,

[1] This is because Banverket is presently in the process of introducing a new generation of interlockings intended for small rail yards.

[2] Not to be confused with the commercial program development tool with the same name.

points, train routes[3] etc. The safety properties are described as a set of invariants (axioms) that express relations between the objects that must always hold in order to maintain a safe situation in the yard. Apart from safety requirements, the specification includes requirements describing the possible layouts of a rail yard. This is needed both when using the specification to reason about arbitrary yards, and to verify that a description of a particular rail yard makes sense.

The specification describes general safety requirements. In order to use the specification to describe requirements for a particular interlocking – e.g. for verification – the specification must be supplemented with a description of the layout and properties of the particular rail yard controlled by the interlocking. This description is given as a set of atomic facts in predicate logic.

The predicates of the conceptual model represent the different states of the interlocking, and track side objects, as well as relations between both. For each type of track side object – such as a point – there is a set of predicates used to express the intended states of the object, such as the possible positions of the point.

The invariants of the specification express the safety requirements in terms of the predicates. E.g. there is a requirement that if a point is occupied (by an engine or car), the point must be locked (in order to prevent derailing after accidentally manoeuvring the point while there is a train moving across it). This requirement can be expressed simply as:

```
ALL pt (occupied(pt) -> point_locked(pt))
```

...where `pt` is a variable that ranges over points. Given predicates to express the locking of train routes and to relate names of train routes to the parts of the rail yard that makes up the route, the requirement that all points in a locked train route must also be locked can be expressed as:

```
ALL pt (SOME tr (locked(tr) AND part_of(tr,pt)) ->
                    point_locked(pt))
```

The rule set of the specification describes how the state of the interlocking (and of track side objects controlled by the interlocking) may change as a result of events from the environment. As an example, the train dispatcher (or traffic management software) may request that a point be moved to a certain position. The interlocking should accept that request only if the point is not locked. This requirement is described using a Delphi rule:

```
WHEN move_right(pt) IS DETECTED
IF NOT point_locked(pt) CONCLUDE right(pt) AND NOT left(pt)
```

[3] A *train route* is a part of the rail yard intended for the movement of a train. When a train is about to use a particular route, all points in the route must be locked in the correct positions, the route must be protected from cars accidentally rolling into it from the side, the route must be free from obstacles etc.

When a request to put a point into the right position (`move_right`) is received by the interlocking, the rule requires that the *precondition* `NOT point_locked(pt)` is satisfied for the state change to be permitted. If that is the case, the state of the interlocking should be changed so that the point is moved to the right position and is no longer in the left position.

Since points are physical objects, the interlocking can not simply assume that the point is in the correct position as described by the predicates `right` and `left`. A separate predicate, `controlled`, represents information about whether the point has actually moved to the intended position.

The specification has been validated in several ways. It has been used to simulate the behaviour of interlockings, and the behaviour has been checked for safety. Several safety properties not directly expressed by the specification has been formally proved to follow from it. E.g. that two signals leading into the same track from different directions can not both show a drive aspect (green light) simultaneously.

Banverket has also inspected and approved a plain text translation of the specification. We consider approval of the specification by domain experts to be crucial for acceptance of formal methods in an application area. If the domain experts do not have sufficient formal methods training to fully understand the formal specification themselves, a plain text version adhering as closely as possible to the structure and concept set of the formal specification should be used for assessment.

3. Case studies

Presently, case studies about using the specification to formally prove the correctness of existing interlockings are in progress. Although the studies have not been completed, preliminary results have been obtained.

Two verification activities are taking place. Under a contract with Banverket, Logikkonsult is verifying a particular instance of a widely used relay interlocking system (SJ model 59). Also, under the supervision of Logikkonsult's staff, a Ph.D. student funded by the Danish State Railway is doing verification work on the Ebilock 850/950 interlockings mentioned in the introduction.

The interlockings are verified by translating the relay circuits and program code, respectively, into propositional logic and then showing that the various states that can be assumed by the interlocking are all among the states permitted by the specification.

The translation into logic presents no particular difficulties in general. Relay circuits can naturally be regarded as networks of logical gates which are trivially translated into logic formulæ. The few cases where the circuitry does not translate readily into propositional logic (timing circuits in particular) are handled as special cases outside the logic with little trouble.

The program code of the Ebilock 850/950 interlockings are divided into a number of modules, each of which are translated into a logical formula describing the input-output behaviour of the module. To accommodate all features of the STERNO²L language, the propositional logic is extended with a limited form of arithmetic.

Since the specification has been written without any particular interlocking system in mind, it should be expected that the same concepts are represented quite differently in the specification and in a realisation. E.g., the specification represents a train route as the set of the track sections making up the route, while the relay interlocking represents a train route using a combination of two relays – one for the beginning and one for the end of the route.

To enable analysis of an interlocking, the different representations of the concepts must be related using formulæ in logic. This can be problematic, as it is not necessarily obvious how concepts are represented in the actual interlocking system. In the case of the Ebilock 850/950 interlockings in particular, this has proven to be a major difficulty, due to incomplete documentation.

The actual verification of requirements is essentially done by proving that the requirement formulæ of the specification are logical consequences of the translation into logic of the interlocking system. Any erroneous state will cause the proof process to fail and gives rise to a counter example, showing a situation where the interlocking fails to fulfil the requirements.

The logical proofs are done using the Stålmarck theorem prover for propositional logic [1,10,11,12]. The specification is written in predicate logic, but as any particular rail yard has a fixed number of objects, the specification supplemented with the rail yard description can be translated into propositional logic by expanding the quantifiers. To accommodate translations of STERNO²L programs the theorem prover provides limited support for arithmetic[4].

The size of the formulæ that need to be proved can be quite large. The relay interlocking under study controls a small rail yard with only two points and 12 signals. The translation of the specification into propositional logic for this rail yard results in a formula with about 10 000 logical operators. The translation of the interlocking circuitry itself results in a formula with about 1000 logical operators. This is well within the capability of the theorem prover in most cases.

So far in the project, only a few requirements have actually been verified although we expect a full verification to present no difficulties. An example of a verified requirement is that a signal cannot show a drive aspect (green light) when there are obstacles in the route behind the signal.

[4] In particular, the theorem prover is not complete when arithmetic is used.

The particular Ebilock system under study is intended for a larger rail yard with 22 points and 33 signals. A translation of the program for this interlocking results in a formula with about 400 000 logical operators. This touches the current limits of the theorem prover used. For all but the most simple properties an analysis cannot be made in reasonable time. The simplest requirements could be analysed in about 5 minutes on a SUN 4/20 computer, while the analysis of more complex requirements needed a day or more of CPU time.

On the other hand, since the Ebilock 850/950 interlocking uses general program modules that are combined according to the layout of a particular rail yard to form a program for a interlocking intended for that yard, it is possible to verify the program modules separately, a few modules in combination, or even entire parts of the rail yard. In this way, it has been possible to analyse many requirements in only a few minutes.

Unfortunately, the analysis of the program code has so far not led to any conclusive results about the correctness or incorrectness of the program. For most of the requirements tried, counter examples have been found. However, it is very difficult to tell if the situations corresponding to the counter examples can actually occur. The reason is that correct operation of each program module depends crucially on it being passed correct data from other program modules during execution. When less than an entire rail yard is analysed, it is necessary to be able to characterise formally what "correct data" from the excluded modules mean, in order to avoid false counter examples. This has proven to be difficult since sufficiently detailed documentation of the software has not been available.

This example shows that in order to formally verify large systems in practise, the systems have to be designed with formal verification in mind so that they can be decomposed cleanly into parts that can be analysed separately.

Although the two case studies concern rail yards of very different sizes, the difference in problem complexity can not only be explained by this. Another important factor is the complicated design of the interlocking program and the fact that the interlocking program is very general – intended to cope with all possible rail yard configurations – while the relay interlocking is tailored to the specific yard. While this brings the advantage that the same software can be used for different yards, it does mean that attempts to verify interlockings for simple yards will still be burdened with having to deal with program code that is unnecessary in that particular instance.

References

1. Stålmarck G, Säflund M: Modelling and Verifying Systems and Software in Propositional Logic. In: Daniels BK (ed) Safety of Computer Control Systems 1990 (SAFECOMP'90). Pergamon Press, Oxford, 1990.

2. Eriksson L-H. Formalisering av krav på ställverk (delrapport fas 1). Report NP-K-LHE-001. Logikkonsult NP AB, Stockholm, 1995. (in Swedish)

3. Eriksson L-H. Formalisering av krav på ställverk (slutrapport). Report NP-K-LHE-003. Logikkonsult NP AB, Stockholm, 1996. (in Swedish)

4. Groote JF et.al. The Safety Guaranteeing System at Station Horn-Kersenboogerd. Logic Group Preprint Series No. 121. Department of Philosophy, Utrecht University, Utrecht, 1994.

5. Hansen KM: Validation of a Railway Interlocking Model. In: Naftalin, Denvir, Bertran (eds.) FME'94: Industrial Benefit of Formal Methods. Springer-Verlag, Heidelberg, 1994. (Lecture Notes in Computer Science no. 873)

6. Morley MJ. Modelling British Rail's Interlocking Logic: Geographical Data Correctness. Technical Report ECS-LFCS-91-186. Department of Computer Science, University of Edinburgh, Edinburgh, 1991.

7. Morley MJ: Safety In Railway Signalling Data: A Behavioural Analysis. In: Joyce, Seger (eds.) Higher Order Logic Theorem Proving and its Applications. Springer-Verlag, Heidelberg, 1993. (Lecture Notes in Computer Science)

8. Höök H. Delphi – A General Description of the Language. Report F 91 0881. Ellemtel Utvecklings AB, Stockholm, 1993.

9. Stålmarck G, Widebäck F. Definition av Delphi. Report NP-FW-001. Logikkonsult NP AB, Stockholm, 1991. (in Swedish).

10. Stålmarck G, Åkerlund O: Formal verification of hardware and software systems using NP-Circuit. In: Malmén Y, Rouhiainen V (eds.) Reliability and safety of processes and manufacturing systems. Elsevier, London, 1991.

11. Säflund M: Modelling and formally verifying systems and software in industrial applications. In: Proc. of the Second International Conference on Reliability, Maintainability and Safety (ICRMS'94). International Academic Publishers, Beijing, 1994.

12. Widebäck F. Stålmarck's Notion of *n*-saturation. Report NP-K-FW-200. Logikkonsult NP AB, Stockholm, 1996.

SIGAV, the Italian High Speed Railway Integrated Management System: Safety and Reliability Overview

G. Aprea°, P. Colantuoni*, P. Firpo+, R. Lido°, D. Pellegrino°, M. Rapone^, F. Senesi°

° Italferr Sis.T.A.V., Roma, Italy
^ Consiglio Nazionale delle Ricerche, Napoli, Italy
* Vector s.r.l., Napoli, Italy
+ Sciro Electra s.r.l., Genova, Italy

Abstract

The possibility of controlling the operation of a wide high speed railway network by a centralised management system, allows to optimise its performance in terms of traffic control and regulation, logistic support management, services provided to the user and integration with the conventional railway and external services. In the specification and design of such a system, safety and reliability cover a fundamental role in that its integrity heavily impact the railway operation. This paper aims at discussing the reliability and safety implications taken into account by the SIGAV, the Italian High Speed Railway Integrated Management System, design group in the SIGAV preliminary design and specification.

1 Introduction

The Italian High Speed System, hereinafter called HS system, is a high performance mass transport system intended as a complex of lines, stations, junctions, rolling stocks, signalling and telecommunications equipment and power supply plants.
The overall design of the HS system has been carried out by means of a "systemic" approach able to ensure, by means of an appropriate development plan, the full integration of the lines, rolling stocks, logistic support, technologies, management systems and customer services design.
This results in *network-oriented* design as the system structure itself is network structure: as shown in Figure 1a , the HS system is organised in two main lines, from west to east and from north to south widely interconnected with the conventional railway in order to provide adequate services also to the cities not located along the HS lines.
HS is intended to provide a *train product* able to harmonise the customer satisfaction with the cost-effectiveness, mandatory features for a system which aims to cover a strategic role in the market. This can be accomplished by:

- improvement of the demand through the improvement of the quality of service;
- improvement of productivity;
- reduction of operational costs.

The above requirements can be fulfilled by means of adequate automated procedures, supported by appropriate informative systems, able to optimise the parameters influencing the HS system operation, in such a way that:

- a high level of system quality in terms of reliability, availability, maintainability and performance can be reached;
- the wide data flows relevant to the railway management process are made available to all the subsystems.

Therefore, it is mandatory to maximise the exploitation of the information flows in such a way that, at each time, it is possible to evaluate the effectiveness of the service and to undertake appropriate corrective actions.

In the context of the HS system, SIGAV represents the global informative system able to continuously monitoring the whole system most significant operational parameters, including those related to the quality of service, and to control them by providing appropriate commands to the peripheral systems, with the main purpose of optimising the management of the High Speed Railway network both in terms of cost-effectiveness and in terms of customer satisfaction.

2 SIGAV Functional Architecture

The SIGAV architecture is organised in three functional levels as shown in Figure 1b. At each of them, a specific operative competency, according to the decisions made at higher functional levels, is assigned.

The link between the three functional levels is realised by means of standard protocols for information exchange and commands dispatching, with duplication of the hardware in accordance to the diversity criteria, which define univocally the relevant functional and operative hierarchic relationships.

The first level, the higher, is called *PSV-PCO level*, where PSV is for *Posto di SuperVisione*, which means Supervision Centre, and PCO is for *Posto Centrale Operativo*, which means Operative Central Location. The role assigned to this functional level is to work out the whole strategies for the service optimisation: all the control, command and optimisation functions of the lower levels are centralised in PSV-PCO; furthermore, a strategic optimisation function aimed at resolving, at a global level, conflicts and degrades interesting the whole HS railway operation, is continuously performed by PSV-PCO.

The second level, the intermediate, is called *PCS level*, where PCS is for *Posto Centrale Satellite* which means Satellite Central Location. The role assigned to this functional level is to directly control and command, according to the requests received by the PSV-PCO or by a local operator when present, a line section (of about 250 km on the average), to manage the link among and with the other functional levels and to optimise the train running. The optimisation is realised by means of operative orders dispatched to trains either as a result of the intervention of PSV-PCO which sends a new traffic operative plan, determined at the whole system level, or as a consequence of local decisions constituting the reaction of the system to problems judged to be solvable in the context of the controlled line section.

The third level is called *PPF level*, where PPF is for *Posto Periferico Fisso* which means Fixed Peripheral Location. Its role is to execute safely the commands received by the higher levels, to transmit and verify the operative orders to be dispatched to the trains and, finally, to manage the communications with the other functional levels.

The scope of this overview is limited to the PSV-PCO functions and to the relevant architecture.

At this level, SIGAV may be subdivided in the following functional areas or subsystems:

- Control and Regulation (CR - Controllo e Regolazione);
- Signalling (SG - SeGnalamento);
- Power Supply (AL - ALimentazione);
- Rolling Stock (MR - Materiale Rotabile);
- Commercial (SC - Sottosistema Commerciale)
- Maintenance (SM - Sottosistema Manutenzione);
- Telematics and Telecommunications (TT - Telematica e Telecomunicazioni);
- Data Recording and Treatment (SEE - Elaborazione dati di Esercizio).

SIGAV has also to exchange information with some external informative systems:

- SIMAV: the integrated system for the HS system maintenance;
- SIF: the Italian conventional railways informative systems.

The proposed PSV-PCO architecture, which will be presented in the following paragraph, allows to evaluate the interconnection characteristics of SIGAV and the design constraints for its realisation.

2.1 Architecture of the HS Supervision Centre PSV-PCO

The PSV-PCO architecture is organised as a *Global Control Room* comprising four operative areas correspondent to four functional groups, as shown in Figure 2a.

2.1.1 Real Time Area

This functional area performs all the control/command functions characteristics of the HS system.

The above functions are those performed, in the conventional Italian railways, by master operators at system level.

The Real Time Area is furtherly subdivided in reason of the "safe" nature of equipment in the following sub-areas:

- Not Vital Real Time Area.
  Groups all the not safety related control/command functions relevant to:
  1. signalling equipment;
  2. power supply equipment;
  3. rolling stocks and crews;
  4. auxiliary equipment;
  5. telecommunications.
- Vital Command.
  Groups all the safety related command functions (i.e. speed reduction orders).

2.1.2 Planning Area

This functional area performs all the *off line* functions relevant to the service global planning, in particular:

- traffic, power supply plants configuration, rolling stocks, personnel;
- equipment maintenance;
- commercial aspects and telecommunication network management.

2.1.3 Interface Area

This functional area includes all the communication functions both at a local network level (LAN) and at a geographical network level (WAN).

All the communications monitoring and managing activities are included.

2.1.4 Training Area

This functional area is equipped with a set of workstations dedicated to the training of the PSV-PCO real time areas personnel.

2.1.5 Global Control Room Layout

In Figure 2b the LAN architecture, which connect the diverse operative areas, is shown.

The Control Room structure (Figure 3a) is based on a *concentric* configuration organised in three hierarchical levels:

- a system wide *Master Plan*, realised by means of retrographic displays, visible to all the operators and displaying all the information about the status of the whole controlled area (Figure 3b);
- *Zonal Operator* work stations intended for the control of bounded sub-areas;
- *Master Operator* work stations intended for coordination purposes.

2.2 General Architectural Requirements

The SIGAV Hardware and Software architecture is defined in accordance to the technologic principles currently adopted by the Italian Railways company in other applications and can be summarised as follows:

- client-server computing
- servers and workstations connected in redounded local network FDDI (IEEE802.3);
- adoption of redundancy for all the critical components;
- fault tolerant mass storage shared by the servers;
- operating system Unix System V for open system architecture, with real-time extension Posix 1003.4;
- relational DBMS Oracle with standard access and management specifications;
- windows graphics facilities X Window OSF/MOTIF;
- software local network communications TCP/IP SNMP.

2.3 SIGAV Vital Functions

The SIGAV architecture is designed for allowing the centralisation of the vital functions which compete usually to the lower hierarchical levels without decreasing the system performance in terms of safety and reliability.
The SIGAV vital functions belong to the Real Time Area and are implemented in the Vital Kernels, constituted by "safe" computer systems, present at each functional level: the *PSV-PCO level* (NV - Nucleo Vitale), the *PCS level* (NVC - Nucleo Vitale Centrale) and the *PPF level* (NVP - Nucleo Vitale Periferico).
The implementation of the vital functions requires that the whole safety targets are always respected within the boundaries characteristics of guided transport systems.
Being the safety, intended as the train running protection functions (spacing, route assignment, etc.), ensured locally by the PPFs and PCSs, the SIGAV Vital Functions, at PSV-PCO level, are those related to the system reactions against the occurrence of dangerous events necessary for maintaining the appropriate level of safety, concerning in particular:

- diagnostic data management;
- control/command of equipment not safely managed at local level (PPF);
- determination and/or request of speed reductions from/to internal functions;
- acquisition and/or request of speed reductions from/to external functions;
- actuation of speed reductions.

The missing or erroneous reaction of the system can cause hazards as explained in § 4.3.

3 Fault Conditions and Subsystems Criticality

The criticality indexes resumed in Table I, are defined for the SIGAV subsystems and functions.

Table I

Subsystems Criticality		Functions Criticality	
A	**High** Subsystem *essential* for SIGAV functionality	α	**Maximum** Function/subfunction *essential* for the relevant subsystem/function integrity.
B	**Medium** Subsystem *critical* for SIGAV functionality	β	**Medium** Function/subfunction *critical* for the relevant subsystem/function integrity
C	**Low** Subsystem *not critical* for SIGAV functionality	γ	**Minimum** Function/subfunction *not critical* for the relevant subsystem/function integrity

The following fault conditions are defined for SIGAV:

1. *Immobilising (I)*: one or more A criticality subsystems are lost.
2. *Service (S)*: one or more subsystems or one or more functions which cause a not negligible degrade in SIGAV performance are lost.
3. *Minor (M)*: one or more subsystems or one or more functions which cause a negligible degrade in SIGAV performance are lost.

All the functions of criticality α belonging to subsystems of criticality A (functions A,α) in addition to all the functions which, independently by their own criticality, result essential for the functionality of functions A,α, concur to the fault condition 1.
As an example, a function belonging to the subsystem SEE, with criticality B,γ, can result essential for the functionality of another function belonging to the subsystem SG with criticality A,α: in this case it will concur, indirectly, to the fault condition 1. and therefore it shall assume maximum criticality at system level.
The other functions concur to the conditions 2. and 3. in reason of the scheme resumed in Table II.

Table II

Function criticality	Subsystem criticality		
	A	B	C
α		S	S
β	S	S	M
γ	S	M	M

4 SIGAV RAMS Requirements

In the following subparagraphs, the approaches adopted for defining the SIGAV RAMS Requirements shall be presented and discussed.
In particular, a novel *function oriented* methodology for RAM apportionment, derived, for the quantitative part, from the Feasibility of Objectives method defined in [1], is presented. This methodology covers also the qualitative requirements apportionment adopted for software and based on a *function weakness* approach.

4.1 Overall RAM Requirements

The SIGAV Overall RAM Requirements are defined, in terms of Inherent Availability, for the fault condition defined as Immobilising (I) in § 3 in that it can be assumed as the only condition which lead to the loss of the whole system.
Inherent Availability is defined as the probability that the system is operational at a specified time, under its rated operating conditions and without taking into account the downtimes due to both preventive and corrective maintenance and the administrative delays.
The relevant reliability block diagram, in terms of SIGAV subsystems is shown in Figure 4.
The Inherent Availability general requirement interests all the computational resources and the interfaces necessary for ensuring:

- the correct performance of the *essential* functions for each *essential* subsystem;
- the correct feeding, both external and internal, for the above functions.

The equipment interested on the basis of their influence on the success of the *essential* subsystems are the following:

- the redounded servers of the Real Time Area and the relevant peripherals;
- the not redounded servers of the Planning Area and the relevant peripherals;
- the LAN Supervision workstation;
- the X.25 and G.703 routers;
- the MUX;
- the switching hubs and the Ethernet links of the Real Time and Planning Areas;
- the FDDI ring.

The following equipment do not contribute to the Overall Availability requirement in that specific reliability requirements are provided for them:

- the Real Time and Planning Areas peripherals;
- the operator workstations.

Furthermore, specific reliability requirements, in terms of MTBF, are defined for all the single equipment for establishing constraints on their service quality and on their maintenance demand.

For taking into account the not negligible contribution of the software failures without defining a specific quantitative availability target for them, a maximum percentage of hardware failures contribution on the overall availability requirement has been stated as the 50% of the total downtime per year. In this way the requirements to comply with are the HW availability requirement and the Overall availability requirement. The SW contribution instead is not quantified and does not require to be quantitatively demonstrated as long as the overall availability requirement is accomplished.

The Inherent Availability Requirements stated for SIGAV are resumed in Table III.

Table III

Subsystem	Inherent Availability A_i	Max tolerable standstill time per year	HW contribution	
			Availability	Standstill Time
CR + MR + SG	0.999886	60'	0.999943	30'

4.2 RAM Functional Apportionment Methodology

The functional RAM apportionment consists in assigning to each *essential* SIGAV function an appropriate quantitative availability target (*quantitative* apportionment for HW), determined on the basis of the hardware involved, and in assigning to all the SIGAV functions an appropriate Integrity Level (*qualitative* apportionment for SW) based on the results of a qualitative functional analysis which provide a measure of the function *weakness* to be combined with the function criticality.

4.2.1 Hardware Contribution Apportionment

The proposed method is an extension of the Feasibility of Objectives (FoO) method defined in [1]. The difference consist in the fact that in the FoO method the apportioned quantity is the system *reliability* while in this case the system *availability* has to be allocated to the equipment and consequently to the functions. The application of the method results in the determination of an *allocation factor* for each equipment as a function of numerical ratings of:

1. *system intricacy* range 1 to 10: 1 for the least complex system;
2. *state of the art* range 1 to 10: 1 for the best developed system;
3. *performance time* range 1 to 10: 1 for the least utilised system;
4. *environment* range 1 to 10: 1 for the least severe environment;
5. *ease of maintenance* range 1 to 10: 1 for the easiest maintainable system;
6. *redundancy* range 100 to 1: 100 for a not redounded system.

The six ratings, determined on the basis of experience in similar systems or of engineering judgement, are multiplied to give the equipment rating which will range from 1 to 10^7.

Once known the population of equipment which concur to the I fault condition the ratings are normalised in such a way that their sum is 1 obtaining the allocation factors for all the equipment.

The functional apportionment procedure may be summarised as follows:

1. let $\{w_1,..,w_N\}$ be the vector of the allocation factors assigned to the N system equipment;
2. let $\{D_1,..,D_{Ni}\}$be the set of Ni equipment involved by the function f_i;
3. let DT_{hw} the tolerable downtime per year due to HW failures (30');
4. the downtime $DT_{hw,i}$ tolerable for the function f_i is given by:

$$f_i \Rightarrow DT_{hw,i} = \left(\sum_{j=1}^{N_i} r_{i,j} \cdot w_j \right) \cdot DT_{hw} \Rightarrow A_{f_i}$$

where $r_{i,j}$ represents the percentage of utilisation of the device j by the function i.

4.2.2 Software IL Allocation

The method is based on the *Failure Risk Matrix* which allows to determine the IL to be assigned to each function by combining:

- the function criticality in terms of its effect on system failures (I, S or M);
- the function weakness in terms of structural and environmental qualitative parameters.

Four classes of weakness are defined and assigned to the functions in reason of the combination of the following three factors:

- expected number of defects *Nd*;
- likelihood that defects become failures *Fd*;
- defects and failure reduction measures *Rd*.

Each of them may assume three discrete values as shown in Table IV.

Table IV		
Factor	Value	Source
Expected number of defects (Nd)	Nd1 High Nd2 Medium Nd3 Low	Structural parameters: • size • flow complexity • data structure complexity
Defect caused failures (Fd)	Fd1 High Fd2 Medium Fd3 Low	Environmental parameters: • time features • human factor • expected occurrences
Defects and failure reduction (Rd)	Rd1 Absent Rd2 Low Rd3 High	Fault tolerance features, developer experience, test effort: • fault tolerance • defect reduction

The appropriate Weakness Class can now be assigned to the SIGAV functions by means of the Weakness Graph shown in Figure 5.
The Class 1 is assigned to the more robust function while the Class 4 is assigned to the more weak one.
Finally, the *Failure Risk Matrix* is applied for determining the appropriate IL to be assigned to the function under analysis as shown in Table V.

Table V - Failure Risk Matrix					
Weakness Class / Criticality	WC0	WC1	WC2	WC3	WC4
1 - MINOR	IL1	IL2	IL2	IL3	IL3
2 - SERVICE	IL2	IL2	IL3	IL3	IL4
3 - IMMOBILISING	IL3	IL3	IL3	IL4	IL4

4.3 SIGAV Preliminary Hazards List

From the safety point of view, the first step to determine the safety requirements is to analyse all the SIGAV vital functions (see § 2.3), on the basis of the SIGAV architecture, for a preliminary assessment of the consequences due to the occurrence of dangerous events (hazards).
The methodology defined in [8] is adopted; the severity levels are classified as shown in Table VI.

Table VI - Hazard Severity Categories		
Severity	Description	Definition
4	**Catastrophic**	Fatalities and/or multiple severe injuries.
3	**Critical**	Single fatality or severe injury. Subsystem loss.
2	**Marginal**	Multiple minor injuries. Major subsystem damage.
1	**Negligible**	Possible single minor injury. Minor subsystem damage.

The results of the analysis are recorded in the Preliminary Hazard List (PHL): among all the listed hazards, the severity level 4 are those related to the following functions/subfunctions:

- SG - speed reduction management (i.e. due to an erroneous acquisition from Control and Regulation function), the worst consequence being the setting of an unsafe speed reduction for the operation;
- SG - line section control and command (i.e. due to a missing/erroneous command acquisition/request or to an erroneous speed reduction processing and management), the worst consequence being respectively the setting/request or the processing of an unsafe speed reduction for the operation;
- CR - operating plan definition (i.e. due to an erroneous test of congruence on the plan or to an erroneous management or request), the worst consequence being, for both cases, the request of an unsafe speed reduction or the missing request of a speed reduction for the operation.

4.4 Integrity Level Assignment to SIGAV Functions

The Supplier of SIGAV shall conduct preliminary safety analyses aiming at assigning a Safety Integrity Level (SIL) to each of the functions listed in the PHL whose severity level is ≥1.

The SIL, once approved by Italferr Sis.T.A.V., shall become contractual safety requirements for the functions/subsystems involved and, therefore, shall be considered as a constraint to develop the safety activities required for that SIL in the Safety Program (see § 4.5).

The SIL are defined as follows:

Table VII - Safety Integrity Levels

SIL	Description	Definition	Characteristic
4	**Higher**	Vital or critical for safety	Fail safe
3	**High**	Not directly critical for safety	Fail operational
2	**Medium**	Essential or not critical for safety	Medium Integrity
1	**Low**	Essential or not critical for safety	Medium Integrity
0	**Not specified**	Not vital	Not relevant for safety

As regards software integrity, the Integrity Levels shall be updated following the assignment of the SILs mentioned above and applying the table VIII as shown:

Table VIII - IL Assignment for Safety

SIL 4 functions	Integrity Level 3
SIL 3 functions	Integrity Level 2
SIL 2 functions	Integrity Level 2
SIL 1 functions	Integrity Level 1

For each first level function, the final Integrity Level shall be the greatest between the Integrity Level originally assigned to it (see § 4.2.2) and the one resulting from the application of table VIII. The Supplier shall apply the same methodology to allocate the Integrity Levels to the lowest level functions.

4.5 RAM and Safety Programs General Requirements

In order to assure that overall reliability, availability and safety goals will be achieved a RAM and Safety (RAMS) Program shall be carried on by the contractor. General requirements of RAMS program make reference to [8], [9] and [10].
In particular for system made of software and hardware, RAMS Program plays a still more relevant role. In fact, the major scope of the program is to detect potential failures causes since the conception phase of the product life cycle. In the case of software this is dramatically important because, as it is well known the cost of bugs fixes exponentially increases along the software development process.
In this context, the approach followed refers to the three important phases of SIGAV:

a) Proposal/ Pre-contractual Phase, Requirements Analysis, Preliminary Design,
b) Development, Detailed Design, Coding,
c) System test stage.

The phase a) has been performed by Italferr Sis.T.A.V.., with the co-operation of Sciro Electra and Vector. The phases b) and c) will be performed by the system Supplier.
To this end the Supplier has to propose a RAMS Program Plan, which identifies all tasks required to accomplish program requirements set by Italferr Sis.T.A.V. and fix milestones and time schedules to achieve the goals.
The work developed in the phase a) assures that system HW&SW requirements fulfil the end-user expectations. It is well known that the major part of software failures derives from the misunderstanding of Purchaser requirements. In this case a deep analysis of Purchaser general requirements for the entire system has been performed, and the description of system has been developed at a level far deeper than the preliminary design, very close to the detailed design, in such a way that RAMS requirements for the entire system could be set.
To define the outlines of the Supplier RAMS Program, a minimum set of tasks from [8], [9] and [10] has been selected in order to ensure that RAMS aspects are properly taken into account by the Supplier at all stages of detailed design and development. In particular the Safety Program specifications require different Safety and Quality management tasks, as well as different level of details for them, according to the Safety Integrity levels assigned to the functions/subfunctions by the Supplier and approved by Italferr Sis.T.A.V.
As far as software is concerned, strong emphasis is given to the RAMS and QA Program to be performed by supplier during the software development process, following the principle that reliability, safety and quality must be designed in from the beginning, more than built in through defects removal. Moreover, a reliability growth process has to be designed and developed by the software developer using quantitative measures to define reliability objectives at different stages of the development and to verify the accomplishment of established reliability goal at the end of each phase, since the first elementary software unit is coded and up to the final system integration and testing.

5 Lifecycle Cost Considerations

In the previous section the general requirements of RAMS Program, as well as underlying philosophy, has been described. Even though the first aim of a RAMS Program is to assure that RAMS requirements are achieved, it cannot be underestimated that it has to be cost-effective. This means that the costs connected with SIGAV shall meet the overall program cost.
In the RAMS context costs are summed up in the Life Cycle Cost (LCC) approach, considering the total cost of the system over its full life. Components of LCC regard acquisition, system operation, maintenance and life-cycle support. Thus following the RAMS contractual specifications considered for SIGAV, the Supplier is given the task to perform maintainability analysis in order to allow the definition of a contractual scenario which could assure the achievement of overall SIGAV availability goals under specified life cycle cost, hence considering not only acquisition costs but still more SW&HW maintenance costs.

References

1. MIL-HDBK-338-1A, Electronic Reliability Design Handbook, Vol. 1, U.S. Department of Defence, Washington, 1991
2. prEN50128, Railway Applications: Software for Railway Control and Protection Systems, CENELEC TC9X SC9XA, Draft, November 1995.
3. Friedman M. A., "Reliability Techniques for Combined Hardware/Software Systems, 1992 PROCEEDINGS Annual RELIABILITY AND MAINTAINABILITY Symposium, 1992, 290-293
4. Perera U.D., "Systems Reliability and Availability Prediction and Comparison with Field Data for a Mid-Range Computer, 1993 PROCEEDINGS Annual RELIABILITY AND MAINATAINABILITY Symposium, 1993, 33-40
5. Keene,S.,Lane,C.,"Reliability Grwoth of Fielded Software", 1993 PROCEEDINGS Annual RELIABILITY AND MAINTAINABILITY Symposium, 1993, 360-365
6. Capasso A., Rapone M., Lamedica R. et al., Valutazione previsionale dei guasti del sistema di alimentazione di una linea ferroviaria per Alta Velocità sull'affidabilità del servizio, 1 Convegno Nazionale PFT2, 1993, Rome, Italy, Proceedings Vol. 1, 369-392.
7. Firpo P., Savio S., and Sciutto G., Fail Safe Digital Systems in Railway Applications: a Comparative Analysis, WCRR '94 Conference, 1994, Paris, France, Proceedings, 1281-1284
8. MIL-STD-882-C, System Safety Program Requirements, U.S. Department of Defence, Washington, 1993
9. MIL-STD-785-B, Reliability Program for Systems and Equipment Development and Production, , U.S. Department of Defence, Washington, 1980
10. MIL-STD-470-B, Maintainability Program for Systems and Equipment, U.S. Department of Defence, Washington, 1980

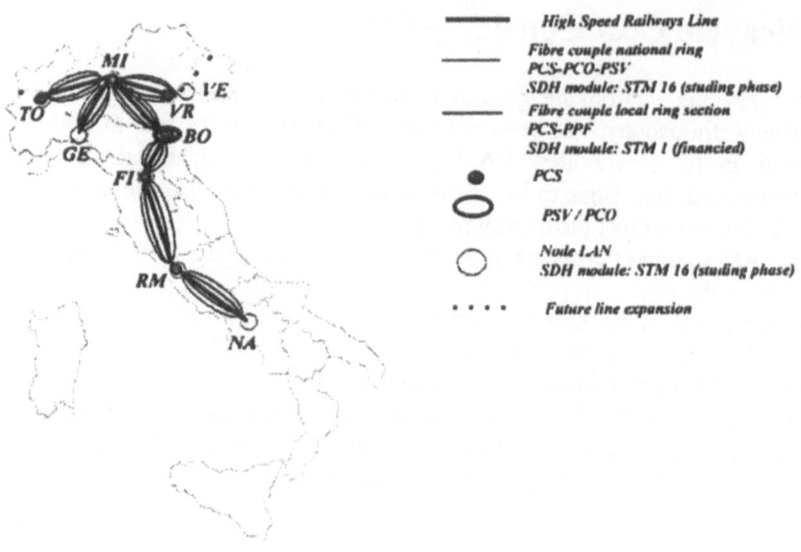

Figure 1a - Italian High Speed Railway Network

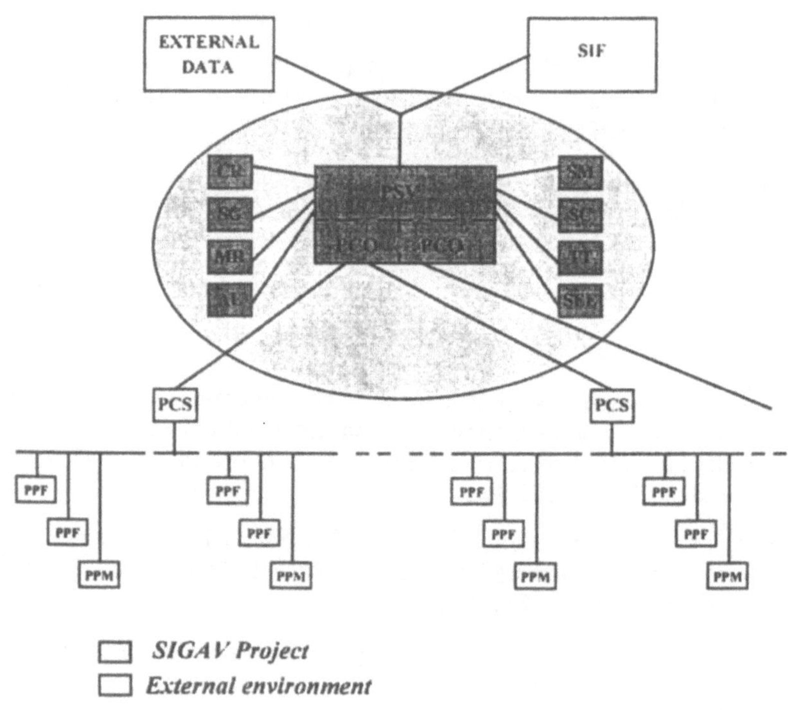

Figure 1b - SIGAV Architecture

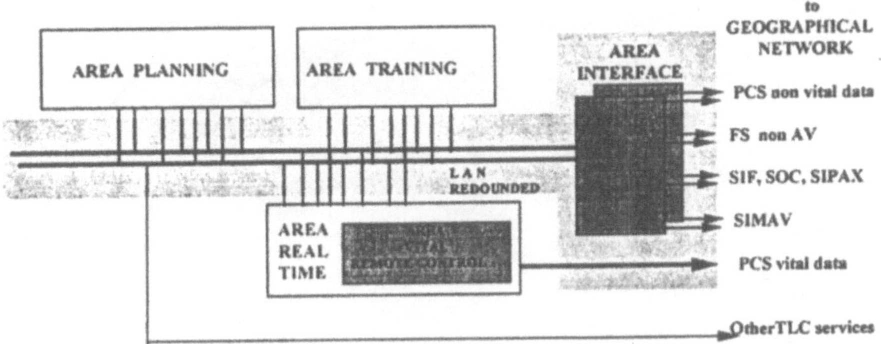

Figure 2a - PSV-PCO Functional Areas

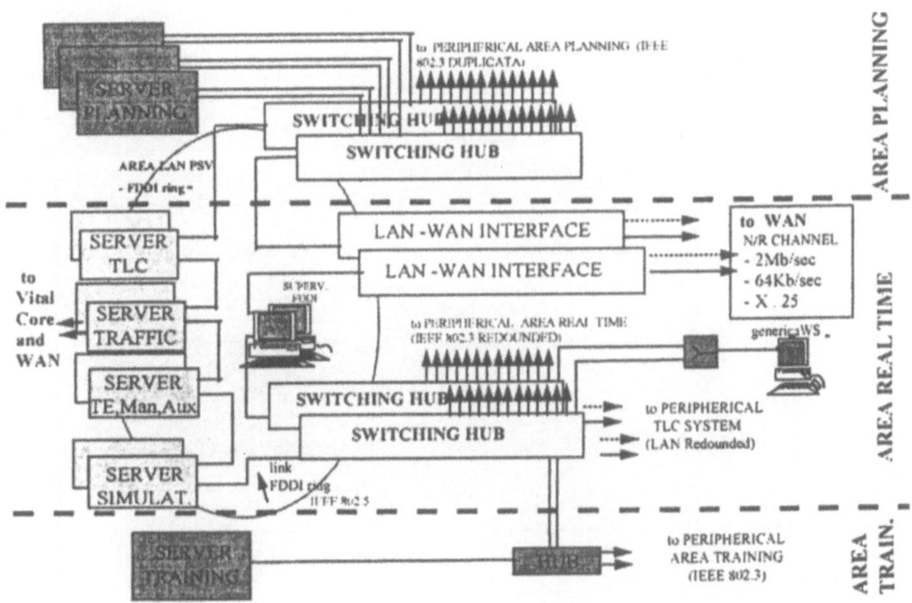

Figure 2b - PSV-PCO LAN and Interfaces

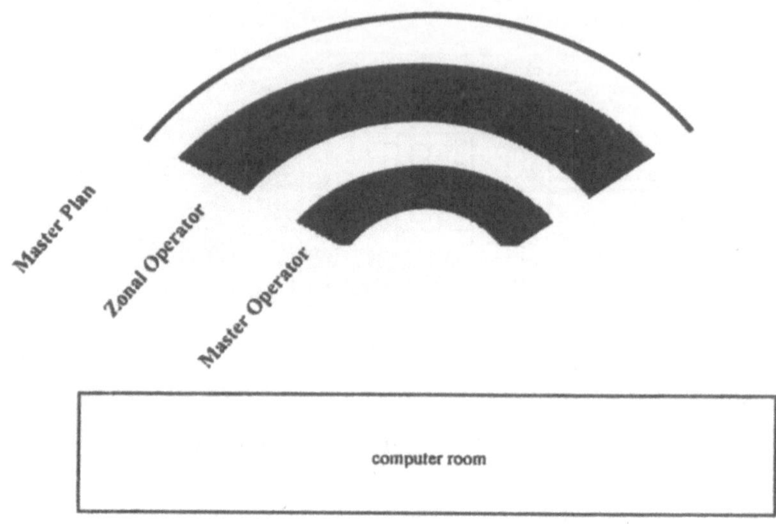

Figure 3a - PSV-PCO Real Time Area General Layout

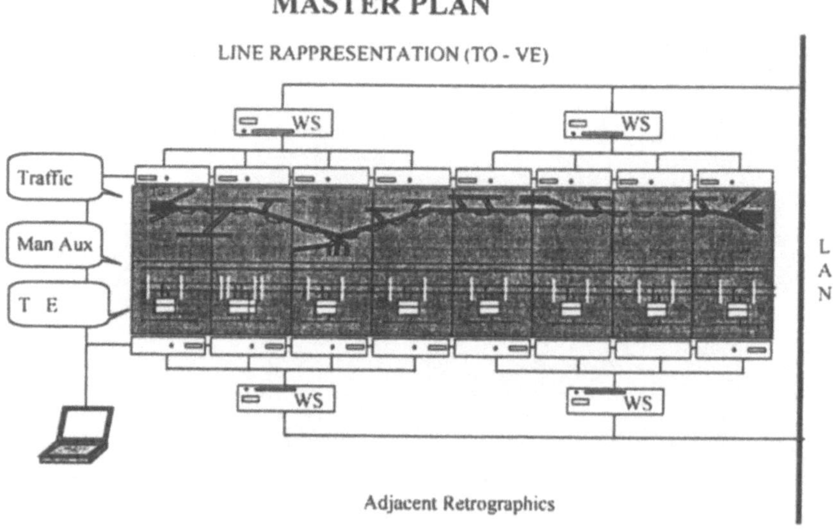

Figure 3b - PSV-PCO Master Plan

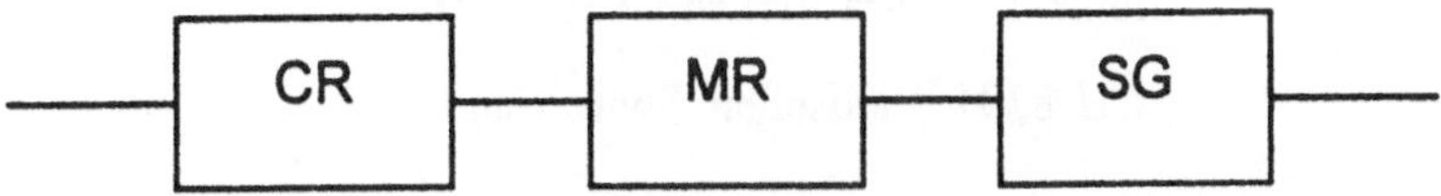

Figure 4 - Reliability Block Diagram for the Immobilising Fault Condition

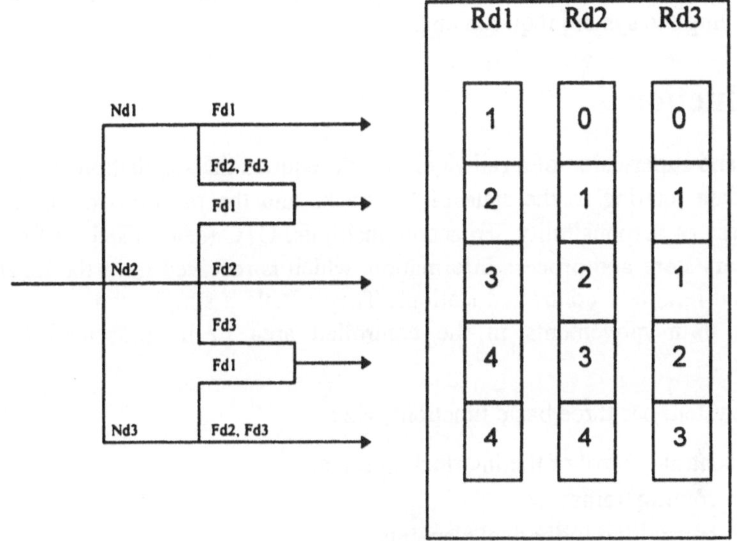

Figure 5 - Weakness Class Assignement by Weakness Graph

A Safe, Reliable Control and Supervisory System for Railway Networks

Dr Neville Rowden

Siemens Integra Verkehrstechnik AG

CH-8304 Wallisellen, Switzerland

Abstract

The design of a modern railway control and supervisory system which uses standard off-the-shelf components is described. Particular emphasis is given to the aspects concerning safety and availability, describing the unique solution the designers utilised in satisfying the system requirements.

1 Introduction

The control and supervision of a railway network requires the acquisition and provision of all data relating to the state of the tracks and the positions of the trains within the area of responsibility. To accomplish this, CTC (centralised traffic control) computers store and process information, which is received from the local interlocking and remotely controlled stations. They enable a simple and exact overview of all train movements in the controlled area to be presented to the station-master.

A CTC system has three basic functions, viz.:

1. the central control of the interlockings, e.g.:
   (a) routing trains
   (b) controlling individual elements
   (c) blocking/clearing routes
   (d) blocking/clearing the operation of points
   (e) critical commands to override the interlocking in the event of a break-down,
   (f) etc.
2. the automation of train traffic:
   (a) tracking trains
   (b) automatic routing
   (c) transferring train-position information to a higher-level management centre
3. the display of train information for passengers

For the most part, the responsibility for the safety of the railway system lies in the interlockings themselves, which can be either relay-based or electronic-based systems. However, certain countries (e.g. Switzerland, Austria, Germany, *et al.*) allow the operator the possibility of overriding the inherent safety of the interlockings by issuing so-called critical commands via the CTC system. For example, a track can be blocked using the CTC system to enable maintenance work to be carried out. By doing this, the interlocking inhibits the routing of any trains over this blocked track. When the maintenance work has been completed, the track needs to be cleared to allow further routing of trains over this track. As this action is potentially life-endangering (e.g. when it is erroneously executed whilst maintenance workers are still actively present on the track), a track-clearing command is defined to be such a critical command.

It is the responsibility of the CTC system to ensure that the execution of these critical commands does not compromise the safety of the railway network.

Another characteristic of a CTC system, which has some bearing on safety, is the general availability of the system. Managing a railway network in many places is a 24 hours a day activity. Large railway networks are so complex that the efficient management of the traffic would be impossible without a CTC system. If the CTC system is out of service, traffic management is only possible through visual verification, which has a detrimental effect on the efficiency and is generally a less safe operation. Great efforts have to be made, therefore, to ensure that enough redundancy is built into the CTC system to enable continuous operation even after partial hardware failures.

In the past, CTC systems have been developed using custom-built hardware to guarantee the prescribed safety requirements. This custom-built hardware not only adds to the cost of such systems, but also restricts the system from taking advantage of the rapid advances being made in off-the-shelf hardware components. In June 1991, it was therefore decided to proceed with the development of ILTIS as an integrated CTC system which would satisfy the necessary safety requirements using standard off-the-shelf hardware components.

2 ILTIS System Design

ILTIS has been designed from scratch as a totally integrated CTC system, based on the company's experience of developing previous systems. It is a multi-computer system where the CTC functions are freely distributable within the available computers (see Figure 1). The number of computers to be used in a particular installation site depends on the size and complexity of the site.

The station-master's workplace is based on a work-station with a single PC-keyboard and a pointing device (e.g. a mouse). Currently, an ILTIS system can handle upto 20 workplaces although this number could be increased at any time, if needed. Depending on the size on the target installation and the number of functions that a station-master needs to carry out, each work-station can control upto 6 full-graphic, colour screens.

All data-flow in the system is over the LAN (local area network), making it accessible to every computer connected to the LAN. Information to and from the interlockings is transmitted directly over the LAN. Similarly, peripheral devices which utilise serial interfaces (such as printers and the system clock) communicate with the LAN through a terminal server. This characteristic enables each computer in the system to take over the functionality of another computer in the event of hardware problems.

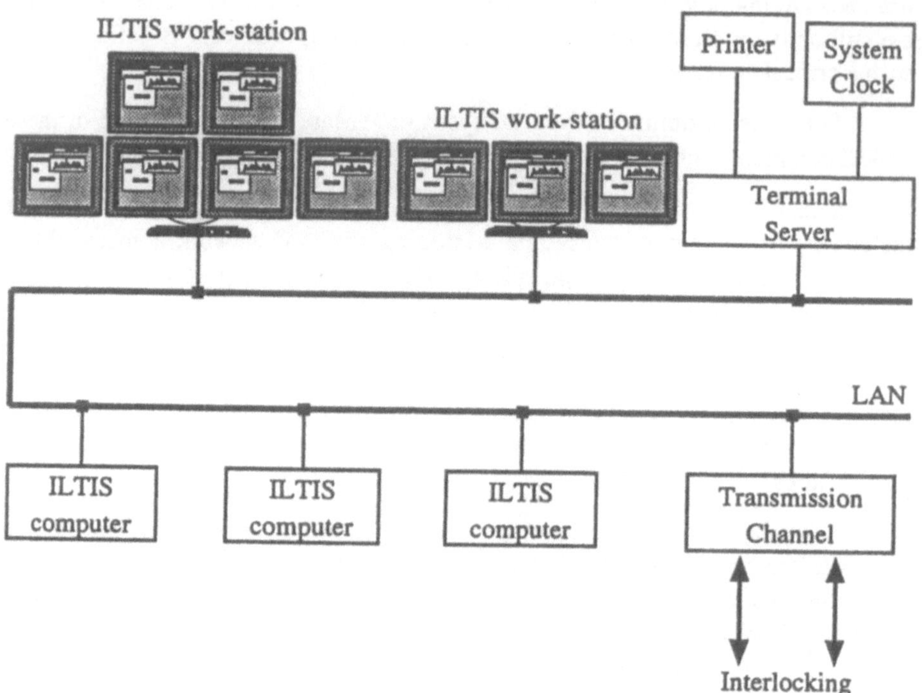

Figure 1 - Typical ILTIS installation

2.1 Standards

In order to facilitate the integration of off-the-shelf components in a system, it is advantageous to adhere to certain internationally recognised standards.

Probably the most important component of any CTC system is its MMI (man/machine interface). It is of the utmost importance that the MMI can provide clear, concise information concerning the state of the interlockings, and enable the station-master to issue error-free commands quickly and efficiently. This has been achieved in ILTIS by conforming to the OSF/Motif™ and X-Window standards. This has also simplified development and ensured that the system is not tied down to a particular hardware architecture. Currently ILTIS is using Motif Version 1.2-3 based on X11R5.

An other import characteristic to be considered in software development is the portability of the software itself, especially when the product is expected to be in service for many years. This need for portability was a significant factor in the choice of Ada (ANSI/MIL-STD-1815A) as the development language for the project. Ada is often highly recommended for the development of safety-critical software, which also made it eminently suitable for the ILTIS project.

2.2 Safety in ILTIS

In order to process critical commands, it is essential that a CTC system is able to fulfil certain safety-related functions, i.e.:

- It has to ensure that the station-master does not make life-endangering decisions based upon faulty information presented to him by the CTC system.
- It has to ensure that critical commands cannot be inadvertently executed.
- Critical commands have to be correctly executed.

In previous generations of CTC systems, it was often necessary to integrate specialised hardware into the system design to ensure that the necessary safety requirements were satisfied. For example, a usual design would be to process the interlocking state information over two independent channels, i.e. using two transmission lines, with two computers processing the input so that it could be presented in a visual format by storing it in two separate video memories. The transmitted data from the interlocking would contain additional information to allow error detection. A specialised computer, which had been methodically proven to function correctly, would then be used to display the two sets of visual information alternately in short cycles onto a screen. Any discrepancies in the data sets would result in a blinking effect on the screen.

Critical commands can only be processed by the interlocking when they are received over two independent channels. Such commands would be issued from the keyboard and, after user confirmation, processed independently by two different CTC computers.

Although effective, this design does suffer from certain drawbacks, viz.

- The user is required to consult the display screen before he issues any critical commands to ensure that all the displayed data is consistent (i.e. not blinking). Although he should not issue a critical command when the system is in an uncertain state, it is not possible for the CTC system itself to control automatically whether the relevant checks have been carried out.
- The system is dependent on specialised hardware which increases costs. Further development in the system is also restricted as the specialised hardware is unable to develop at the same rapid pace as other hardware because of the safety constraints placed upon it.

Off-the-shelf work-stations could never guarantee to the required level of safety a display needed for the design of a CTC system. Work-stations incorporate too much complex, single-channelled hardware, which could never be certified as safe.

The software used to drive this hardware is complex and was never conceived to be applied in a safety-critical environment. Although the occurrence of an error which could jeopardise the safety of the system is only a remote possibility, it has to be accepted that it could nevertheless happen. This does not mean that work-stations cannot be used in safety-critical applications, but it does mean that the design of the system has to be tailored to permit their use.

The ILTIS system designers were therefore called upon to solve this problem with an original approach if they were to satisfy the system requirements. The design, which has been patented (see Pixley, Rowden and Aepli [1, 2]) has been rigorously controlled and accepted by the Swiss Federal Railways.

2.2.1 Acquisition of State Information

The information pertaining to the state of each interlocking is always processed via two hardware independent transmission channels (see Figure 2). This information is transmitted to the TCM-software (Transmission Channel Manager) which runs simultaneously in two different ILTIS computers in either a so-called Main or Mirror state, each TCM state handling the information from one channel. The information is transmitted in the form of data packets which contain additional error detection information (Hamming distance 4). The data packets also contain a sequence number which is used by the TCM to detect any data packet loss.

Lost data packets can only be detected with the successful reception of the next data packet. To confirm that the flow of data packets is not being completely lost, watchdogs are sent over the transmission channels in the form of a data packet containing a sequence number. Watchdogs are normally sent every 60 seconds.

The interlocking information is then passed from the TCM to its partner ILM (interlocking manager) which is the next stage in the ILTIS chain of processing. The two ILM states (Main and Mirror) have to run on different computers to ensure independent processing. Again a similar system of error detection and lost data packet detection is used between TCM and ILM. The ILM process generally runs in the same computer as its partner TCM process, but an installation site could also be configured so that TCM and ILM execute on different computers.

The ILM maintains a data base containing all the dynamic state information from the interlocking. This data base is controlled for consistency using checksums. The ILTIS system software issues watchdogs every 20 seconds which will only be positively acknowledged when the checksum corresponds with the information in the ILM data base.

To control that the ILM software is not functioning incorrectly due to hardware faults or corrupted code, each ILM regularly compares the information contained in its partner data base with its data base to guarantee consistency. This flow of information is also transmitted in the form of data packets with additional error detection information.

2.2.2 *Display of Interlocking Information*

Up to this point, the processing of the interlocking information is completely dual-channelled. In order to display the information on the work-station screen, the Main-ILM process transmits its data to MX (Motif and X-window process) using the same system of error detection and lost data packet detection as between TCM and ILM. MX displays the information on its screen in a graphical format.

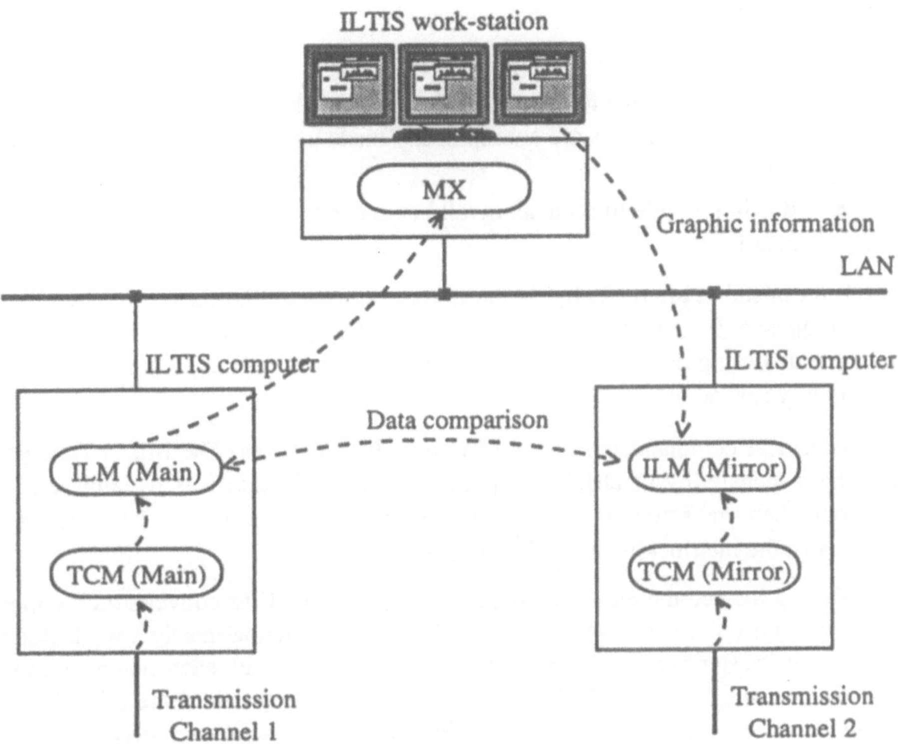

Figure 2 - Data Flow of State Information

As a first step in ensuring a safe system, information from the display is "read" by the Mirror-ILM process at regular intervals and compared with the data in its data base. Processing all the data from the display would place an unacceptable load on the network and the processing power of the computers but it is possible to reduce the data being processed by targeting certain critical types of elements (points, tracks, etc.) and within these elements only a selected few of pixels. The pixel characteristics (i.e. colour) can be compared with the information in the Mirror-ILM process.

If there were any unexpected discrepancy between the data, the station-master would be informed and the system would enter a safe state until data update requests had been successfully executed.

Further preventative measures are incorporated in the execution of critical commands themselves to complete the steps necessary to ensure safety.

2.2.3 *Execution of Critical Commands*

A critical command is only carried out by the interlocking when it has been transmitted over two independent transmission channels within a fixed time frame.

In ILTIS, all commands are entered using a mouse by selecting target elements from the screen which has advantages compared with entering text commands from a keyboard, e.g.:

- The station-master's attention is concentrated solely on the target element.
- It is only possible to input valid commands with a mouse. Invalid combinations of parameters are automatically made unavailable for selection by the system.
- It is impossible to issue an invalid command as no text is input from a keyboard.

As fewer mistakes are made by the station-master issuing the command, there is, as a consequence, a marked improvement in the safety of the system. The fewer mistakes made during input, the less chance of an incorrect critical command being inadvertently carried out.

A critical command is sent by MX to both ILM states. The Mirror-ILM now has the responsibility of checking the given command against its dynamic data to find out where the safety of the system is being compromised (i.e. where the safety features of the interlocking are being by-passed).

Having located the elements in question, the Mirror-ILM converts their current states into a text-format which is then displayed on the station-master's work-station The station-master has to confirm that he is in agreement with each statement in the text window on the screen (achieved with a mouse click) before the critical command can be executed (see Figure 3). If there were any discrepancy between the information displayed graphically and in text, the critical command has to be aborted as that would imply that the displayed state information could not be trusted to be accurate.

The critical command can only be transmitted to the interlocking from the ILM (Main and Mirror) when all the relevant steps have been successfully implemented by the station-master.

By using diverse processing (i.e. displaying information in graphical and textual formats), the execution of a critical command can be safely carried out although the software and hardware used for the display is complex and only single channelled.

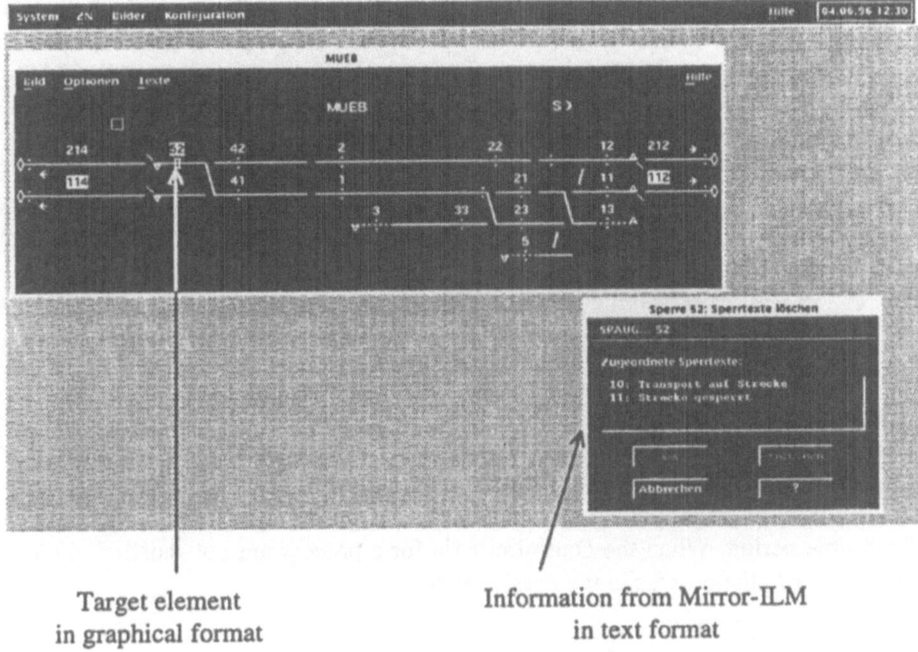

Figure 3 - Processing a Critical Command

3 Availability in ILTIS

The availability of a CTC system is becoming an increasingly important factor as railway networks become more and more dependent upon them. Current systems usually solve this problem by having redundant computers, which can be manually switched into service whenever they are needed. From experience, it has been seen that this solution suffers from a number of drawbacks, viz.:

- The redundant computers are not in active service. When a situation arises where they need to be brought into active service, they themselves could have developed technical problems which, as the redundant computers have been dormant, have gone unnoticed.
- When there is a problem in a distributed system of computers, it is not always clear to the user, which of the computers is causing the problem. The manual switch-over to the relevant redundant computer therefore leads to a process of trial and error.
- As each computer in the distributed system needs to have a similarly-configured redundant computer within the system, this effectively doubles the total number of required computers.

In ILTIS, a different approach has been taken to this problem. Here the necessary redundancy has been built into the software design. In addition to the Main and

Mirror states seen with the ILM and TCM processes (see Section 2.2), each process has a Passive state. Processes in this state are always prepared (either as a hot-standby or warm-standby) to take over the processing from the corresponding Main- or Mirror-state. Naturally when the process is a safety-critical process (e.g. ILM or TCM), the Passive-state may not execute on the same computer as either the Main- or Mirror-state.

The switch-over procedure is controlled by the ILTIS system software and is processed on two levels, viz.:

- control of each ILTIS process, and
- control of each ILTIS-computer

3.1 Process Control

This is standardised for each process running in ILTIS. Watchdogs are sent cyclically to the target processes which have to be positively answered within a defined time period. When the control criteria for a process are not satisfied, there a switch-over to the process in the Passive state.

3.2 Computer Control

There is no master computer in ILTIS which is responsible for the general well-being of the system. The responsibility for watching over the computers is distributed throughout all the computers in the system. Each active computer in the system is responsible for another computer in the system.

Again, by using a system of watchdogs between all computers in the system, the ILTIS system-software in each computer can control if a particular computer is causing problems. When this is the case, the problem is reported to the responsible computer for that computer. When the responsible computer has received three such reports, it can automatically force the defective computer to be taken out of service. This again forces a switch-over for all processes running on this defective computer to the relevant processes in the Passive state running on other computers.

3.3 Design Advantages

With this design, it is no longer necessary to have redundant computers but redundant computer power. There has to be enough reserve computer power configured into the system to handle the additional work-load caused by a switch-over. The distribution of the Passive states is fixed when the installation is configured. This means that the redistribution of the load after a switch-over is not dynamic, but follows a predefined program.

The consequence of this design is that it is no longer necessary to double the number of computers needed in the system to provide the necessary stand-by capabilities. Usually, only an additional 20%-30% computers are needed.

4 Conclusions

The first ILTIS with safety-critical features has been running since July 1995 in Berne, Switzerland. The users quickly adopted to their new work environment and the feedback from them has been positive. Currently there are 6 ILTIS systems in operation in Switzerland and Hungary and several more are in the pipeline.

The portability of the system has also to a certain extent been tested. ILTIS was originally conceived to run on Vaxes (from Digital Equipment Corporation) running on a VAXELN operating system. During the development, it was decided to switch to the OpenVMS operating system, as it was felt that OpenVMS offered more security for the future. The porting from VAXELN to OpenVMS was achieved within a week.

After the delivery of the first ILTIS system to Olten, Switzerland (a system without safety-critical features) in December 1993, the decision was made to target the ILTIS development to AXP-computers (also from Digital Equipment Corporation) using the OpenVMS operating system. This again was achieved without any significant problems.

Thought is currently being given as to whether it makes sense to port ILTIS to other platforms and/or operating systems. It has been estimated that a transfer to UNIX, for example, should not cause any significant problems and could be accomplished within a few months.

Abbreviations

CTC	centralised traffic control
ILM	interlocking manager (ILTIS software)
LAN	local area network
MMI	man/machine interface
MX	Motif and X-window process (ILTIS software)
OSF	Open System Foundation
TCM	transmission channel manager (ILTIS software)

References

1. Pixley, D., Rowden, N.T., Aepli A. Verfahren zur Gewährleistung der signaltechnischen Sicherheit der Benutzeroberfläche einer Datenverarbeitungsanlage, German patent - DE 43 06 470 C2
2. Pixley, D., Rowden, N.T., Aepli A. Verfahren zur Gewährleistung der signaltechnischen Sicherheit der Benutzeroberfläche einer Datenverarbeitungsanlage, Swiss patent - CH 683953 A5

Assessment and Certification Requirements in the European Railway Industry

S Mitra
Lloyd's Register
London, UK

I D R Shannon
Opal Engineering
London, UK

Abstract

This paper discusses the current and future requirements for the assessment and certification of software based safety-critical systems used within the European railway industry. It outlines the CASCADE project and reviews the findings of a pan-European survey carried out under its auspices.

1 Introduction

In the past, and to some extent at present, the European railways have followed diverse policies towards new technology, standards and safety. As a result, different national or regional railway organisations have developed different safety cultures, i.e. safety requirements and safety measures. In recent years, a strong trend towards a common, harmonised approach to safety has emerged. The following factors are responsible for this change:

- greater concerns for safety balanced against the desire for a cost-effective and efficient rail service that can compete with road and air.
- technical innovations, such as modern information technology and telecommunications, especially those based on computers, which although complex in nature and expensive to develop, are cheap when mass produced.
- regulations emerging from the European policies on the single market economy and mobility of people and goods.

CASCADE is developing a Generalised Assessment Method and a certification framework which will facilitate harmonisation of working practices and provide a mechanism for the mutual recognition of certification across the European Union.

The CASCADE certification framework will only succeed if it gains widespread support and acceptance within the safety critical community. Currently, CASCADE is targeting the railway industry. To achieve this support it must be demonstrated that the assessment and certification objectives are relevant to the railway industry. For this purpose CASCADE has undertaken a survey within the railway sector to determine the current and future requirements for a safety assessment and certification service.

2 CASCADE

The CASCADE project focuses on the assessment and certification of software intensive safety-critical systems. It is part funded by the CEC under the ESPRIT III programme and includes collaborators from the UK, Germany, France and Denmark. The need for assessment is driven, in CASCADE's view, by the increasingly sophisticated usage of computer based control systems and the economic and technical advantages of systems incorporating such components. Where computers are used in applications with safety implications there is an emerging requirement either to develop systems which comply with the safety regulations, for instance LOTI [1] and MbBO [2], and internationally recognised standards, such as ISO/IEC 1508 [3], or to purchase systems that comply with those regulations and standards.

In particular, procurers require independent assessment of conformance that is cost effective, objective and which constrains liability in the event of failure. The suppliers of assessment services must ensure that their offerings be objective, repeatable, cost effective, technically sound, generic and risk-limiting in both a technical and legal sense.

The CASCADE project views assessment and certification as an integral part of the system safety case. The safety case is required to present the evidence for the justification of those safety functions that the system must exhibit. The technical work of the project has focused on how to articulate these properties and on determining how current methods, techniques, tools and management practices provide evidence to support the assessment judgement.

CASCADE has produced a generalised framework for the assessment of systems that can be deployed by the assessors and trusted by the users. From the experience gained in the application, the project will also produce guidelines for the development and procurement of safety-critical systems to be deployed by the users.

3 Survey Procedure

The survey began with desk research to identify appropriate organisations to interview, to gather relevant information in preparation for the interviews and to satisfy the survey objectives. It was considered important that the interviewer could demonstrate a reasonable understanding of the interviewee's own industry. Ideally those undertaking the desk research should conduct the interviews relevant to that research.

In order to examine the total picture that was to be covered by the limited number of interviews that could be carried out and to base any predictions from the survey on strong evidence, the market structure was examined to determine which companies are recognised as market leaders. Before undertaking face-to-face interviews the objective of the interview was made clear to respondents and the form of feedback explained. To achieve the most effective results the survey was conducted to the following procedure:

- important issues of concern to assessment and certification services, such as inter-operability, harmonisation, technology and standards, were identified by reviewing the technical and business literature including trade journals, selected industry overviews published by market research firms, Financial Times reports, Economic Intelligence Unit reports, and Union Internationale des Chemins (UIC) reports;
- each partner identified six to eight organisations in their region and established direct contact with them to ascertain whether they would be willing to participate in the survey; organisations contacted included regulatory authorities, the systems/software suppliers, influence groups, other certification bodies, and transport authorities;
- for those who agreed to participate, the interviewer provided them with a letter of invitation explaining the purpose of the survey and a Synopsis of CASCADE;
- the interviewer contacted the interviewee and agreed a date and venue for the interview;
- during the interview the topics identified as of concern were discussed;
- after the interview the interviewer prepared a report which was then sent to the survey leader.
- the survey leader collated all the reports and prepared a summary.

4 Survey Population

Altogether, 44 organisations from 12 countries contributed to the survey (Belgium, Denmark, Eire, France, Germany, Italy, Netherlands, Norway, Portugal, Spain, Sweden and the United Kingdom). Of these, 20 have been interviewed. Twenty-nine organisations, such as UIC, Railway Industry Association (RIA), European Commission (EC), Institute of Transport studies at Leeds University, European Railway Research Institute (ERRI), have provided the material for the desk research.

The distribution of the survey population is shown in Table 1. Operators included infrastructure and rolling stock owners and service providers. Regulators included the approval authorities, transport ministries, and European Union (EU) directive making bodies. Standards bodies included the research and development organisations developing the technical specification for inter-operability, such as UIC and EC.

Type of Organisation	No. Interviewed	No. Surveyed
Operators	10	26
Suppliers	6	11
Regulators	2	5
Standards Bodies	2	2
Total	**20**	**44**

Table 1: Distribution of Survey Population

5 Results

The following facts and areas of concern emerged from the survey.

5.1 Procedures and Standards

Most organisations surveyed recognised the importance of quality management systems in the production, procurement and maintenance of safety-critical software. Most organisations are EN29001 [4] certified, and most procurers require their software suppliers to comply with EN29000-3.

Many procurers, especially the infra-structure owners, have well-developed procurement procedures which include requirements for safety-critical software. These de-facto standards have a lot in common, however, there are several major differences mainly attributable to differences in the safety approaches of the organisations.

There are many published national standards, for example: AFNOR [5], RIA [6], which have been used by procurers in many other countries which have influenced the international and European standards.

The present survey found that most suppliers have started to follow the Comité Européen de Normalisation Electrotechnique (CENELEC) (prEN50126, prEN50128 & prEN50129) [7, 8, 9] and ISO/IEC 1508 [3] standards, even though they are still in their draft form. Many procurers expressed the view that they would require their suppliers to comply with the CENELEC standards once adopted by the industry.

5.2 Regulations and Directives

Often public sector industries are required by law to follow safety regulations designed to protect the public and workers from common hazards. Such regulations also cover technical, economic and social issues, such as inter-operability, non-restrictive trade and free-flow of goods and services between different trading nations. The European railway industry is subject to many different statutory regulations which can be divided into two categories:

- national regulations, designed to meet the railway safety policy of the member states;
- the EU directives on railways, designed to harmonise the safety policies of the member states.

The white paper on Community Transport Policy (CTP) recommends:

- development and integration of community transport systems to provide an efficient and cost-effective means to meet the mobility requirements of citizens and goods.
- safety in transport for users, third parties and the environment.
- social protection and external relations to promote social efficiency and equity.

The Community Railway Policy proposes the following legislative actions:

- establish a transparent and efficient relationship between railways and the State and to allow a certain amount of free access. (Directive 91/440/EEC)[10].
- create a contractual framework for the state to buy railway services for social reasons.
- develop a European network of high speed trains. (Directive 94/112/EEC) [11].
- develop combined road/rail transport system.

In the wake of these developments many member states have taken steps to devise new regulations for their railways. The UK Government's rail privatisation proposals are contained in the Railway Act 1993, which is designed to meet the requirements of directive 94/440/EEC [10] and improve efficiency of the railways.

The scope of council directive 91/440/EEC [10] includes all publicly owned and private railway organisations active in the sub-urban and regional transport services. It requires that member states make their railways efficient and adaptive to market needs by achieving the following measures:

- managerial autonomy,
- realistic financial structure,
- separation of infra-structure from operations, with the minimum requirement being of an accounting separation,
- granting of the right of access to the rail networks for international groupings of railway organisations and for the individual railway operators.

The reorganisation and the privatisation of the railways can only be successful if the services are based on new cost effective technologies and at the same time are safe and reliable. The EC directive requires standards to ensure the safe implementation of technology and the harmonised approach to safety and efficiency throughout the European railway system or service.

A new directive for an operator's Licence is under preparation. This is expected to specify the essential requirements for operating a railway system or service.

5.2.1 High Speed Train Network Directive

For a long time it has been recognised that high-speed rail transport is a new mode of transport that offers substantial advantages to users, and hence needs special attention. The commission has proposed a draft directive [94/0112(SYN)] [11] to facilitate the introduction of High Speed Trains (HST). The HST directive will establish inter-operability within the European High Speed Network. This refers to all of the regulatory, technical or operational conditions which must be met in order to ensure, without interruption, movements by high speed trains on infrastructure within the network which accomplish specified levels of performance.

The HST directive describes six essential requirements. To facilitate the description and technical specifications, the high speed train system is divided into eight sub-systems. Each sub-system will be covered by a mandatory Technical Specification for Inter-operability (TSI). The standard component constituents of the sub-systems will be specified according to the requirements of the European standards, such as prEN 50128 [8].

The directive requires that compliance with the TSI be assessed by independent Notified Bodies (NBs). NBs will authorise the use of appropriate CE markings for the systems or products which will be recognised by all the member states. The HST directive has been reviewed by the social council and will then be presented to the Council of Ministers.

The survey identified the following concerns:

- there is too much emphasis on the sub-systems, it does not address whole-system issues;
- it is not clear how the different safety policies of the member states will be reconciled when implementing the directive;
- the monitoring of accreditation of NBs should be performed by experts from the railway industry;
- there is a need to address the safety approval procedures for the existing systems which have a long history of use.

5.3 Certification and Approval Policies

Many state railways and urban transit operators used to certify their own systems. Such second party certification is still employed by many member states.

Given the composition of the present systems, many would like them to be certified in a hierarchical manner. Where systems are composed of standard components the use of certified components could simplify the assessment of the whole system.

Increasingly, the onus is placed on suppliers and operators to demonstrate safety by making a safety case which satisfies the regulator. This has the advantage of being less restrictive than a system based on prescriptive rules. However, there is some evidence that the technical difficulties of dealing with software in safety cases are a deterrent to using software, even where it might have safety advantages [12].

For Off-the-Shelf Progammable Electronic Systems certification against accepted safety standards would, in most instances, be accepted as supporting evidence in the Proof of Safety.

Union Internationale des Chemins (UIC) is developing a conformity assessment scheme for railway equipment. The scheme will be applicable to all types of equipment: mechanical, electrical, and electronic. UIC expects that there will be an approved set of standards for this purpose.

5.4 Computing Techniques and Tools

Structured development methods, such as ASA and SADT are commonly used. Many organisations now use mathematical formal methods, such as Z or B, at different stages of the development. B has been used to develop Automatic Train Control and operation systems for the high speed and metro trains such as TVM and SACEM and is presently being used by MATRA Transport International to develop a new fully automated metro train for Paris transport authority (a part of which is used as one of the CASCADE case studies).

Different safety architectures are in use, some of which are very specific to the railway sector. A recent proposal made to the Commission, named ACRUDA, plans to develop assessment criteria for railway safety architectures.

Commonly, ADA and PASCAL are used as the implementation languages for SIL 3 & 4 software. Some developers also use C for this purpose. In the USA, Hughes are developing high integrity systems in C for the San Francisco Metro. Some codes, such as the Fast Fourier Transform routine for Texas Instrument chips, are written in assembly language.

Use of tools in Validation and Verification processes are becoming widespread, for example, tools to perform static analysis, emulation, and simulation.

5.5 Legal Issues

There is very little information on this subject in the way of case laws at the moment. However, many regulators and infra-structure owners believe that safety is primarily the responsibility of operators. However, suppliers are responsible for the damage caused by their products. Some suppliers believe that they are not responsible for the systems developed under contract for procurers.

Duty of Care applies to all those who help to develop, install, operate or approve the systems. Most suppliers, operators and certifiers take insurance cover for negligence claims against them.

The UK Railway Act 1993 requires all operators to insure against the worst-case catastrophes, such as fatalities or injuries to passengers.

However, a great majority of interviewees refrained from discussing the legal liability issues because they thought:

- the legal and actuarial analyses focus more on identifying a principal factor for liability reasons, e.g. identification of `proximate causes' of mishaps. This does little to improve the overall safety of the system.
- the legal approach to causality is directed more towards establishing guilt and liability rather than the underlying cause and ways and means to prevent accidents.

5.6 Market Factors

Safety assessment and certification are important issues that must be resolved before too much investment is made in the development of new European networks and the modernisation of state railways. Any assessment approach and certification scheme must be cost effective. It was found that many of those surveyed so far (70%) believe that for new systems the cost of assessment should be no more than 20% of the development cost and for products, such as an interlocking equipment, this should be no more than 5% of the product cost.

In general the cost of certification (or approval) for safety critical systems should be in the range of 5-10% of total cost (i.e. development, installation, operation and maintenance).

6 Conclusions

The survey revealed that the European railway industry urgently needs a pan-European assessment and certification scheme for software intensive safety critical systems. The scheme should be based on an approach that requires demonstration of safety for such systems. A large section of the European railway industry is using the draft CENELEC standards, prEN 50126 [7], prEN 50128 [8] and prEN 50129 [9] mainly to improve their development and procurement processes. It is widely believed that compliance with these standards will be mandated for all new systems. However, compliance with the standards will not be sufficient for the safety demonstration, it should address safety requirements which are system engineering issues. The CASCADE project is working towards such a fully integrated method.

Acknowledgements

The authors would like to acknowledge the help provided by the following individuals of the CASCADE consortium: Tim Moltzen (DSB), Andy Harrison (Railtrack), El Miloudi El Koursi (INRETS), Bao Letrung (INRETS), Heinrich Krebs (TÜV), Ann deGoyese (MATRA) and Jean-Marc Meynadier (MATRA). The co-operation received from the survey population is also thankfully acknowledged.

References

1. Loi d'Orientation des Transports Intrieurs (LOTI), 1942 & modific, a renouvel le cadre d'organisation des transports en France, 1982.

2. Magnetschwebebahn-Bau-und Betriebsordnung (MbBO), Entwurf June 1995.

3. ISO/IEC 1508 Functional Safety: Safety-related systems, part 1-7, 1995.

4. EN 29001 Quality Systems - Model for quality assurance in design, development, production, installation and servicing (1987).

5. AFNOR: Nf - F71 011, Nf - F71 012 & Nf - F71 013: Installation fixes et materiel roulant ferroviaries - - Surete de fonctionnement des logiciels - generalie, December 1990.

6. Railway Industry Association (RIA) Safety Related Software for railway signalling, BRB/LU LTD/RIA Technical Specification NO 23, London UK, 1991.

7. prEN 50126 Dependability (RAMS) for Guided Transport Systems - Part 2: Safety (draft 1993).

8. prEN 50128 Railway Applications - Software for railway Control and Protection Systems (draft February 1994).

9. prEN 50129 Railway Applications - Safety-related Electronic Railway Control and Protection Systems (draft 1994).

10. Council Directive 91/440/EEC *On Development of Community's Railways*, August 1991.

11. Council Directive 94/0112/EEC *On the interoperability of the European high speed train network*, COM(94) 107, April 1994. (Draft)

12. Reason James *Latent Errors and Systems Disasters* in Human Error CUP 1990.

Session 6
Management and Development

Failure Classification Schemes for Analysing System Dependability

Chris Loftus, Fred Long, Dave Pugh and Ian Pyle
Department of Computer Science
University of Wales, Aberystwyth

Abstract

It is recognised that the dependability of systems can be improved by enhancing the processes used to build them. However, to ascertain the effectiveness of methods and techniques of reducing or masking software design faults requires the analysis of historical data. For such analyses to be effective, appropriate kinds of data must be recorded and structured using a suitable classification scheme. We explore the requirements that a classification scheme should meet by looking at the four kinds of data (causal, process, product and failure effect) that need to be recorded to enable an assessment of process and product on system dependability. We then propose a framework classification which encompasses these four kinds of data and their interrelationships.

1 Introduction

It has long been recognised [1] that recording data associated with system failure is an essential part of improving the dependability of future systems by allowing systems engineers to compare existing or proposed systems with others of known characteristics. If a comparison can be made based on the nature of the products and/or the processes used for developing and checking them, then the engineer may be able to assume that certain quality attributes are also similar (for example, reliability). This data will help predict the dependability of systems and help in the selection of suitable techniques for development, verification, validation and fault-tolerance so as to meet required integrity levels.

The two key aspects to achieving this goal are the use of a suitable data classification scheme and the establishment of failure databanks which provide the mechanisms for industry to submit and analyse such data. This paper focuses on the former, recognising the specific problems of software-based systems with significant systematic failure modes as well as random faults. The use of databanks has a precedent in other engineering disciplines, for example, FACTS (Failure and ACcident Technical information system) in the process industries.[1]

The intended uses of a classification scheme place requirements on the kinds of data that it should cover. The main purpose of the scheme is to record data to be used to assess the dependability of existing or planned software-based systems. We argue that both the process and the product need to be analysed in order to achieve this goal. Therefore, the classification should enable the recording of both process and product related data. The classification must, of

[1] Descriptions on several databanks can be found in [2].

course, also enable the recording of failure data (the fault/error/failure causal chain) and relate this to both data about the product and the processes which led to the development of that product.

In all, four classes of data must be covered by the scheme: process, product, the fault/error chain (which we term *cause*) and the effect of failure on the behaviour of the system. We have modelled these classes, and the relationships between them, using a classification framework. This framework provides a very abstract view of the data. We have not attempted to specialise these classes. Instead, it is our hope that these more detailed sub-classifications will be developed and pieced together within our framework. We are aware of only one standard that presents a classification which partly conforms to our framework: IEEE 1044 [3]; discussed in [4].

Our motivation for a fault/failure classification scheme stems from our research interests in software failure modes and effects criticality analysis [5]. This technique assesses the probability and severity of different kinds of latent fault (called failure modes) which might occur in the operational system. Determining the failure modes possible and their probability requires knowledge of the development, checking and fault-tolerance techniques used, and their effectiveness at preventing, detecting or making unlikely various kinds of failure mode. It also requires knowledge of the final system itself (for example, code complexity). A careful analysis of development/checking techniques can provide some insight into their effectiveness. However, recording field data — which includes data on the techniques used, the faults discovered, and data about the system — should both validate our qualitative analyses and provide quantitative evidence of technique effectiveness.

This paper begins by providing terminology regarding faults, errors, failures, and failure modes. Section 3 provides a more detailed discussion of the requirements on a classification scheme. Section 4 then proposes a classification framework which begins to meet these requirements. Finally, Section 5 presents conclusions, and directions for further work.

2 Terminology

This section defines the terms fault, error, failure, and failure mode, since many papers use these terms synonymously. For fault, error and failure we base our definitions on those found in [6] and agreed by most researchers and practitioners in the systems engineering community. However, for the purposes of this paper, we extend the scope of these terms to the software development and checking process.

Fault, error and failure are different parts of the cause and effect chain (causal chain). This chain (see Figure 1) can be viewed from two directions:

1. Given an identified or hypothesised failure effect on system behaviour we can examine the design of the system to deduce what aspects of the system or its construction might cause that behaviour.

2. Alternatively, by examining the system design and construction process, we identify potential causes of anomalous system behaviour and, by a process of induction, discover whether the causal chain leads to actual anomalous behaviour.

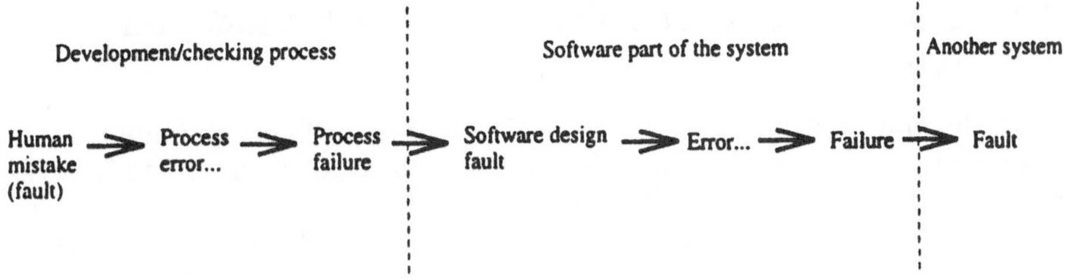

Figure 1: The fault/error/failure causal chain.

Consequently, a failure (at the visible effect end of the chain) is caused by an error (an incorrect system operational state) which, in turn, is caused by either one or more physical faults or design faults which have been activated randomly (e.g. ionic radiation on integrated circuits) or as a consequence of one or more data inputs. For a software design fault, we can take the chain back further to discover faults in design, requirements documents or in the transformation process (for example, between design and code). Within the chain there may be any number of intermediate error state links between the 'root' cause or causes and the observable system failure. By failure mode, we imply a classification of possible faults.

3 Requirements on Classification Schemes for Representing Software Based System Failure Data

We now examine the requirements a classification scheme should meet if it is to be used in assisting the process of improving the dependability of future systems. Some of these requirements are general, and pertain to most, if not all, classification schemes. Other requirements can be derived by analysing the purposes of the classification.

3.1 General Requirements

There are a number of general requirements that classification schemes should address. Whatever its purpose, any classification scheme should partition the universe concerned by providing a complete set of non-overlapping classes. It may be hierarchical, with each class divided into smaller subclasses, or it may be multi-faceted, where classes can be combined as required [7]. In general, each level in a hierarchical classification scheme should be based on a single guiding principle, by which the classes are distinguished with sharp boundaries.

There are two antagonistic general requirements, namely, ease of use and richness. The success of a classification scheme may depend on its usability: a simple scheme is likely to be easier to use than a rich, complex scheme, but if the scheme is too simple then its use will provide few benefits.

3.2 The Purposes of a Fault/failure Classification Scheme

Since the mid 1970s it has been known that classifying faults and failures alone is not sufficient, especially where the purpose is to record data for later analysis; for example, to avoid, mitigate and/or recognise potential fault reoccurrence in future projects. Myers [8] suggested that failures should not be classified according to manifestations but rather by causes. By associating causal data with the failures observed, it should be possible to use such data to improve engineering practices in future projects. Recording the cause of the fault, the nature of the fault itself and the effect of an activated fault on system behaviour are all essential parts of a good classification scheme constructed to improve the quality of future software-based systems.

The response to this has been a move towards classification schemes which also classify causes. Bowen [1] recommends a minimal set of three software sub-classifications: cause, severity and source (from hereon called *process* to avoid confusion with *cause*). The *cause* captures a causal description of the failure rather than describing the symptoms, that is, the causal chain. *Severity* records the effect of the failure on mission performance. *Process* details the lifecycle phase in which the failure occurred. We extend this classification in two main ways: to include data about the nature of the artifact (called *product* data), and to include data not only about the lifecycle phase but also about the methods and techniques used during each phase.

These extensions are required to enable the recording and structuring of both detailed process and product data. We discuss the need to include both process and product data in more detail in Section 3.2.1.

The following sections discuss the benefits of analysing data recorded using a classification scheme and where the overriding benefit is to improve system dependability. These benefits are based on five suggested by Bowen [1]. Given the intended uses of a classification scheme, we derive requirements about the kinds of data that a scheme should cover.

3.2.1 A Requirement for Process and Product Data

We argue that both the software development/checking process and the nature of the final system influence the types and probabilities of residual faults.

The use of development/checking methods and techniques provides a set of constraints on the kinds and probabilities of faults which might reside in the final system. For example, the use of Ada will eliminate a large number of faults associated with weak typing in other languages. However, we contend that although knowledge of the development/checking process will identify the potential presence of particular kinds of fault, this data may be too 'coarse grained' and of little use for a particular system under inspection. We may discover, for example, that using a particular set of methods and techniques can identify classes of faults which are highly unlikely to survive (A); another set of fault classes for which their survival probabilities is considered greater (B), and a final set of fault classes for which we really cannot estimate their probabilities based only on knowledge of the methods and techniques used (C).

Therefore, to further differentiate the nature and probabilities of latent faults in set C, and to build confidence in sets A and B, it is also necessary to study the system itself, for example, using code metrics, and scrutinising the software for atypical patterns.

3.2.2 Evaluation of Modern Software Engineering Practices

Data recorded and structured using a classification scheme may be used to enable an evaluation of all aspects of the systems and software engineering process. The evaluation should determine the efficacy of software engineering practices at improving the quality of software-based systems. We believe that the quality attribute of most concern is system dependability rather than other attributes such as cost effectiveness. Consequently, it is important that a candidate classification should cover process data which includes details about the programming practices used to develop a system which has failed. This is justified since there is a clear relationship between software engineering practices and dependability (e.g. the relationship between the strength of a language type system and run-time failure rates).

Indeed, the classification should enable the capture of systems' engineering process histories which we define to be the development methods, techniques, programming languages, testing techniques and fault-tolerance techniques used. Also, it must, of course, classify the fault or faults which have caused failure and record the effect and severity of that failure on the system or application.

3.2.3 Provide Guidance to Software Engineers and Managers

Software engineers and project managers will be one group of users of historical data. They will want to know the effectiveness of a particular technique at preventing, masking or detecting particular failure modes. To determine which failure modes are likely, and need checking for, they will need to be able to query historical data on particular development techniques (presumably the ones used in their projects) and the kinds, probabilities and typical severity of faults possible after using these techniques. For example, designers will want to know which design techniques best meet required integrity levels.[2]

For reasons discussed in Section 3.2.1 knowledge of the methods, techniques and engineering processes alone is not sufficient when estimating the kinds and probabilities of residual fault. The systems engineer will also need to understand the influence that particular components have on system dependability. This will entail understanding the characteristics of these components (such as complexity), how these correlate to kinds and probabilities of faults, and the the relationships between components and the rest of the system in order to understand error propagation and severity. Much of this data will be system specific and provided by design documents.

A future approach may be to classify components, subsystems and systems according to architectural styles: typical characteristics of components including their interconnections and configuration rules. Investigation into architectural styles is an active area of research [10]. Researchers have begun examining the link between quantitative quality measures, such as reliability and performance, and architectural style [11].

[2] The standard IEC 1508 [9] provides qualitative recommendations on the combinations of techniques to use to achieve required integrity levels.

3.2.4 Evaluation of the Effectiveness of CASE Tools

Methods and techniques are often supported by CASE tools. Determining the effectiveness of tool support may therefore be an important part of evaluating the effectiveness of the methods and techniques themselves. Therefore, recording whether automated support is available for a particular technique is worthwhile.

3.2.5 Validate and Support Quantitative Dependability Models

Perhaps the most important benefit required of a candidate classification scheme is that historical data, recorded and structured using it, can be used to validate and support quantitative dependability models. These models are used to help predict the reliability, safety and availability of existing or planned high-integrity systems [12].

The most common models include Continuous Time Markov Chains, Markov Reward Nets [13, 14] and Stochastic Petri Nets [15] and their variants. These models require calibration and have parameters which need instantiating from empirical data. Some data can be obtained by system prototyping, simulation, testing and experimentation (e.g. fault injection [16], and empirical comparisons of techniques [17]). However, most must be obtained from field data, especially where ultra-reliability is required. Therefore, a suitable classification scheme should cover many of these parameters, including most importantly:

Failure rates of hardware and software components. These rates will often be domain specific. For hardware, this data is often well known. Failure rates for styles of software components are limited and inconsistent.

Time to system failure and/or loss. Failure rates must also be considered at the system level. Many models attempt to calculate or to use mean time to failure and mean time to loss figures (the latter is concerned with unsafe system loss, the former both unsafe and safe system loss). Associating such figures with particular styles of system may help validate future estimates and indicate whether systems are becoming more dependable.

Persistence of fault. The classification should also capture whether the hardware or software fault is transient, permanent or intermittent.

Severity of system failure. The classification should enable engineers to record the impact of the failure on the external environment: life-threatening etc. Existing standard classifications should be used (for example [9]).

Timeliness. For non safety-critical, but, nevertheless real-time systems, it is useful to record missed soft real-time deadlines. The percentage of missed deadlines gives an indication of performance and where larger workloads could be handled.

3.2.6 Provide Feedback to Standards Concerned with System Dependability

The analysis of historical data could be used as feedback to standards concerned with system dependability. The proposed standard IEC 1508 [9] makes recommendations regarding development, testing and fault-tolerant techniques which should be used to meet required integrity levels. However, little rationale (quantitative or qualitative) is provided as to why particular techniques are recommended in order to meet given integrity levels.

4 A Fault/failure Framework Classification Scheme

Given the requirements identified in Section 3, we now present a fault/failure framework classification scheme. We start by characterising the four kinds of data which must be covered by any classification suitable for meeting the intended uses identified in Section 3.2.

For causal chain data a comprehensive fault sub-classification needs to be defined. This would cover fault classes such as *logic problem*, *incorrect function called*, *conflicting requirements* and so on. The sub-classification would be hierarchical to allow grouping of related classes of faults, for example, all classes of logic fault. It should also capture the causal chain as depicted in Figure 1.

Product data should be captured as a model of the system architecture in terms of stereotypical patterns of component and connector types, together with metrics such as code complexity. The purpose of this is to allow the definition of the context within which faults occur. However, for both the architectural model and the process history, anonymity is of utmost importance.

Process data consists of a detailed process history including development and checking phases undertaken to date and the types of methods, techniques, and languages used during those phases. This would form an audit trail linking architectural elements with the process histories of those elements. The classification scheme should allow users to indicate whether tools (and perhaps their characteristics) are used to support identified methods and techniques.

Data on system failure effect and context of use should include: the severity of failure on the environment (based on existing classifications [9, 18]); an indication of how the system was being used operationally (the kinds of hazard or danger possible will be influenced by the behaviour of the system just prior to its failure); and a means to allow users to record the elapsed time since a particular system last failed (safely or unsafely), or missed a soft real-time deadline.

Figure 2 represents these four kinds of data diagrammatically, presenting a framework which a complete candidate classification scheme would conform to.

In Figure 2 causal data is represented by the entity *fault class*. The *fault class* entity has a relationship with itself called *caused by*. The purpose of this relationship is to capture the causal chain such that each fault (or error) is related to the fault(s) or error(s) which led to its activation and ultimately the failure of the system. Each fault class instance is related to the component in which it resides by the *associated with* relationship. Faults can also be associated with the transformation process (e.g. misinterpretation of requirements during design).

Product data is represented by the entity *component style*. Instances of this entity are used to provide a stylised and abstract view of the system. We provide a broad definition of what a component can be. It can be any system object which is considered to have its own identity; examples include, the

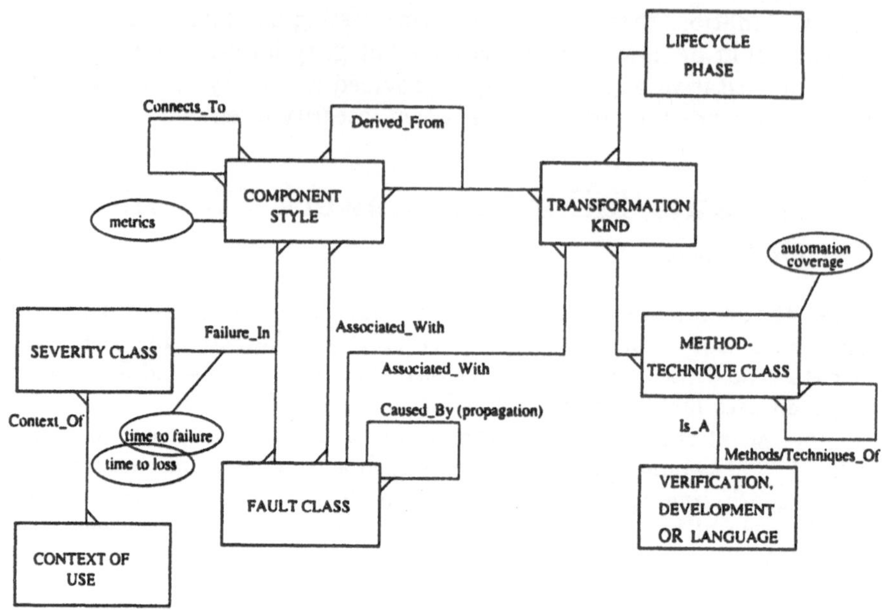

Component: Anything from system down to smallest indivisible unit. Can be software, data, document or hardware.

Method: Set of heuristics and procedures which explain when to use a set of techniques or subsidiary methods to create or check a set of products and how to check that the information content is complete/consistent.

Technique: Set of heuristics and procedures that explain how to create or check a set of one or more products.

Figure 2: Diagrammatic representation of a framework for a failure classification scheme (an entity-relationship style of notation is used).

smallest indivisible software unit, a communication channel (such as a unidirectional Unix-style pipe), documentation etc. Not all components represented will necessarily contain detected faults since we believe it is useful to capture a representation of the system as a whole. Connectivity amongst components within the architecture is represented using the relationship *connects to.* The nature of this relationship (such as asynchronous calling) would be specified using the *nature of* attribute. The derivation of one aspect of the system software product state from another is represented by the *derived from* many-to-many ternary relationship that relates the system component states[3] and a transformation process (such as text edit, design edit, compile and so on).

Process data is represented by the entities *transformation kind*, *lifecycle phase* and *method/technique class* and its subclasses. The transformation process relates stylised system components with process data. Based on their intended use, we have identified three basic kinds of methods and techniques. Some attempt has been made elsewhere to classify methods and techniques further [19].

Failure effect and context of use data is represented by the entities *severity class* and *context of use* and the relationships *failure in* and *context.* The *failure in* ternary relationship relates the kind of component(s) containing the

[3]These may be versions of the same component or related, but very different, components (for example, a requirements specification and the design documents derived from it) .

top-level error(s)/fault(s) with the class of fault(s)/error(s) causing the failure. Attributes are used to capture failure times. The *context of use* entity records the nature of relevant hazards (for example, loss of aircraft, or overheating in a chemical reactor).

5 Conclusions and Future Directions

To improve the dependability of systems in the future pertinent data about the performance of existing systems will need to be recorded. By an examination of the potential uses of such data in order to meet this goal, we derived a set of requirements on the classes of data and their interrelationships that need to be recorded, and hence covered by a classification scheme. Four classes of data were identified: causal, process, product and effect (severity). Based on these requirements, we proposed a framework classification (Figure 2) providing an abstract view on the characteristics of each class of data and how they interrelate. Further specialisation is required for each of these abstract classes.

One of the most serious barriers to the effectiveness of any future scheme is the unwillingness of organisations to publish failure data. This may be because they believe that such data may undermine their perceived competence or the quality of their products, or because the effort involved in in publishing their data is simply too great. These issues must be addressed by both those developing the failure classification (for example, to keep it simple) and those operating failure databanks (for example, by ensuring data is recorded confidentially and is non-attributable).

Regarding classification usability, although the framework appears complex (since it is composed of many sub-classifications), we contend that all four classes of data are required if accurate predictions of the dependability of future software-based systems are to be achieved.

We have begun to address some of the issues identified in this paper, for example, problems associated with IEEE 1044. There are three main directions for our future work: first, definition of failure modes representing causal data, based on investigating existing field data, and capturing new failure data; second, investigation of whether the product (design and code) can indicate the presence of possible failure modes by the recognition of atypical patterns; third, developing automated intelligent tools for capturing failure data leading, potentially, to the establishment of a failure databank service.

Finally, we strongly recommend the establishment of a failure databank service, reflecting the examples provided by other areas of engineering. Such a databank should use a classification scheme conforming to a framework similar to ours. Moreover, the service should place emphasis on solving the non-technical problems identified above.

References

[1] Bowen J. Standard error classification to support software reliability assessment. Proc. of the National Computer Conference, vol 49, AFIPS, 1980; pp 367–705.

[2] Cannon A, and Bendell A (eds). Reliability data banks. Elsevier Applied Science, 1991.

[3] Institute of Electrical and Electronics Engineers. IEEE standard classification for software anomalies - IEEE Std 1044-1993. 1993.

[4] Loftus C, Long F, Pugh D, Pyle I. Failure classification schemes for analysing system dependability. Tech. report, University of Wales, Aberystwyth, Department of Computer Science, January 1996. UWA-DCS-95-020.

[5] Loftus C, Pugh D, Pyle I. Software failure modes, effects and criticality analysis — position paper. Tech. report, University of Wales, Aberystwyth, Department of Computer Science, Aberystwyth, January 1996. UWA-DCS-95-018.

[6] Laprie J. Dependability: from concepts to limits. Proc. of SAFECOMP '93 (Poznan Poland), Springer-Verlag, 1993; pp 157–168.

[7] Hunter E. Classification made simple. Gower, 1995.

[8] Myers G. Software Reliability — Principles and Practices. Wiley, 1976.

[9] IEC. IEC draft 1508 - Functional safety: safety-related systems. Geneva, 1995. 65A/179-185.

[10] Garlan D. Research directions in software architecture. ACM Comp. Surveys 1995; 27(2):257–261.

[11] DeMillo R, Young M. Quantitative aspects of software architecture (position statement). Proc. of the first Int. Workshop on Architectures for Software Systems, in cooperation with ICSE-17, Seattle, Washington 1994; pp 72–79.

[12] Tomek L, Mainkar V, Geist R, Trivedi K. Reliability modeling of life-critical, real-time systems. Proc. of the IEEE, 1994; 82(1):108–121.

[13] Muppala J, Woolet S, Trivedi K. Real-time systems performance in the presence of faults. IEEE Computer 1991; 24(5):37–47.

[14] Heimann D, Mittal N, Trivedi K. Availability and reliability modeling of computer systems. In Yovitts M (ed) Advances in Computers, vol 31, New York Academic Press, 1990; pp 175–233.

[15] Molloy M. Performance analysis using stochastic petri nets. IEEE Trans. on Computers 1982; C-31(9):913–917.

[16] Arlat J et al. Fault injection for dependability validation — A methodology and some applications. IEEE Trans. on Software Engineering, 1992; SE-16(2):166–182.

[17] Shimeall T, Leveson N. An empirical comparison of software fault tolerance and fault elimination. IEEE Trans. on Software Engineering 1991; 17(2).

[18] The Motor Industry Software Reliability Association (MISRA). Development guideline for vehicle based software. Motor Industry Research Association, November, 1994.

[19] Pyle I, Lissandre M, Hruschka P, Jackson K. Real-time systems: Investigating industrial practice. Wiley, 1993.

Session 7
Human Factors

Human Factors in High Integrity Software Development: a Field Study

J. Griffyth
Mathematics/Computing Department,
Open University
Milton Keynes, UK

1 Introduction

Human factors are critically important during the development of software, and nowhere more so than for safety critical applications. This paper reports a study of human factors in the development of such systems. The particular aim of the study was to consider how human factors affect the productivity of the development process and the integrity of the developed product. The background to this focus is explained and the methods used to investigate this difficult area are outlined.

Preliminary findings from the study are reported, focusing on how competence, communication and motivation within the development team are related to the efficient production of high integrity software. The paper concludes by considering how the findings can lead to process improvement.

2 The PRICES context

The study was undertaken as part of the PRICES[1] project. The aim of the project is to enhance productivity and integrity in the development of safety critical software through the evolution of a new Code of Practice. The fundamental premise of the project is that the main barriers to productivity and integrity are associated with human aspects of the development 'system' rather than with technical aspects.

The project has sought to understand the influence of human factors on the development process and product by obtaining quantitative and qualitative evidence of their effects. This data has been used to develop a object based model of the process and the product which makes these relationships explicit. The model is focused on a measure in which productivity of the development process is seen as a function of:

---

[1] PRICES (IED4/1/9202) is a UK DTI/EPSRC SafeIT project, in its third year at the time of writing. Project partners include Lloyds Register, Rolls-Royce, Open University, BAe SEMA, Analysis International, City University and GP- Elliott.

a) the number of errors introduced at the start of the process

and

b) the efficiency with which these errors are detected and corrected until the density of residual errors is sufficiently low for delivery.

By tracking a development project from the beginning of the lifecycle and measuring defects found and corrected against time, we can obtain a measure of team productivity i.e. how long does it take a team of *x* people to produce *y* lines of code to *z* level of integrity. Correlation of this productivity measure with human issues is provided by metrics describing the status of human factors attributes within the team, such as motivation, and collected at the same time as the defect data. The model will not be discussed in detail here - it merely provides a framework for comparison and interpretation of data sets.

3 Preliminary Work

The first task of the study was to gather basic information about the way safety related software is currently produced to determine those elements of current practice which are human intensive, those which are sound, and those which offer scope for improvement. The study drew on material collected during the first PRICES activity which sought to describe current development practice in some detail. This was achieved primarily through a postal survey which attracted 105 responses, and through face-to-face interviews with relevant development groups. The salient points which emerged are reproduced here - for a full discussion see [1].

3.1 The PRICES Survey

The survey confirmed a view of the safety related development process as one in which the vast majority of development groups adhere to a conservative waterfall model or variant, and tried and tested procedures, methods and tools. There is little to distinguish these projects from conventional development groups other than a few specialised analysis and testing activities and a general strategy of more rigorous application of typical software engineering methods.

The survey showed that safety related development remains an essentially human centred process. Very few development activities are automated - even where tools are used to mechanise tasks, the human user makes the decisions, finds solutions to problems and takes responsibility for the form and quality of the end product. Thus, the skills and performance of the people involved have a critical effect on the success of the project - it is in relation to these human issues that productivity and integrity seem to be vulnerable. It follows then that they are perhaps open to improvement through optimisation of these human factors.

3.2 Classification of Human Factors

Having established the significance of human issues, the next objective was to both classify and quantify the human factors which have particular influence in the development process. This was a crucial activity to constrain the scope of the work since development is a human intensive activity, and humans have innumerable characteristics and behaviours. With the overall project goal in mind, it was felt important to concentrate on factors which are of significant effect and are open to measurement and adaptation. From the survey and interview data it was clear that several human factors were perceived, by the developers and the researchers, to be influential. These are now briefly described.

3.2.1 Competence

Competence, i.e. developer knowledge, training and experience of the application domain, of the characteristics of the overall system, and of the development environment, tools and methods was seen as essential to project success. The competent application of knowledge is a crucial attribute of software engineering, because software systems are abstract entities which can only evolve through a process in which the characteristics of the application domain problem are abstracted, manipulated and represented as a program.

3.2.2 Communication

The communication of knowledge and requirements between the client and the developer, amongst the development group, and between the software and system developers was a process which could impact strongly on productivity and integrity. The importance of communication in the development process reflects the fact that commercial software development is a group activity, in which the social functioning of the group structure contributes significantly to group performance.

3.2.3 Motivation

Motivation is what arouses people to action, and directs the nature of the action over time [2]. Psychology has tended to distinguish between biologically and socially/individually motivated behaviour and here it is the latter kind which is of interest. At an individual level, motivation and satisfaction amongst the software developers surveyed was not high due to pressures within, and acting on, the organisations in which they worked.

3.6 The Human Factors Study

These three factors emerged most frequently and consistently[2] from the data gathering activity and were carried forward as the main focus of the human factors study. However, these 'person related' factors[3] rarely act directly, or in isolation, on productivity or integrity; more typically there is an interaction with many other factors.

For example, while requirements negotiation can be seen as an event dominated by communication issues, the quality of the requirements specification is also determined by the analyst's ability to recognise and understand the client's problem, (which in itself is a product of the novelty and complexity of the task), and by the tools and facilities with which he works. Thus there are factors at work in the interaction which are a product of the individual characteristics of the developers, the organisation, the task and the environment. Within the PRICES project, we recognised the complexity of the interaction and complementary work packages investigated other aspects of the development system such as tools and methods. Within the human factors study we elected to concentrate on person related factors.

The next task of the study was to collect both quantitative and qualitative data about these factors in the development process to add detail and depth to the model. We chose to derive this data from live projects and perceived a need to try to elicit a richer perspective on the process than that typically obtained by project management. By this, we mean that many conventional metrics and monitoring techniques offer little insight into why and how defects occur and persist. We used interview, observation and metrics collection activities combined in what we refer to as the field study.

4 The Field study

The field study was bound by many pragmatic constraints. Ideally, the study sample would have included groups from all of the major safety critical industry sectors, working on software products of varying size and integrity. In practice, we were limited by our own project resources to a small number of organisations who had active development groups. We drew on the contacts made in the survey activity and were able to study three development projects in detail.

---

[2] % of interviews where issues were raised:
Competence 100 Communication 100 Motivation 90

[3] The FASGEP project [3] has suggested that human factors can be classified as person, task or environment related - examples of these classes might be an individuals' experience of the development environment, the complexity of the algorithm to be coded and the available hardware and tools.

A detailed account of the organisations and projects in the study can be found in [4] - a summary of characteristics will suffice here. Organisations 1 and 2[4] were small companies owned by larger corporations. Their interest in software arose from the increasing reliance on software components within their application domains to replace manual and mechanical control and monitoring systems. Both 1 and 2 were involved in the design and production of the whole system. Organisation 3 was a much larger company with a long history of involvement in safety related systems. The organisations differed most in size, in the maturity of their high integrity development process and in their managerial structure and style. Table 1 summarises these differences.

	Organisation 1	Organisation 2	Organisation 3
Size	Small	Small	Large
Maturity	Low	Low	High
Style	Informal	Formal	Formal
Team Structure	Flat	Hierarchical	Hierarchical

Table 1 Summary of Organisational characteristics

Access to a live development project was negotiated with each organisation. Each project was visited on a regular basis for up to a year, with visits timed to coincide with the end of development lifecycle phases. The objective of the visits was to gather evidence about the development process.

The field study was intended to capture two very different kinds of evidence - both quantitative and qualitative. On the one hand, we needed to collect measurements to flesh out our rudimentary productivity model. On the other, we recognised that a metrics oriented approach could only confirm that certain relationships existed and were significant - it could not tell us why.

We needed to explore the underlying causal relationships in a more insightful way if we were to be able to offer process improvement advice rather than merely describe the status quo. The study was therefore composed of two main strands i.e. a metrics collection activity and an interview/observation activity.

---

[4] Field study collaborators were guaranteed anonymity and data confidentiality

4.1 Metrics Collection

Conte et al. [5] offer some useful criteria for factors for inclusion in productivity models. They propose that factors must be measurable, generalisable, significant and independent. The first of these, i.e. measurability, must be satisfied for a metrics collection activity to be feasible.

Measurability is problematic where human factors are concerned - apart from physical characteristics which are of no interest here, attributes are not directly observable. We can not look inside a designers head and measure the amount of knowledge he has, but we can infer its' presence in relative quantities from other indicators such as his length of experience in an application domain, or a programming environment.

The difficulty inherent in finding metrics for human aspects of systems is sometimes used to undermine the value of this kind of research. However, we found that the objectivity and relevance of commonly accepted metrics for more 'concrete' aspects can be as arguable e.g. code complexity.

Table 2 shows examples of the factors which were 'measured' in the study. In addition to the human factors, a number of other factors were deemed important - these factors are clustered around the top level objects in the development 'system' e.g. development tools and tasks.

For each organisation:			
Process Maturity	Stability		
For each project:			
Team Size	Effort	Duration	
For each team member:			
Competence	Motivation	Communication	Role
For each lifecycle task:			
Complexity	Concurrency	Support from method	
For each product:			
Size	Defects	Time to Produce	
For each method:			
Understandability	Coverage	Proof of Completeness	
For each tool:			
Value for money	Adaptability	Supports communication	

Table 2 Examples of Monitored Factors

Many factors were decomposed into components, or attributes, and for each factor or attribute, a metric was selected. The factors, attributes and metrics are described more fully in [4]. In themselves however, these metrics could only tell us about the relative productivity of a development group, and very little about the organisational and industrial context in which the group worked. To gain any insight in to these issues, we needed to adopt a more sociological approach to the problem.

4.2 The Interview/Observation Activity

There has been a growing recognition in recent years that the most important components of the context for software development are the human and organisational capabilities available and that conventional research strategies are ill equipped to recognise and explain such phenomena. There are now a number of ecological research methods which are able to elicit a much richer kind of data based on an 'ethnographic' [5] paradigm. This involves setting aside 'the empirical positivist' perspective on the developer and acknowledging that it is the developer who knows what is important and that development events only have meaning within the wider relevance of his task and environment [6].

In PRICES we felt it important to give full recognition to these aspects of the development process - thus, for every set of metrics data acquired, there was a programme of interviews and observation sessions in which the people involved in the development process were asked about their work, and about the metrics data collected from them.

5 Findings

The study generated an enormous amount of data, much of it in the form of tape recordings and transcripts. The analysis of this data is still in progress but broad trends can already be detected. Among the more interesting issues to emerge are those discussed below.

5.1 Productivity and Integrity

Analysis of the defect introduction/correction rate for the three projects suggests that teams 1 and 3 had similarly productive development processes whereas team 2 fell somewhat behind. The introduction of a new tool and quality management system for team 1 had compromised their typical productivity rate to an extent which probably deprived them of first

---

5 Context and naturalness of data important, no preconceived ideas about what to look for.

place. However, the qualitative evidence showed that these productivity levels were achieved in very different ways.
In software engineering, and especially at the higher integrity levels, conventional wisdom dictates that the key to productivity and integrity is formalisation of the process - Pressman [7] suggests that we need to enlist the help of 'comprehensive methods...better tools...more powerful building blocks...better techniques...and an overriding philosophy for co-ordination, control and management'. If we were to rank the field study organisations according to the degree of formality and control of their development practices we would see a ranking of 3, 2, 1 - thus, a productivity ranking of 1, 3, 2 is somewhat at odds with what we might have predicted.

We can hypothesise about the reasons for this by closer inspection of the metrics data, supported by the qualitative evidence obtained by interview and observation. The quantitative data showed that the productivity of all teams suffered most at the same points - primarily when completed work was reviewed and revised. These revisions were prompted either by detection of errors or changes in resources. It also revealed large differences between the groups in terms of the match between human factors and the task, and the environment. It showed little difference between the teams in terms of the size and complexity of the development product. The qualitative data showed how the teams managed change, and how successful their different approaches were.

Team 1 had good support for negotiating and implementing changes in that they were a small group, working in close physical proximity to each other and to the team manager. They had good interfaces with those responsible for system/software boundaries, access to the end user environment, and worked within a culture which encouraged their participation in planning.

Team 3 was much larger, and structured in a much more formal hierarchy. Their development process was highly proceduralised and relied heavily on formal communication methods such as memoranda. In this mature process, formal monitoring and control procedures had evolved to track work progress and ensure that activities were synchronised and completed. This team could not respond as easily to change as team 1 and were not able to reach consensus on decisions as quickly but were relatively productive.

Team 2 was much the same size as 1, but structured similarly to 3. Interfaces between system, hardware and software engineers and between engineers and managers were poor and restricted to formal channels. Although the development process was heavily regulated by procedures, there was little support within the methodology for negotiating and meshing the interests and activities of team members. Thus, development activities tended to carry on even when problems had materialised upstream, leading to more rework than was necessary.

5.2 Other Issues

Other issues became visible as the study progressed. It was noted that the introduction of new tools, methods and procedures was detrimental to both productivity and integrity. Team 1 suffered from the negative effects of moving their development to a new operating system and tool environment, incurring a steep learning curve for their engineers. In this case, the changes were necessary to keep pace with technical developments in the industry, and were supported by training and evaluation activities. However, the study uncovered a number of cases where expensive tools and methods had been imposed on teams with no provision for training or support and which were eventually discarded as being too difficult to learn to use or to integrate into the development process.

Another issue concerned the differences in perception of the project between managers and their subordinates. Managers tended to report their projects as being relatively smooth running while project engineers reported their work as stressful and unpredictable. In the field study, many engineers admitted to delivering documents to meet scheduled milestones while fundamental aspects of the work reported in the deliverable remained unresolved. There were many complaints that managers had little idea of what actually went on in their projects and were not sympathetic to their teams' difficulties.

5.3 Human Factors

In terms of the three factors of particular interest to PRICES, we will need to do a great deal more analysis before we can offer a detailed account of their role in the development process. However, we can summarise the results so far as follows:

Competence was confirmed as a significant contributor to team performance. Knowledge of the interfaces between the engineers own domain and others was a particularly valuable characteristic. Unfortunately, competence tended to be undermined by a lack of adequate training and support, and by the often inappropriate introduction of new work practices.

Communication was revealed as essential to prevent problems gathering momentum and incurring increasing costs in rework and stalled activities. Many formal communication methods were exposed as ill suited to the activities they supported, and in some cases, actively hindered the resolution of difficulties.

Motivation issues came up repeatedly, often focused around the engineers' need for appropriate levels of involvement in work planning and decision making. The tendency to work to fixed price contracts, leading to ever shortening deadlines when recovering from mistakes and defects, seemed a particularly demotivating practice.

6 Conclusions

One of the primary implications of these results concerns the relationship between process improvement and process control and formalisation. It seems that many of the problem solving activities which are critical to productivity and integrity can not be formalised without impinging on the human behaviours and characteristics which we currently rely on to successfully negotiate solutions. In the foreseeable future, it seems unlikely that the problems of changing and misunderstood requirements will simply cease to exist - yet simply applying more control and more procedures is unlikely to have any beneficial effect. This argues that the way forward is to understand these human factors rather better than we do, and seek ways to support them more effectively.

In terms of our three person related factors, all seem open to enhancement through relatively simple practices such as improving training and support in new techniques, choosing tools more carefully and with proven benefits in view, opening up and supporting communication channels across domain boundaries and fostering an environment in which individual team members are encouraged to take responsibility for their own tasks, for the achievement of the teams' goals and to participate in the setting of those goals.

Finally, we can consider the value of the research methods we employ in pursuit of process improvement - although the collection of quantitative data is a necessary and desirable way to increase our understanding of the development process, we need also to acknowledge the social context of software engineering. If we choose to ignore these human aspects because they are complex and difficult to measure we may find that little further progress will be made.

References

1. Templeton SP. Baseline of Current Practices used in the Development of Software for Dependable Systems, D17, PRICES Consortium, 1994

2. Maslow, A. Motivation and Personality, Harper Row, 1954

3. Cockram T, Salter J, Mitchell K, Cooper J, Kinch B & May J. Human Error in the Software Generation Process. Technology and Assessment of Safety-critical Systems, Springer-Verlag, 1994

4. Griffyth J. The PRICES Human Factors Study, D21, PRICES Consortium, 1996

5. Conte S, Dunsmore H & Shen V. Software Engineering Metrics and Models, Benjamin/Cummings, 1986

6. Crellin J & Preece J. Assessing the Usability of Different Evaluation Techniques in a Small Company, Journal of Information and Software Technology, 35, 6, 1991

7. Pressman R. Software Engineering: A Practitioners Approach, McGraw Hill, 1994.

Human Factors in Safety-Critical Systems: An underestimated contribution?

S.J. Westerman & G.R.J. Hockey
Department of Psychology, University of Hull, Hull, HU6 7RX, England

Abstract

This paper discusses current and future prospects for the application of human factors techniques in the assessment and development of safety-critical systems. It argues that HF issues are given insufficient priority within this context, considers reasons for this situation, and makes suggestions for developing the partnership between human factors and engineering.

1 Introduction

This paper considers the role of Human Factors (HF) in the assessment and development of safety-critical systems (SCSs). It is intended as a document which stimulates discussion within the engineering and HF communities, and consequently seeks to present a range of perspectives on the interface between engineering and HF, rather than a single prescriptive model. Within the present context, HF is taken to pertain to all human involvement in the development, maintenance and operation of a SCS; it is particularly concerned with the appraisal of human capabilities and preferences as they relate to human error (although see [1], for a less context dependent definition). The views expressed here are based on a review of the literature (including internet newsgroups), responses to a semi-structured questionnaire (administered over the internet), and a series of semi-structured interviews with engineering and human factors specialists.

Assertion: Although there appears to be an increasing recognition of the importance of HF to SCSs in some quarters of the engineering community [2] [3], generally the contribution which HF can make within this context is underestimated [4].

2 Human error and system failures

Although the potential for human error within the operation and maintenance of SCSs (e.g. the Bhopal or Challenger disasters) is increasingly widely recognised, the number of system failures that are the result of design error is probably underestimated. Many errors that are attributed to the human operator within a SCS might also be more properly attributed to inadequate foresight during design [5]. Similarly, in addition to system failures that are directly attributable to errors in the software design process (as appears to be the case with the recent failure of the Ariane rocket launch), it can also be argued that hardware failures that have a safety-critical outcome are the result of human error on the part of the designer, in that the potential for particular circumstances to cause failure had not been foreseen, and appropriate design contingencies not prepared.

A range of engineering methods are available to structure the design process of SCSs in such a way that the effects or the frequency of human error are reduced (e.g. formal methods, independent testing, n-version software). Although these methods are generally considered to fall outside the remit of HF (and for this reason we do not presume to discuss them in detail), it is worth noting that they rely on established cognitive heuristics in order to provide additional levels of system reliability. Each reduces the probability of system failure by forcing, in a more (e.g. formal methods) or less (e.g. n-version software) systematic way, designers to adopt different cognitive perspectives to problem solution. While each of these methods are applied widely and have much to commend them, a number of limitations associated with their use have also been identified:

- They are insufficient to ensure reliability. Generally the scale of SCSs is such that complete coverage using these methods is not possible [6], and/or complexity is so great that errors also occur in the applied prevention methods. For example, n-version software does not guarantee that the same error will not occur in each version [7]. There are a number of reasons why human designers may be prone to applying the same misconceptions about an SCS [8].
- None of these methods adequately addresses the difficulties of initial problem description. "Showing consistency between the requirements and the code is not adequate to raise confidence in safety, since most safety problems stem from flaws in the requirements" [5], p. 496.
- The application of these methods to one SCS does little to enlighten designers about the causes of error within other SCSs.
- These methods are very restricted with respect to the identification of limitations of human components within a SCS.

For these reasons, it is argued that the application of these methods must be supplemented by the use of HF techniques for error reduction, and that this should occur throughout the design lifecycle [9] [10]. HF expertise is frequently only brought in at a late stage of the design process [11], sometimes as part of developing a safety case. The railway signaling industry provides an example of SCS development based on little HF support, but with subsequent impetus for HF input given by the need to generate safety cases. The use of HF in this way severely restricts the effectiveness of any input [9] [11]. Large design decisions have been taken, budgets have been set, and the scope for radical alteration to a system is limited. HF specialists are consequently restricted to 'tinkering at the margins'. Even in industries where effort has been made to increase the availability of HF expertise throughout the design lifecycle (e.g. the nuclear industry), the integration of HF may be restricted by the difficulties of communication between engineers and HF specialists (see below).

3 The application of human factors to SCS design

There are a wide range of HF methods which can serve to increase system reliability. A complete review is beyond the scope of this paper, however, some broad areas of HF application will be considered (see [1] for a brief review of HF methods).

Several HF techniques exist for the purpose of describing the component elements of jobs and the sequence in which they are performed (e.g. cognitive task analysis, hierarchical task analysis). A further set of HF techniques exist for the appraisal of job performance. These range from primarily qualitative (e.g. observation, verbal protocols, critical incident techniques) to primarily quantitative (e.g. in-situ data logs, laboratory-based simulation studies, work samples, questionnaires and tests) methods of assessment. Techniques may relate to the direct assessment of performance, or to the associated demand experience by the operator (e.g. self-report workload scales, secondary task). They may include, or be supplemented by, additional techniques designed to examine the effects of what may be termed Performance Shaping Factors (PSFs: [12]); motivation, noise, fatigue, etc. These may exert an influence on the quality of task performance or the level of perceived effort and other 'costs' associated with performance.

These techniques are well established and can be expected to provide reliable data relating to task description and performance. Their application, in various structured forms, has been demonstrated to result in substantial increments on performance efficiency [13] [14]. The difficulty, from an SCS design point of view, is that these methods are time-consuming to implement. Further, although behaviour within the confines of a particular system, once analysed can be reliably predicted, the generalisability of results to different systems can be poor. This is because human behaviour is the result of an immensely complex set of interacting processes. Although HF techniques can be used to investigate the role played by many of these, and such investigations have enabled a number of useful models of human performance to be developed (e.g. relating to sustained attention; multiple-task performance; problem solving), HF still struggles to meet the quantitative requirements of many engineering applications. Accurate prediction of physical interaction with systems is generally possible (using anthropometric data), however, a comprehensive quantitative appraisal of the cognitive model that will be applied to task performance is beyond the current state of the HF art (see Section 5.3). This has implications for the use of various HF standards and guidelines that have been developed on the basis of HF research and application, e.g. [15] [16] [17], and which serve to inform the system design process. In attempting to generalise across systems, guidelines and standards run the risk of becoming so bland as to be useless.

Within the SCS engineering community, perhaps the most well known and most widely applied HF techniques come under the umbrella of Human Reliability Assessment (HRA). This refers to a group of methods designed to identify and quantify potential sources of human error (see [18] for review). These are generally founded on principles of Probabilistic Safety Assessment (PSA), and seek to decompose task performance into a number of components to which error probabilities can be attached. In order to achieve this goal some methods rely upon databases of human error statistics, while others rely on expert judgment. However, absolute human error probability quantification has been demonstrated to be unreliable (see [19]). The reasons for this include problems of generalisability (when using human error probability statistics contained within error databases), as discussed above (see also [20]). Also, a basic assumption of HRA is that the probabilities of failure of different system components (including the human) are independent. This is demonstrably not the case (e.g. the Challenger disaster), and furthermore the nature of this non-independence may be complex and difficult to specify (see [4]). Finally, some errors occur with such low frequency that they are difficult to analyse [21].

Two further areas in which there is a substantial literature and experience of application are selection and training. Individual differences have been shown to account for substantial variance in the performance of computer based tasks [22]. A wide variety of tests are available to the HF specialist for examining individual differences in performance or preferences. The advantages of using such information are well documented, with a good deal of research evidence to support its utility. Similarly, training has also been the focus of sustained HF research and application [23]. There is a large literature describing the acquisition, retention, and transfer of job skills; and many investigations of new methods of training delivery (e.g. computer-based training). Simulation (including the use of virtual reality techniques) of potentially dangerous system states can be a particularly important element in training for SCS operation (see [24]). The utility of training research looks set to increase in line with technological innovation and the increasing complexity of SCSs.

4 The under-utilisation of human factors

Given the range of HF techniques and the potential benefits to SCS design, why is it that the application of HF is not more widespread? There are a number of potential reasons why this situation might arise.

There is a perception within some parts of the engineering community that HF expertise is not really necessary for the development of safe systems (e.g. see [25]). Responses from engineers, to questionnaires and interviews, suggest a number of possible reasons for this. The problem may arise because engineers believe that they understand HF well enough, and feel that they have a better grasp of how a system will be used than a HF specialist. Given that the system operator is an engineer it can be argued that an introspective approach will be sufficient, though there is a good deal of evidence to suggest that it is not a reliable method of identifying the potential for human error. An example can be seen in the case of recent trends in aircraft cockpit automation. Within the aircraft industry engineering capabilities were overriding human factors requirements with respect to the use of automation on the flight deck. Increasingly the pilot was becoming a 'monitor', i.e. acting as a fail-safe for the automated system that was flying the plane. Unfortunately, this automation encouraged a state called 'automation complacency' in which the pilot no longer felt engaged with the task and tended to make increasing numbers of monitoring errors. The solution was provided by an HF investigation into the management of control between the human and the automatic controller [26]; as a result, 'my turn' systems are being designed to enable the control of the aircraft to be periodically switched to the pilot.

Another, perhaps more basic reason for an under valuation of HF is that, despite a potentially high level of interest in the application of HF methods, there is a lack of understanding in some parts of the engineering community about what the HF approach is, and what it can do. This may be related to a tendency for engineers to use software development models which focus upon the evolution of an artifact rather than the human behaviour underlying it [27], and be further compounded by the fact that the significance of many design decisions may be temporarily masked by the adaptability of the human operator.

Several sources (respondents) stressed the importance of costing information (or the lack of it) as a determinant of the application of HF expertise to SCS development. Given that finances are limited, HF must be able to make a case for inclusion within a project on financial terms, and must compete with engineering for project resources. However, there are a number of fundamental difficulties associated with expressing the benefits of HF interventions in financial terms. First, unless the system already exists, and previous performance data is available, and the system undergoes no other changes, then an accurate assessment of the benefits of an HF intervention will not be possible. These constraints necessarily rule out gathering data from new systems (in which HF has been incorporated in the early stages of design), unless it parallels an existing system. Second, the outcome of a successful HF intervention may be the avoidance of an error which would occur with very low frequency. The 'pay off' may occur only after a number of years. However, the avoidance of such a safety related error may more than justify the 'up-front' costs. Nevertheless, it will be very difficult, if indeed possible, to demonstrate this association, particularly bearing in mind the difficulties of generalisation for HF, as discussed above. These difficulties will be compounded by the fact that budgets are frequently controlled by engineers.

Coupled with these factors is a demand, on the part of engineers, for fast, 'straightforward', and precise answers to questions. This encourages the development and use of guidelines and standards which are held to generalise across situations (perhaps with the application of some correcting factor such as PSFs).

Engineers are often uncomfortable with the level of precision that accompanies human factors solutions. However, the intrinsic variability of human system components is substantial [22], and the current state of the art is such that the performance of these components cannot be described, particularly accross systems, as precisely as that of non-human components. The consideration of human factors issues adds additional layers of complexity to the situation and is sometimes perceived by engineers as uneccessarily slowing down project completion. However, the quality of HF intervention may also be influenced by specialists pandering to the requirements for fast solutions, resulting in 'quick and dirty' implementations with no life beyond the immediate application. Such evaluations are of limited effectiveness for error detection [28], perhaps because the most significant system problems are inherently resistant to detection by over-simplified methods.

Results of a previous study indicate that the quality of the HF specialist is a major determinant of favourable outcome for HF interventions in software development projects [9]. A number of points have been made by respondents which bear upon this issue. HF specialists need to convey information in a meaningful way to engineers (and vice versa). Communication difficulties can arise between engineers and HF specialist because of the use of different jargon. Further, when a HF specialist is brought in to assess an existing system, a major difficulty relates to the acquisition of sufficient knowledge about the task domain. This may bear upon a broader issue concerning the background of HF specialists. Nevertheless, given a situation in which engineers have spent many years conceiving a complex SCS, it is not realistic to bring in an external HF consultant for a brief period of time, and to expect their input to be based on a full appreciation of all system properties (unless they have prior knowledge).

5 How can the uptake of human factors be improved?

We regard it as axiomatic that the potential of HF methods should be fully realised within the development of SCSs, as within the broader engineering community. A full-scale investigation of how this may be achieved is planned for the near future. For the present, we address the issues raised in the previous section, by providing a provisional framework for debate.

5.1 Education and training

The most pressing need, and probably the most effective in the long run, is for better education and training of both HF specialists and engineers. While the occurrence of large-scale disasters raises awareness of the need for HF at all levels (the public, project managers, engineers, and government agencies) their impact is relatively short-lived. More fundamentally, we need to be proactive in promoting such awareness. There may be advantages with more direct education management, and more widespread incorporation of HF concerns in the training of engineers and SCS professionals (within both degree courses and professional training). Equally, we should consider the increased emphasis on engineering issues in HF training. For example, HF specialists could be encouraged to frame problems identified by an HF analysis in terms of engineering solutions. Only when the two professional groups converse in something like the same language would we expect a rapprochement of attitudes and priorities

There may be some advantages to promoting the use of 'in-house' human factors specialists/teams who can develop the necessary system specific expertise, as occurs in the nuclear industry in countries like US, Canada, or Sweden [3]. Lundell and Notess [9] found that favourable HF interventions tend to be those in which the level of involvement between HF specialists and engineers is greater. Although in their survey it did not seem to matter whether an HF expertise was provided on a consultancy basis or in-house, this research relates to all types of software development, and it may be that the complexities of SCSs makes them a special case.

5.2 Dissemination

Wider dissemination of HF products is required. HF needs to be making more of its success stories, given the difficulties, described above, in generating costing information, much of this may have to be done in a qualitative manner. HF should look for opportunities to present model case studies showing the richness of the approach in different application areas. Where possible it should provide examples of the costs and benefits. It should be noted that the benefits of HF intervention are not restricted to error reduction. They can also include: greater performance efficiency; reduced operator training times; reduced maintenance times; increased operator job satisfaction; etc.. In the June, 1996, issue of the Ergonomist there is a call for this type of HF outcome information which can be used within a database resource for future justifications of HF interventions. Such initiatives must be welcome.

In addition, focussed dissemination of HF is required. Instead of standing on the fringes of the engineering community HF specialists need to be more actively demonstrating to engineers the advantages that are associated with the application of HF techniques. At the same time engineers could improve the process of mutual understanding by addressing HF audiences, perhaps relating to the effectiveness of HF interventions or research (providing feedback), or describing engineering requirements from HF and proposing methods by which they may be achieved.

5.3 A framework for human factors

The importance of developing a general framework for HF has been identified by a number of authors [25] [29]. The requirements of HF in relation to SCSs will be a subset of these, and could be addressed directly. This framework should provide a detailed description of areas of application for HF, the techniques available; information as to their utility; and demands for research. In addition, it should address the association between HF methods and the software development lifecycle (see [30]).

Within this framework, research funding is required which encourages investigation of broad (see below for examples) human factors problems. A good deal of unifying research, with strong theoretical and applied roots, must be done if increased precision and generalisation are to be achieved. It is imperative that this research is designed for wide ('horizontal') dissemination. Too much HF data is lost through an overly 'vertical' research design and outlook. For example, if HF investigations of the development of SCSs within the nuclear industry are reported only in nuclear industry journals, these results may be overlooked by HF specialists and engineers working in other areas of SCS application, and therefore not contribute to the wider HF perspective that is required. In contrast, if HF research is established which addresses issues that relate to more than one industry, greater benefit may result with respect to both dissemination and the development of a HF framework.

More specifically we would identify the following key areas within HF which would benefit from immediate research input:

- HF modeling tools: There is increasing opportunity to simulate performance of systems and their components. This can be used as part of an HF analysis. Increasingly, the potential of virtual environments is being explored within this context, e.g. [31], to assist system designers to visualise the outcome of design decisions. Alternatively, several projects have developed computable models of different aspects of human cognition, e.g. [32], although these tend to be somewhat restricted, and have not yet provided an applied model of human error (see [33] for a theoretical account of a system for modeling human error).
- Creativity within design: Although there is a substantial literature within psychology relating to creativity (e.g. see [34]) consideration of these issues within the context of system design is very limited. The importance of addressing human error within the design process is increasingly recognised, as discussed above. However, perhaps because of the complexities involved, this area has been comparatively ignored within research circles. Very little progress has been made in transferring and extending this work within the context of systems design.
- Quantitative assessments of mental models: A mental model refs to "... a rich and elaborate structure, reflecting the [designer's/operator's] understanding of what the system contains, how it works, and why it works that way" [35], p. 51. However, while the mental model has been an important construct within HF for a number of years, HF is poor at making quantitative assessments. Further research is urgently required.
- Team composition: The importance of the composition of teams (design, operation, between phases, etc.) is an important determinant of performance efficiency, and perhaps susceptibility to error [8]. A comparatively small literature exists, in other fields, which bears upon these issues, however it can be valuably extended within the context of SCSs.

6 Conclusions

As technology advances and systems become more complex the need for an effective partnership between engineering and HF will be accentuated. This will be evident in a number of areas: developing correct specifications will become more problematic; testing/verification will become more difficult; improved methods of code generation will be required; there will be an increasing need for co-ordination of project components between individuals and groups; and it will become increasing difficult to maintain a safety culture.

Action in this area should be seen as being at the cutting edge of SCS design issues. Given that intervention is possible at such a wide variety of levels - ranging from individual project decisions to government legislation and funding - it would seem that progress is definitely possible.

Acknowledgments

We are very grateful to our interviewees and questionnaire respondents for their valuable input. We hope that in summarising contributions we have retained the essence of their replies.

The work described here forms part of the project "Human Factors in the Design of Safety-Critical Systems" (Grant no. IED4/1/9311) from the UK Safety-Critical Systems Research Program, supported by the DTI and EPSRC.

References

1. Hockey, G.R.J. & Westerman, S.J. Human factors engineering: An introduction. Report No. SCS-02. University of Hull, 1994.
2. McDermid, J. Issues in the development of safety-critical systems. In F. Redmill & T. Anderson (eds.). Safety-critical systems. Chapman & Hall, London, 1993, pp 16-42.
3. Herman, L. An international survey of human factors involvement in the nuclear industry. In E.J. Lovesey (Ed.) Contemporary ergonomics. Taylor & Francis, London, 1993, pp. 215-220.
4. Wickens, C.D. Engineering psychology. HarperCollins, New York, 1992.
5. Leveson, N.G. Safeware: System safety and computers. Addison-Wesley, Reading, MA, 1995.
6. Rees, C. & Oddy, G. Safety critical software for defense systems: Requirements of interim Defence Standard 00-55. GEC Journal of Research, 1995; 12: 43-49.
7. Littlewood, B. & Strigini, L. The risks of software. Scientific American, 1992; 267:62-75.

8. Westerman, S.J., Shryane, N.M., Crawshaw, C.M., et al. Cognitive diversity: A structured approach to trapping human error. In SAFECOMP '95: Proceedings of the 14th International Conference on Computer Safety, Reliability and Security. Belgirate, Italy, 11-13 October. Springer, London, 1995, pp. 142-155.
9. Lundell, J. & Notess, M. Human factors in software development: Models, techniques, and outcomes. In Proceedings of CHI '91, 1991, pp. 145-151.
10. McCafferty, D.B. Successful system design through integrating engineering and human factors. Process Safety Progress, 1995; 14: 147-151.
11. Lim, K.Y. & Long, J. The MUSE method for usability engineering. Cambridge University Press, Cambridge, 1994.
12. Miller, D.P. & Swain, A.D. Human error and human reliability. In G. Salvendy (Ed.), Handbook of human factors, Wiley, London, 1987, pp. 219-250.
13. Nielsen, J. Usability engineering. Academic Press, London, 1993.
14. Miller, M.A. & Stimart, R.P. The user interface design process: The good, the bad, & we did what we could in two weeks. In Proceedings of the Human Factors and Ergonomics Society 38th Annual Meeting, HF&ES, Santa Monica, CA., 1994, pp. 305-309.
15. Defence Standard 00-25: Human factors for designers of equipment. Parts 1-12, Ministry of Defence, 1984-1992.
16. Defence Standard 00-56: Hazard analysis and safety classification of the computer and programmable electronic system elements of defence equipment. Ministry of Defence, 1991.
17. Smith, S.L. & Mosier, J.N. Design guidelines for designing user interface software. Technical Report MTR-10090. The Mitre Corporation, Bedford, MA, 1986.
18. Kirwan, B. A guide to practical human reliability assessment. Taylor & Francis, London, 1994.
19. Hollnagel, E. Human reliability analysis: Context and control. Academic press, London, 1993.
20. Taylor-Adams, S. & Kirwan, B. Human reliability data requirements. International Journal of Quality and Reliability Management, 1995; 12: 24-46.
21. Kirwan, B., Martin, B.R., Rycraft, H., et al. A. Human error data collection and data generation. International Journal of Quality and Reliability Management, 1990; 7: 34-66.
22. Egan, D.E. Individual differences in human-computer interaction. In M. Hellander (Ed.), Handbook of human-computer interaction. North Holland, Amsterdam, 1988.
23. Patrick, J. Training: Research and practice. Academic Press, London, 1992.
24. Welch, D.L. Ergonomics in nuclear power and process safety. In E.J. Lovesey (Ed.) Contemporary ergonomics, Taylor & Francis, London, 1993, pp. 240-245.
25. Meister, D. Conceptual aspects of human factors. John Hopkins, Baltimore, 1989.
26. Parasuraman, R., Mouloua, M., Molloy, R., et al. Adaptive allocation reduces performance costs of static automation. In R.S. Jenner & D. Neumeister (Eds.). Proceedings of the Seventh International Symposium on Aviation Technology. Columbus, OH. 26-29th April, 1993, pp. 178-181.

27. Curtis, B. & Walz, D. The psychology of programming in the large: Team and organisational behaviour. In J-M. Hoc, T.R.G. Green, R. Samurcay, & D.J. Gilmore (Eds.), Psychology of programming. Academic Press, London, 1990.
28. Kerr, K.C. & Jordan, P.W. An investigation of the validity and usefulness of a "quick and dirty" usability evaluation. In S.A. Robertson (Ed.). Contemporary ergonomics, Taylor & Francis, London, 1995, pp. 128-133.
29. Dowell, J. & Long, J. Towards a conception for an engineering discipline of human factors. Ergonomics, 1989; 32: 1513-1535.
30. Hefley, W.E., Buie, E.A., Lynch, G.F., et al. Integrating human factors with software engineering practices. In Proceedings of the Human Factors and Ergonomics Society 38th Annual Meeting. HF&ES, Santa Monica, CA, 1994, pp. 315-319.
31. Smets, G.J.F., Stappers, P.J., Overbeeke, K.J., et al. Designing in virtual reality: Perception-action coupling and affordances. In K. Carr & R. England (Eds.), Simulated and Virtual Realities. Taylor & Francis, London, 1995, pp. 189-208.
32. Roth, E.M., Woods, D.D., & Pople, H.E. (1992) Cognitive simulation as a tool for cognitive task analysis. *Ergonomics*, 35, 10, 1163-1198.
33. Reason, J. Human error. Cambridge University Press, Cambridge, 1990.
34. Smith, S.M., Ward, T.B., & Finke, R.A. The creative cognition approach. MIT Press, Cambridge, MA, 1995.
35. Carroll, J.M. & Olson, J.R. Mental models in human-computer interaction. In M. Hellander (Ed.) Handbook of Human-computer interaction. North Holland, Amsterdam, 1988, pp. 45-65.

PERE: Evaluation and Improvement of Dependable Processes

Authors

Robin Bloomfield†, John Bowers‡, Luke Emmet†, Stephen Viller*

†Adelard,
London, UK
+44-(0)181-983-1708
{reb | loe}@adelard.co.uk

‡Department of Psychology
Manchester University, UK
+44-(0)161-275-2599
bowers@hera.psy.man.ac.uk

*Computing Department
Lancaster University, UK
+44-(0)1524-593793
viller@comp.lancs.ac.uk

Abstract

In the development of systems that have to be dependable, weaknesses in the requirements engineering (RE) process are highly undesirable. Such weaknesses may either introduce undetected system weaknesses, or otherwise significant costs may arise in their correction later in the development process. Typically, the RE process contains a number of individual and group activities and thus is particularly subject to weaknesses arising from human factors. Our work has concerned the development of PERE (Process Evaluation in Requirements Engineering), which is a structured method for analysing processes for weaknesses and proposing process improvements against them. PERE combines two complementary viewpoints within its process evaluation approach. Firstly, a classical engineering analysis is used for process modelling and generic process weakness identification. This initial analysis is fed into the second analysis phase, in which those process components that are primarily composed of human activity, their interconnections and organisational context are subject to a systematic human factors analysis. In this paper we briefly describe PERE and provide examples of the application experience to date.

1 Introduction

Requirements engineering (RE) is the process within the earlier phases of the system lifecycle that concerns the discovery, analysis, negotiation and definition of system requirements, resulting in a specification of what the system must do in order to satisfy user needs, integrate with other installed systems, satisfy commercial demands, meet safety regulations and so on. The importance of the RE process is generally recognised and it is acknowledged that problems originating in the RE process are hard to detect and expensive to put right later on in the system's development. Furthermore, in the context of dependable systems, getting the requirements wrong may have disastrous consequences.

Within the REAIMS[1] project, we have been developing a number of improvement strategies to address problems in RE, particularly focusing on the development of dependable systems. In this paper we report on one aspect of the

---

[1] Requirements Engineering Adaptation and IMprovement for Safety and dependability

REAIMS work that has considered the safety and reliability of the RE process itself. PERE (Process Evaluation in Requirements Engineering) is a method for assessing requirements processes, examining them for weaknesses and proposing protections against those weaknesses. Although PERE has been specifically developed for the evaluation of requirements processes, the analysis and process improvement techniques employed are applicable to problems within the broader process improvement domain. Process evaluation and improvement may be necessary in any domain where the process is required to be dependable.

In this paper we briefly present an overview of the PERE method and its background, and give examples of its application. Further details on PERE can be found in [1].

1.1 Dependable systems

Dependable systems are conventionally those in which failure of one or more RAMSS (Reliability, Availability, Maintainability, Security and Safety) attributes would have critical consequences. Within Safety Engineering, numerous techniques have evolved to aid engineers in safety analysis and risk reduction for safety critical systems. Such techniques include Fault Tree Analysis, Event Tree Analysis, Failure Modes and Effect Analysis, and Hazops [2]. Recent effort has considered the application of such techniques to computer-based systems [3, 4], although there is little work in the application of such techniques to the development process itself. Furthermore, existing quality improvement frameworks such as SEI CMM [5], Bootstrap [6], and those associated with ISO 9000 [7], do not focus on the particular problems of the *development* of dependable systems.

1.2 Process weaknesses due to human factors

Classical hazard and safety analyses of processes focus on the mechanistic aspects of that process. However, it is increasingly recognised that RE processes need to be understood in social as well as technical terms [8, 9, 10, 11], and thus may be subject to process weaknesses arising from human factors. A comprehensive process analysis must therefore give consideration to the human factors of the process by taking a more human centred view of the process than is traditional.

The human factors literature is diverse and impacts on RE activities from multiple angles. Due to current space restrictions we shall only touch on the wealth of relevant literature; more comprehensive reviews can be found in [1, 12, 13].

1.2.1 Errors and violations in individual work

A large amount of research into "human error" [13] has emerged from cognitive approaches to the understanding and modelling of individual failures. This work has generated important distinctions such as that between skill-based, rule-based and knowledge-based "levels" of cognitive activity [14]. Skill-based errors—otherwise known as *slips* and *lapses*—occur in the execution of routine skilled work, typically characterised by "strong but wrong" error behaviour [13]. Rule-based behaviour, where previously generated "if...then" rules are applied, can be subject to rule-based mistakes if those rules are "bad" or "misapplied". Knowledge-based mistakes arise when an individual is working in a novel situation and is not able to reuse a prepackaged solution, or previously generated rule. *Hindsight biases, confirmation biases* and *availability biases* are examples of how solutions to current problems can be distorted through the misapplication of prepackaged solutions.

Furthermore, hazards may also be caused by procedural *violations* if the procedures are overly prescriptive, poorly defined or do not support the processes actually followed.

1.2.2 Group process losses

Social psychological research has concentrated on the effects of working in social and group settings. This wide and diverse body of research includes work on phenomena such as *social facilitation* [15], *group performance* [16], *group leadership* [17], *conformity and consensus* [18], the effects of *minority opinion* [19], *group polarisation* [20] and *groupthink* [21].

Such phenomena have been studied in a vast variety of settings, from political decision making to industrial shopfloor work, in both naturalistic and laboratory-based studies. While RE processes have not been an explicit topic of extensive study for social psychologists, these group weaknesses associated with how teams perform together and coordinate their work can confidently be assumed to be potential sources of process losses for socio-technical development processes such as RE.

1.2.3 Organisational context

It is increasingly recognised that the organisational context and safety-culture surrounding a process is a further determinant of the safety of that process. For example, *latent organisational failures* [13] may lie dormant until some active trigger event coupled with insufficient defences precipitates an accident. Debates between the *Normal Accident* [22, 23] and *High Reliability* [24] theorists, on the scope for optimism concerning safety for organisations operating in hazardous conditions, have resulted in recommendations aimed at improving organisational reliability. These include strong "safety culture", high levels of technological and personnel redundancy, decentralised authority, organisational learning, and so on.

2 Method description

PERE is an integrated process improvement method that combines two complementary viewpoints onto the process under analysis:

1. *Mechanistic viewpoint*—an analysis of the process in mechanistic terms, as a number of interconnected process components. This analysis uses techniques adopted from classical safety analysis, adapted for a consideration of the RE process.
2. *Human factors viewpoint*—an analysis based on the application of human factors and social scientific principles to assess weaknesses and protections at an individual, group and organisational level using the results of the mechanistic viewpoint to scope the analysis.

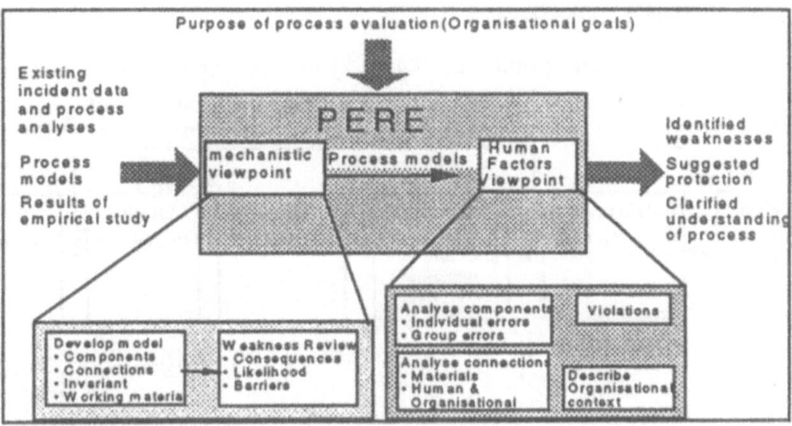

Figure 1: Overview of PERE[2]

An overview of the method can be seen in Figure 1. This dual viewpoint approach has been adopted since it has the following advantages:

- *Structured, usable approach*—PERE enables human factors considerations to be presented in a usable manner, through the application of a structured grounded checklist. This checklist is *grounded* in that each item contains references to human factors review documents and *structured* since the user is guided through the checklist by means of navigational questions. This navigation is guided and scoped by the results of the mechanistic viewpoint analysis. As a result, a manageable subset of the checklist is used, preventing the combinatorial explosion of having to consider each checklist item for each component.
- *Sensitive to actual RE process improvement needs*—since RE processes in practice combine human and automated processes, it is appropriate to combine two complementary viewpoints within the method, each concentrating on different aspects of the process. PERE exists within the process improvement paradigm and combines both "hard" and "soft" process improvement approaches.
- *Knowledge dissemination*—PERE integrates classical engineering analysis and human factors analysis. This structured, usable, yet technically defensible approach means that engineers in the process and safety domains will have access to the relevant social scientific research and broader human aspects that determine process dependability and which would not typically be within their domain.
- *Enhanced coverage*—since each viewpoint comes from a different research tradition, there is a certain amount of redundancy in the PERE process, resulting in increased coverage of the process under analysis as process weaknesses are trapped under different guises. This redundancy further improves the dependability of the PERE process itself.

---

[2] Other REAIMS modules such as PREview-PV and MERE can be used to supply process data or complement any existing process documentation.

2.1 Mechanistic viewpoint

PERE's mechanistic viewpoint (see Figure 2) has its origins in the classical safety analysis technique, Hazops [2], and Object-Oriented inspired analysis.

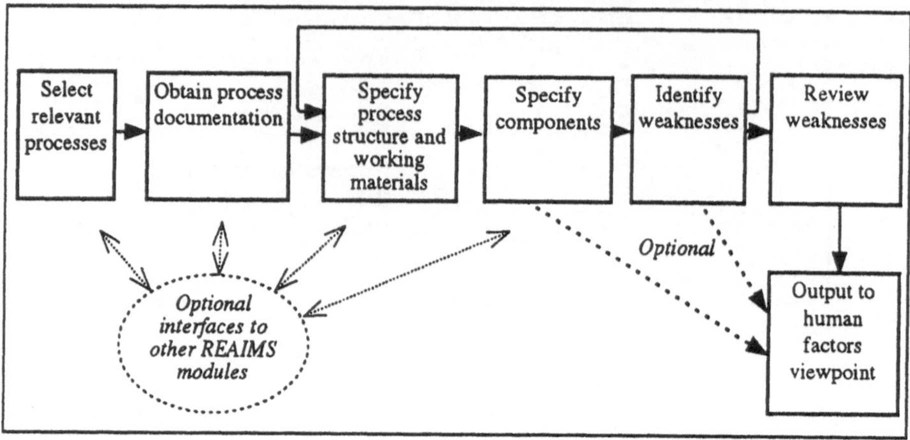

Figure 2: PERE's mechanistic viewpoint

For this viewpoint it is assumed that both human and machine activity in the process are analysable into components. The model we describe is based on the principles of using modularity and abstraction to describe systems, considering generic component classes (process, transduce, channel, store and control) as subject to generic component weaknesses, and explicitly considering the "working material".

Once the process structure and working material is described, the PERE analyst completes a PERE component table (a row of which is shown in Figure 3) to describe the process model. This process model is then reviewed for weaknesses by considering the generic weaknesses associated with each component and also the specific weaknesses associated with the components attributes.

Component name	Component Class	Interfaces and working material	(Optional) State	Invariant	(Optional) preconditions and resources	(Optional) external control

Figure 3: PERE Component Table (PCT)

In documenting this analysis a PERE Weakness Table (see Figure 4) is completed. The weaknesses identification and review steps are iterated until no more weaknesses are identified. The results of the mechanistic analysis are then passed on to the human factors viewpoint, although provisional results may be fed forward if, say, one component is considered to be particularly vulnerable to human error.

Actual Weakness	Weakness Class	Likelihood	Consequence	Possible protections	Possible secondary weaknesses

Figure 4: PERE Weakness Table (PWT)

2.2 Human factors viewpoint

For the human factors viewpoint (see Figure 5) the top level analysis shares the perception of the mechanistic viewpoint of processes in terms of interconnected components. As a result, the human factors viewpoint builds on the mechanistic analysis.

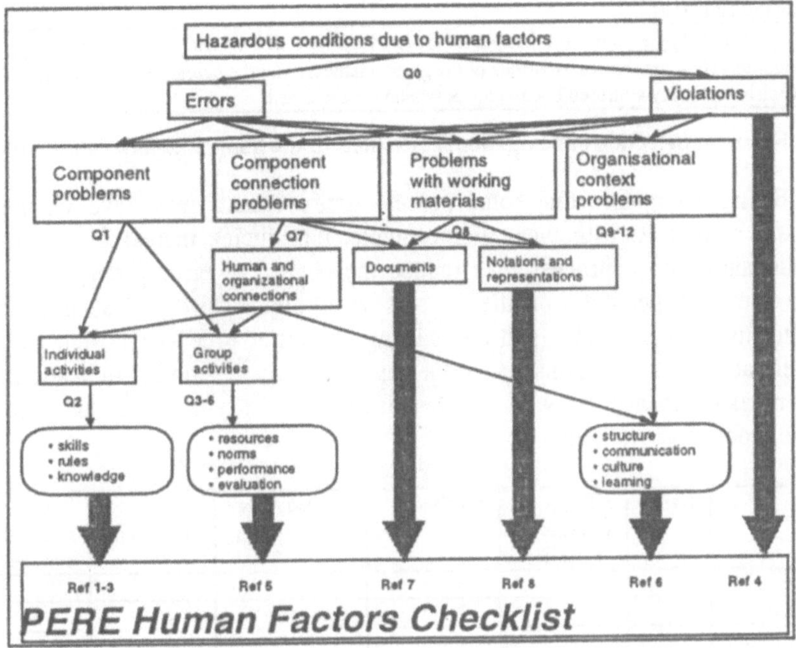

Figure 5: PERE's human factors viewpoint

In this phase we consider those components that are composed primarily of human activity, their interconnections and working material, and organisational context. The analysis proceeds by means of a series of structured questions (see Figure 6), which enables the analyst to search for only those human factor weaknesses that are relevant for the particular process under consideration (e.g. it is not generally necessary to consider knowledge-based component weaknesses for a skill-based component such as typing).

The application of the human factors viewpoint concludes with a completed PERE human factors table (see Figure 7), which includes suggested protections against the identified weaknesses. Of course whether they should be actually implemented for a particular application depends on factors such as the *reason* for investigation, an assessment of the *risk* associated with the weakness, and considerations of *prioritisation* and financial *cost* of the protections.

Q0 Is it suspected that this process component, its connections or its working materials can be vulnerable to error and/or violation?
Q1 Is the component principally characterised by individual or group activity?
Q2 Is the component principally characterised by skill-based, rule-based or knowledge-based activity?
Q3 (resources) What are the available human resources for the group to fulfil its function?
Q4 (norms) How is the function of the group presented to group members and what are the norms (specifications of what the group and its members should do) that govern the activity of the group in executing this function?
Q5 (performance) How are the contributions of group members produced and coordinated?
Q6 (evaluation) How are the contributions of the group members and the overall products of the group (decisions, jointly authored documents or whatever) evaluated?

Figure 6: Selection of navigational questions in the human factors viewpoint

PERE also recognises that some process improvements may have *secondary weaknesses* associated with them. For example, introducing increased monitoring and redundancy into a process may reduce the chance of error propagation, but decrease the manageability of the process at a higher level. If such process improvements are more risky than the existing weakness, process redesign may be more preferable than evolutionary incremental improvement. Another indication that process redesign is necessary is if the existing process encourages or requires extensive procedural violation.

	Name	HF weakness analysis	Likelihood	Consequence	Possible protections	Possible secondary weaknesses
Component problems						
Interconnections and working material problems						
Organisational context problems						

Figure 7: PERE human factors table (PHT)

3 Applications and future work

PERE has been applied in different contexts by different users (see also [25]). These have included applications to RE processes in the aerospace and railway industries. In this section we give an overview of how PERE has been applied to the development of a system for enhancing corporate memory (MERE), to a software engineering process, and to a typical standards making process.

3.1 Corporate memory process

MERE (Managing Experience in Requirements Engineering) is another module in the REAIMS family that concerns the capture and reuse of experience within an organisation, such that previous incidents and good practice—often dubbed "corporate memory"—can generate requirements for other similar projects and products. The MERE process defines the lifecycle of these generated requirements from the collection of incident data, through elaboration and validation, to application and verification for a new product.

The application comprised an initial site visit and process observation from a human factors viewpoint, followed by the development of a process model and a full application of PERE. The PERE analysis aided the ongoing development of MERE by providing some insights and suggestions for how the MERE process might be improved through simplification and redesign.

3.2 Software engineering process

PERE can also be applied in a single pass "fast-track" approach. One such application on a REAIMS partner's formal specification process was conducted over two days and involved a number of meetings between PERE analysts and process stakeholders. The analysis was primarily aimed at evaluating a process guide that had been written, by means of constructing and analysing a process model built from stakeholder interviews and the process guide. Although it would be normal for a PERE analysis to involve more in-depth process capture and analysis, there was some payback even for this "fast track" application, in terms of different types of weaknesses identified in the process as described by the process guide.

3.3 A standards making process

A standards making process can be considered to be an RE process in which the user requirements of the industrial standardisation participants are captured, negotiated and developed into a standards document. Typically, the process is primarily constructed from complex human centred activities, such as document production, group meetings and document review activities.

PERE was applied to the standards process [26] in order to gain increased understanding and identify possible options for process improvement. This increased understanding is needed, since the current standards process may be long and protracted (possibly up to 10 years per document), and standards making as a result may seriously lag behind technological and market developments within the industrial sector that the standard was designed to support.

The initial standards process model was built from official and supporting standards documentation, resulting in a standards model that was considered to be typical of many of the processes by which standards emerge in national and international standards organisations. This initial modelling was augmented with field work, which involved interviewing and observing standards makers. The aim of this field work was to improve the process model that had been constructed and to elicit information on what standards makers considered to be the *actual* problems they faced. The results of this application have included:

- *increased process understanding*—in the form of process models and identified weaknesses of the actual, rather than idealised, standards process. For example, it was seen that although the *document production* aspects of standards production are typically well supported by the current process, other more human centred aspects, such as *consensus building*, may only be implicitly supported.

- *process improvement suggestions*—in terms of possible process protections and process redesign options to safeguard against the identified weaknesses;
- *preliminary validation of PERE*—PERE was able to pick up many of the actual weaknesses identified by standards makers.

3.4 Future work

PERE is being exploited by REAIMS partners in the course of their everyday work, and is under continuous development. We are investigating tool support for PERE, and a "shareware" version of PERE is available on the world wide web[3].

4 Acknowledgements

The authors would like to thank project members and reviewers for helpful comments throughout the development of PERE. This research has been conducted as part of the CEC ESPRIT project number 8649, REAIMS. The project partners are GEC Alsthom, Adelard, Aerospatiale, Lancaster University, Manchester University, RWTÜV, Apsys, Digilog. The order of authors is alphabetical.

5 References

1. Viller, S., Emmet, L., Bowers, J., Bloomfield, R. (1996) PERE: A Method for Requirements Process Dependability, Submitted to IEEE Symposium on Requirements Engineering-RE'97, 5-8th January 1997, Annapolis MD (also available as technical report CSEG/4/96, from Cooperative Systems Engineering Group, Computing Department, Lancaster University).
2. Kletz, T.A., Hazop and Hazan—Identifying and Assessing Process Industry Hazards. Institution of Chemical Engineers, Rugby, UK, 1992.
3. Kletz, T., Chung, P., Broomfield, E. and Shen-Orr, C., Computer Control and Human Error. Institution of Chemical Engineers, Rugby, 1995.
4. McDermid, J.A. and Pumfrey, D.J., A development of hazard analysis to aid software design. In: Proceedings of COMPASS'94 (Gaithersburg, MD, 1994) IEEE Computer Society Press.
5. Paulk, M.C., Curtis, B., Chrissis, M.B. and Weber, C.V., Capability maturity model, version 1.1. IEEE Software 10, 4 (1993) pp. 18-27.
6. Bootstrap Project Team, Bootstrap: Europe's assessment method. IEEE Software 10. 2 (1993) pp. 93-95.
7. Huyink, D. and Westover, C., ISO 9000. Irwin Professional Publishing, New York, 1994.
8. Bowers, J. and Pycock, J., Talking through design: requirements and resistance in cooperative prototyping. In: Proceedings of CHI'94 (Boston, MA, 1994) ACM Press, pp. 299-305.
9. Goguen, J.A., Social issues in requirements engineering. In: Proceedings of RE'93 (San Diego, CA, 1993) IEEE, pp. 194-195.
10. Quintas, P., Ed., Social Dimensions of System Engineering: People, Processes, Policies and Software Development. Ellis Horwood, London, 1993.
11. Westrum, R., Technologies and Society: The Shaping of People and Things. Wadsworth Publishing Company, Belmont, CA, 1991.

---

[3] http://www.comp.lancs.ac.uk/computing/research/cseg/projects/reaims/

12. Bowers, J., Viller, S. and Rodden, T., Human Factors in Requirements Engineering, REAIMS Deliverable D1.2, REAIMS/WP1.2/LU004, Lancaster University, 29th August 1994.
13. Reason, J., Human Error. Cambridge University Press, Cambridge, UK, 1990.
14. Rasmussen, J., Skills, rules, knowledge; signals, signs and symbols; and other distinctions in human performance models. IEEE Transactions on Systems, Man and Cybernetics SMC-13, 3
15. Manstead, A.S.R. and Semin, G.R., Social facilitation effects: mere enhancement of dominant responses? British Journal of Social and Clinical Psychology 19, (1980) pp. 119-136.
16. Steiner, I.D., Task-performing groups. In: Contemporary Topics in Social Psychology Thibaut, J.W., Spence, J.T. and Carson, R.C., Ed., General Learning Press, Morristown, NJ, 1976.
17. Hemphill, J.K., Why people attempt to lead. In: Leadership and Interpersonal Behaviour Petrullo, L. and Bass, B.M., Ed., Holt, Rinehart & Winston, New York, 1961.
18. Van Avermaet, E., Social influence in small groups. In: Introduction to Social Psychology Hewstone, M., Stroebe, W., Codol, J.-P. and Stephenson, G.M., Ed., Basil Blackwell, Oxford, 1988, pp. 350-380.
19. Maass, A. and Clark, R.D., Hidden impact of minorities—15 years of minority influence research. Psychological Bulletin 95, 3 (1984) pp. 428-450.
20. Isenberg, D.J., Group polarization: a critical review and meta-analysis. Journal of Personality and Social Psychology 50, (1986) pp. 1141-1151.
21. Janis, I.L., Victims of Groupthink. Houghton Mifflin, Boston, MA, 1972.
22. Perrow, C., Normal Accidents. Basic Books, New York, 1984.
23. Sagan, S.D., The Limits of Safety: Organizations, Accidents, and Nuclear Weapons. Princeton University Press, Princeton, NJ, 1993.
24. La Porte, T.R. and Consolini, P.M., Working in practice but not in theory: theoretical challenges of 'high reliability organizations'. Journal of Public Administration Research and Theory 1, 1 (1991) pp. 19-47.
25. Märtins, F., Schippers, H., Viller, S., Bowers, J. PERE: Process Evaluation in Requirements Engineering with special consideration of human factors, In: Design for Protecting the User, 13th Annual CSR Workshop, 11–13th September, Bürgenstock, Switzerland 1996
26. Emmet, L. and Bloomfield, R. Application of PERE to the standards process; REAIMS Deliverable D5.2a, REAIMS/WP5/AD/W/103: Adelard; 1996

Safety and Technology Transfer

Zdzisław Żurakowski
Institute of Power Systems Automation
Wroclaw, Poland

Abstract

The concept of the term 'Safety culture' based on the paper [4] is presented together with a discussion of the factors influencing the safety culture and its meaning for safety of potentially unsafe high advanced technologies. Majority of such technologies and threats involved refer to applications of programmable electronic systems. Trends of social changes due to technology transfer, which may have an unfavorable impact on safety culture are considered on examples of changes in Poland since beginning of 70-ties when government and management staff of companies aimed to base the development of national industry on foreign technology. Conclusions are drawn, useful when the range and consequences of technology transfer are considered.

1 Introduction

Difficulties in keeping pace with the developments in science and technology caused by different reasons, existing in most of the countries in the world, on the one hand, and the increase in the people's expectations in relation to their living standards on the other, have caused a heavy pressure for technology transfer.

On the grounds of existing experience, it is obvious that technology transfer is a very strong factor – or tool – influencing – consciously or not – almost all areas of life in a given country. This influence may be advantageous or not.

The advantageous effects include the following:
- assistance in filling the fundamental needs of a country's population,
- assistance in rising some industries to the higher level in a given country.

The disadvantageous effects – depending on the way transfer is made and on its scope – are likely to numerous. The effects connected with the influence of technology transfer on social development are known to some extent (for example [2]). Less known are however – as it seems – problems of the influence of technology transfer on Safety Culture which is currently considered as one of

---

This paper is partially done within the European Union Research Project COPERNICUS CP'94 No. 1594 *Integration of Safety Analysis Techniques for Process Control Systems* (ISAT). It is revised and improved version of paper presented in Polish at the Polish conference Real-Time Systems'95 [12].

fundamental factors in maintaining operation safety of potentially dangerous complex technologies. The number of such technologies is on the increase at present. It results from creation of new technologies on the one hand, and on the other from introduction of programmable systems with much more complex structure, compared to the traditional one, to the existing technologies. The assessment of programmable systems – mostly due to their software – creates hitherto unknown problems. At the same time, the assessment is required not only before putting a system into operation but also after each alteration made during operation.

At present, such potentially dangerous technologies include not only nuclear power engineering, but also for example:
- computer-controlled chemical processes,
- computer control systems for railway and air traffic,
- computer control systems for railway engines, cars etc.
- robotics,
- computer-controlled medical devices.

Potentially dangerous technologies include also, for example, application of computers to perform critical functions in regard to results any functional errors are likely to produce in social life, e.g. telecommunications, power systems control, booking systems.

From the existing experience it appears that implementation, functioning and certification of critical systems, in regard to their safety, decisively depend on integrity and qualifications of individuals, as well as on Safety Culture in the environment they work in.

Many publications, e.g. [6], put the emphasis on assessment of human dependability that is essential in all stages of design, improving and operation of systems. What is concerned is quality assurance in all stages of system existence, as well as taking into account all data concerning human factors at the assessment of system reliability. Recently it became obvious that without any assessment of a human factors, evaluation of reliability in programmable systems cannot be complete. Other issues stressed include the urgent need for adaptation of specialists who handle critical systems in regard to such questions of human factors as the above-mentioned influence of individual dependability. These issues have not yet been analysed in Poland. Such approach to system assessment has not been used as well.

This paper presents the concept of the therm "**Safety Culture**", the influence of technology transfer on social development from the point of view of widely considered safety, and concluding comments about prevention methods concerning disadvantageous results of technology transfer. Though part of discussion and conclusions has a general nature, the paper is based mainly on observation of processes that take place in Poland where since the beginning of the 1970s the very strong desire for basing industrial growth on technology transfer from abroad was the programme clearly expressed by the political leadership of the country, and later since the 1980s until now by the management of state-owned industrial companies. The paper deals only with technology transfer from abroad to a given country. The technology transfer inside a country, for example from universities to industry, is considered as a separate phenomenon, generally related to the normal development of the country, however – as is evident by experience – lack of partnership on an industry side to cooperate in such a transfer has also disadvantageous results.

2 Safety Culture

Such terms as quality, safety and security are without doubt the values conditioned by culture, and perhaps even more – they have defined meaning only within given culture. The term **"Safety Culture"** was introduced by INSAG* in its "Summary Report on the Post-Accident Review Meeting on the Chernobyl Accident", IAEA, Vienna (1986), in the publishing series dedicate to safety. In its subsequent report, "Basic Safety Principles for Nuclear Power Plants", published in 1988, the term was expanded and highlighted as a fundamental management principle. Since these reports were published, the term **Safety Culture** has been in use more and more often in documents concerned with safety of nuclear power plants. Currently this term is widely used in practice but the meaning remains open and there are no, commonly accepted, simple and convenient methods allowing to state the safety culture in practice. These problems are given particular attention in the subsequent report [4] in the same series. The first part sets out INSAG's views on the nature of Safety Culture. The purpose is to provide clarification and to develop a commonly shared understanding. The second part and the Appendix seek to give practical value to the concept, identifying characteristics that may be used to judge the effectiveness of Safety Culture in a particular case.

The report presents the following definition of **Safety Culture**: *it is that assembly of characteristics and attitudes in organization and individuals which establishes that, as an overriding priority, nuclear plant safety issues receive the attention warranted by their significance.*

The definition relates Safety Culture to personal attitudes and habits of thought and to the style of organizations. As a consequence, it is stated that Safety Culture has two general components. The first is the necessary framework within an organization and is the responsibility of the management hierarchy. The second is the attitude of staff at all levels in responding to and benefiting from the framework.

The attitudes of individuals are this component of Safety Culture that is especially difficult to verify. As a principle, such verification is impossible or extremely difficult in separation from organizations. The report's authors assume that these elusive attributes lead however inevitably to perceptible aspects which can serve as indicators of Safety Culture in different kinds of organization. The report presents major components and characteristic features of Safety Culture in

---

* INSAG – is an advisory group in IAEA, which serves mainly as a forum the exchange of information on nuclear safety issues of international significance and formulates, where possible, common safety concepts.

form of concise descriptions. These components and their connections are illustrated in Figure 1.

The Appendix comprises comprehensive sets of questions witch can be used to aid judgement of the effectiveness of Safety Culture in a particular case, to aid forming opinion about the effectiveness by organization outsiders, or by organization insiders in order to make a self-assessment.

The report gives special attention to the fact that good working practices and correct procedures, when they are an element of safety, are not sufficient, if they are utilized in a mechanical way. The requirements in this particular case exceed correct implementation of good working practices. It is required that all responsibilities essential for safety are exercised carefully, watchfully, with appropriate consideration, full knowledge, sound judgement and acceptable sense of responsibility.

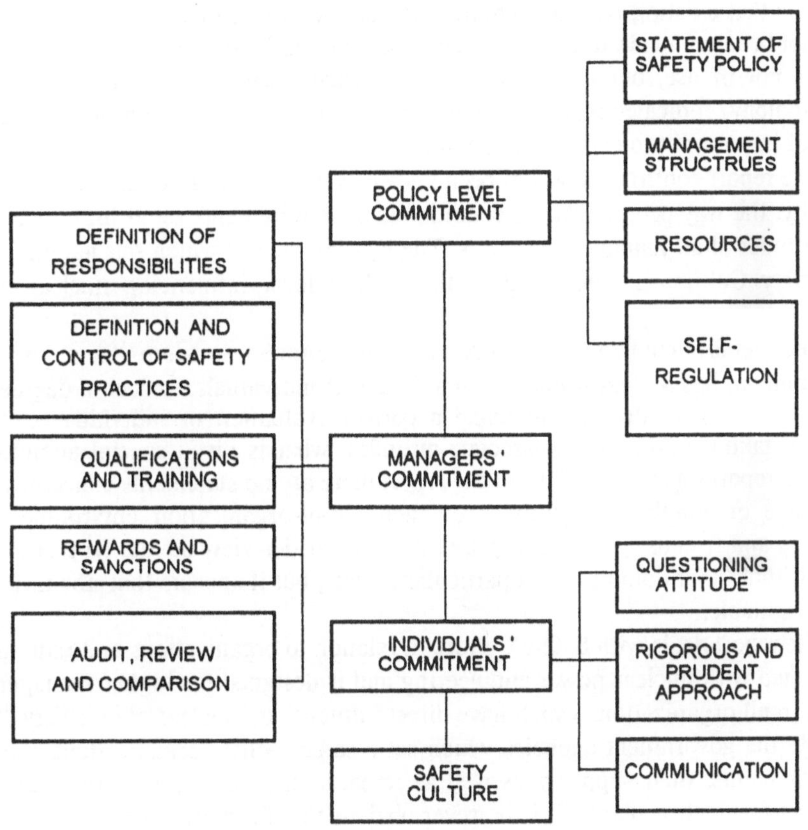

Figure 1. The major components of Safety Culture [4]

The most important component of Safety Culture is of course staff. As it was mentioned in the introduction, the existing experience in the area of programmable systems caused rising interest in human factors. In [6], similar as in other publications, it is stated as follows:

„*Nowadays the insight has grown that dependability of electronics depends on three factors:*

– Hardware.

– Software.

– Human factor.

All three of these have to be examined in order to arrive at a sound assessment of a system" (p. 86). „*Numerous publications discuss the matter and try to find solutions to the problem. The results are promising*" (p. 87).

To some extent, the human factor has been always taken into account in quality planning, even when it was not termed quality planning. But the above-mentioned approach seems to appear as a depersonification of the human factor (what hitherto was said about only in relation to blue-collar). Unfortunately, it appears to be obvious that existing realities require just that. As it was mentioned, such analysis has not been yet made in Poland, such approach – at least in open formulation – is rather not in use, but observation of the human factor in Poland (using above terminology) indicates that such approach must be seriously taken into account in case of important projects also in Poland.

The report confirms what has been known even from folklore: in each important activity, the way people work is conditioned by requirements set at high levels and by attitudes of all management levels. The legislative level which sets national basis of Safety Culture is considered as the highest level influencing nuclear plants safety.

The report confirms also one of the common wisdom facts: working environment has a great influence on attitudes of individuals, to such a degree that it is actually very difficult to make a correct evaluation of individual attitudes without taking into account associate attitudes, what is given special attention in IAEA's report on software [3]. In the report there are no statements concerning the influence on employers' attitudes of such out-of-organization environments as families and friends (peer groups), and also in broader view of the type of culture and attitudes that dominate in a particular society, but it appears that the influence is indisputable.

The report deals with Safety Culture in relation to organizations and individuals connected with nuclear power engineering and is designed for higher management levels in all organizations which have direct influence on nuclear plant safety. They include the government agencies which set a safety policy, organizations running nuclear plants, their suppliers as well as research and development organizations working for nuclear plants. It is however obvious that formation procedures are the same in all areas. It is confirmed in the report [3] dedicated to software and prepared by large international staff with collaboration of the most eminent experts. Its subsection on staff attitudes states that the attitudes required from engineers working on software for safety systems do not differ from the attitudes required from engineers working on other issues. It is also confirmed to some degree by the same report at the beginning of its Section 3 "Universal Features of Safety Culture".

The first impression a reader has been reading this report is that it contains ideas rather well-known and intuitively obvious, at least for someone who had something to do with industrial safety issues even concerning very old technologies. There is nothing surprising in it, nothing that would create hopes for quick solving of existing problems. Then however comes a thought that technologies change considerably as well as the word, and if once it was sufficient to create attitudes on the grounds of experience accumulated rather intuitively, then now new technologies – for example using electronic programmable systems – require on **higher management levels** such an approach that is based on a methodical analysis of reality.

3 Consequences of Technology Transfer for Safety

Depending on transfer range and many other factors (some of them will be presented further on), the consequences of technology transfer from abroad may be very serious and expanding outside the industrial branch which imports technology and outside industry in general. The fact that since the beginning of the industrial era, technicians, managers and technology know-how have been migrating across the borders of the whole word (especially of Europe), can create an impression that technology transfer is something normal that have been existing almost always. But the fundamental change in technology level and the role technology can play in social development force to look at technology transfer in a different way.

Examples of results that can be brought about by incorrect approach to technology transfer from abroad to a particular country, as experienced in Poland, are presented below.

1. Decrease in activity or even complete desistance from own efforts in implementing new technologies due to the fact that the activity of politicians and managers was switched to technology transfer, and not to the development based on own resources.
2. Impoverishment of world's diversity, being a great achievement of humankind, due to copying solutions and not to harmonize different ones allowing to enrich variety and maintain the . tendency towards unification, required by economic and operation considerations.
3. Degradation of considerable part of human resources, especially those with university education, due to:
   - halting professional development,
   - reduction in professional and living ambitions,
   - deepening misunderstanding of scientific and technologic development of world and of development of word in general,
   - reduction in feeling of self-worthiness,
   - increase in cosmopolitism,
   - halting development of personality.
4. Forming of individual attitudes characterized by decreased skill of risk calculating and willingness to take risk, reduced resistance to frustration which is inevitable at facing difficulties in implementation of new solutions etc. Such

attitudes can reach the highest country's management levels resulting in reduction in self-dependence and internal controllability of a particular society. Such a society becomes more and more susceptible to variable concepts and expecting outside help in taking decisions and in realizing them. Such phenomenon can be illustrated on the example of one of large Polish factories. First, its managers were trying very inefficiently for a dozen or so years to develop its own product. It took next several years to apply for help from Western countries in preparation of documentation, in supply of components and in financing prototype manufacturing. In the end, tenders were invited for the position of a strategic foreign investor who would buy 51% of stock because the management came to the conclusion, according to its privatization manager, that even if the company had obtained the documentation and money, it would have required an executive partner to take part in company management in order to successfully implement the new product. The company's name has been not revealed because the author does not want to make an impression that this example was given for any critical or intervention reasons, and because it is, in some degree, the typical case.

5. Reduction in attractiveness of technical university education and of careers in industry, science, technical design and management. As a consequence less and less talented and ambitious young people decide to study technical sciences what eventually deteriorates the work force quality in a very short time according to the principle of positive feed back.

6. Reduction in demand for journals and books of science and technology what as a consequence decreased the attractiveness of writing to these journals and books, loss of interest from authorities, and eventually financial problems and quality deterioration of such journals.

7. Disadvantageous changes in the language of a particular country and, in its vocabulary resources, due to deterioration or extinction of the process of creating and assimilating terms connected with the most important achievements in technology. It is a very interesting topic for psychologists, but even in the light of our rather superficial knowledge it is evident that semantically hazy and incomplete world image presented by such a language is not likely to not influence the world image taking shape in a particular society, its mental development etc. It is a well-known example of African languages which have failed to develop linguistic tools for mathematics and technical sciences as a historical process. Taking into account, the proportion of the phenomenon, it should be stated that some symptoms of such an incident in regard to the Polish language may be observed in informatics (computer science). It concerns also the terms used more commonly. For example, often used the Polish term *technologia*, taken from English publications as the translation of word "technology", in press publications is mostly used in the way that is not clear semantically, and rather differently from traditionally used in Polish technical educational establishments and factories.

The above-mentioned results of technology transfer from abroad concerns phenomena which with various intensity can be observed in Poland due to the above mentioned strong attitudes among government and industry leaders to transfer technology from abroad that have existed since the early 1970s. It applies then to the country whose education system, science and general experience were at rather high levels before this tendency came to being, in spite of technical surrealism in industry. Thus, it can be observed as policy of industry support on technology transfer resulted in deterioration of the above-mentioned areas, even if some modernization of many products based on foreign solutions appears to make an impression of industrial progress. These observations and their conclusions apply directly only to Poland, and to a considerable degree probably also to other East European ex-communist countries. However as review of Western publications suggests, similar negative effects related to improper transfer of technology occur either in highly developed countries, for example in USA, and in the Third World countries [2,7].

The examples mentioned do not constitute a complete list of results that can be produced by incorrect approach to technology transfer, since the purpose was not to make the list but to present directions of changes that take place in a particular society, and to demonstrate a type of attitudes, and in a broader view, a type of culture which takes shape under the influence of these results. Some of the more important reasons for unfavourable results of technology transfer in Poland within the last 25 years are given in the publication [10].

Already now, human intelligence, creative skills, ingenuity and, capability to undertake scientific research and to utilize them in order to fulfil one's own needs are considered as the most important resources of a particular country [1]. Taking it into account and analysing the above-mentioned directions of social changes connected with technology transfer in regard of safety, it should be stated that actually all the above-mentioned phenomena infringe the safety of a particular country, especially in the long run, by creating circumstances favourable for deterioration of the country in the world's community with all consequences of this fact. Even more serious doubts might be expressed, doubts that the proliferation of technology transfer practices from the outside as a foundation for economic growth in most countries of the world, might unfavourably influence the direction of civilization evolution on the Earth, the culture type, the human type currently taking shape and as a consequence it might have an effect on **civilization safety in general**, but these reflections go beyond the scope of this paper. It is possible only to quote a joke which was allegedly known in the West and was recalled not long ago in one of radio programmes, "the Poles are too intelligent for this part of Europe where they live". This wisecrack is rather painful for the Poles, because among others agents some of the above-mentioned results of the technology transfer policy in Poland within the last 25 years have caused a deterioration of intelligence in the Polish society and significant "normalization" of the situation in this respect.

There is no doubt however that also in the strictly technical respect of the safety of installations and technical devices, the above-mentioned directions of changes caused by technology transfer have a considerable effect on safety, among other things through degradation of Safety Culture understood according to Section 2, especially in the long run.

Considering, for example, the Safety Culture in the field of safety - related real-time systems, the following facts should be taken into account:

1. Software is designed – in any case in state-run civilian companies – without any use of software engineering methods, standards etc.

2. No standards or regulations have been issued in Poland regarding such systems.
3. There is no certification establishment for such systems – at least civilian one – or even no articulated needs in this respect. There is also no elementary knowledge about it among engineers, managers, designers, and civil servants up to the highest management levels and the state level.
4. Problems of real-time systems have even not been included to the report on Development Strategy for Informatics in Poland – Status, Prospects and Recommendations, issued by the 1st Congress of Polish Informatics, Warsaw, 1995. Members of the government, including the Prime Minister and his plenipotentiary for informatics, do not talk about it in their speeches. The government supports and deals with such domains as the development of computer networks or the application of computers in management etc. It is right indeed that these fields are managed but applications of computers do not end here.
5. There are no journals for above problems. A special section on real-time systems in the "Informatyka" journal has been published for a year. But the journal is in bad financial condition, so it turns itself to whatever it can, deteriorating the quality. The only one Polish journal on computer science and technology cannot get allowance in order to work calmly.
6. There are no valuable books on real-time systems development and assessment. Polish publishers have not yet published any books on fundamental problems of software engineering.
7. The existing application scope of safety-related real-time systems in Poland is rather impossible to determine, because such systems are not subject to registration in Poland.

In the existing situation, illustrated by the above-mentioned facts, it is difficult to analyse the Safety Culture at all. Fortunately – it should probably be considered so – the application scope of this type of systems in Poland is still rather small. But it will be rising and should be rising, because it is necessary if Poland want to keep in touch with the developed world.

4 Concluding Comments

1. Technology transfer should not be considered as a way of comfortable living, of avoiding troubles connected with efforts concerning staff training and promotion, scientific research, designing consecutive versions of a particular product, taking risk as well as with day-to-day doubts as to whether the correct decisions were made or a venture will be successful or not. As a side note, it should be emphasized that after the disillusioned optimism of the 1950s and '60s over possibilities of economical growth of the world [5], it is well-known now that **development is not a natural state** and it is not enough to give it a chance. **The natural state is stagnation** – "*Africa can break a man's heart. Economic development by no means can be treated as granted, it requires strenuous efforts*" [5]. The way to new answers leads everywhere through risk, difficulties, doubts, learning and often long approach to mature solutions. In Polish industrial circles such attitudes that could evidence the consciousness thereof hardly catch sight. What can be commonly

caught is handling papers and not technology, and rather light-hearted (unconcerned) vegetation.

2. It seems that it is possible to put a thesis that the Safety Culture of a particular country cannot surpass of certain, rather low level, without conducting and implementing own scientific research as well as without any independent testing and evaluating of existing solutions of high-advanced technology what makes it possible to gain experience in a particular field by wide circles of experts and other persons indirectly connected with the field, including members of highest governing bodies setting the safety policy.

3. It seems that is possible to accept that as it used to be said in the past that the transfer of particular technologies requires a particular technical culture, then now it is possible to say that in order to secure the safe utilization of state-of-the-art, potentially dangerous technologies, a particular level of Safety Culture is necessary. In case of European countries, including Poland, it was hitherto assumed that due to long-term experience in industrial growth, the level of technical culture implicitly taking into account Safety Culture in these countries is sufficient. In the present however, when on the one hand more and more potentially dangerous technologies are being introduced, and on the other hand the stagnation in the area of development of technologies of that type exists in such countries as Poland, it may be necessary to change this assumption. Perhaps it will be unavoidable in future to establish some kind of international commission of experts for assessment whether the level of Safety Culture in a particular country makes it possible to transfer a specific technology. In any case, the situation should be observed in this respect. Concerning above-mentioned potentially dangerous technologies it is worth to remark that for instance in complex programmable systems even difficulty in determination whether a failure was caused by an error in operation or by a fault in the system [9] may be a reason for a failure (it is known from experience that it is easier for an accident to happen in an unlighted place).

4. The conclusions as to what activities should be undertaken in Poland in order to keep the Safety Culture at the appropriate level arise from this paper clearly. It should be however added that both public opinion and politicians have currently more influence on the course of events than experts in Poland and perhaps not only in Poland. Taking into account this fact and a significant speed of technology development, it would be perhaps profitable to establish in Poland a journal dedicated to problems of science and technology, as well as social context and consequences of their development. The purpose would be not to rise the level of erudition, but to improve the understanding of the world and to supply the fundamental knowledge to wide circles of politicians, journalists, managers, businessmen, financiers etc., which need it in their daily professional activities. Almost every contact beyond little circles of experts, proves the fact that understanding of the world is very weak and based rather on novelty and fashion.

Acknowledgements

The author wishes to thank the International Atomic Energy Agency, which has kindly given permission for the use of copyright the figure from its publication.

References

1. Colombo U. The technological revolution and the future of the Third World. IEEE Technol Soc Mag 1991;10:25-32
2. Das A, Jedlicka A. Social impact analysis on technology transfer. J Technol Transf 1993;18:49-54
3. Software important to safety in nuclear power plants. International Atomic Energy Agency, Technical Report No. 367, Vienna, 1991
4. Safety culture. International Atomic Energy Agency, Vienna, 1994 (Safety series No. 75-INSAG-4)
5. Landes D. Develop on other light (in Polish), Ameryka, reprint from The New Republic, 1989
6. van der Meulen MJP, Stalhane T, Cole B. Programmable Electronic Systems Analysis Technique in Safety Critical Applications. In: Proceedings of the 12th International Conference on Computer Safety, Reliability and Security, SAFECOMP'93, Poznań-Kiekrz, Poland, 27-29 October 1993, pp 85-95
7. Petrella R. Science and technology in the interest of eight billion people: is it possible? Sci Technol Innov 1995; 8:21-28
8. Weizenbaum J. Human authority, responsibility and accountability in large-scale real-time systems. In: Real-Time Data Handling and Process Control, Proceedings of the First European Symposium held in Berlin (West), 23-25 October 1979, North-Holland Publishing Company, 1980, pp 623-628
9. Who is here pilot? (in Polish), Forum No. 22/1995, reprint from The Economist from 8.04.1995
10. Żurakowski Z. Is industry support in Poland on technology transfer a chance to Poland? (in Polish), Biuletyn Stowarzyszenia Polski Rynek Oprogramowania (Bulletin Polish Software Market), No. 7, Warszawa, 1994, pp 28-29
11. Żurakowski Z. Safety and technology transfer (a short setting of the problem). Typescript on Spring Meeting TC7 EWICS, Siena, April 1995
12. Żurakowski Z. Safety and technology transfer (in Polish), In: Proceedings of the II Conference on Real-Time Systems, Szklarska Poręba, Poland, September 20-23, 1995, pp 67-77

Invited Paper

SAFETY CASE FOR THE NERC AIR TRAFFIC CONTROL SYSTEM

Shoky Visram, NATS UK
Werner Artner, Frequentis Austria
Philip Marsden EDS UK

ABSTRACT

1 OBJECTIVE AND SUMMARY OF THE PAPER

1.1 OBJECTIVE

Our objective is to show, in a current large safety related system implementation (the New EnRoute Centre (NERC) system) how the customer established his safety assurance requirements and expectations, and how suppliers met these.

Part 1 explains how safety is taken into consideration in a large Air Traffic Control system, the customer expectations of his suppliers, and how he provides the Safety Case.

Part 2 shows how requirements were addressed both at system level and in software, and how this related to the customer expectations.

Part 3 discusses the changes in a small company to meet the customer requirements.

1.2 SUMMARY

The rapid expansion of air traffic in the UK air space demanded a significant upgrade to the Air Traffic Control capability. To meet this demand, a purpose built building and large safety-related computer system is being installed in the South of England. Both customer and suppliers must be able to show that this new system assures the safety of the flying public. The end product of the safety assurance programme is a comprehensive Safety Case, which is offered to the statutory regulation authority. Endorsement must be given before live operations start in late 1997.

Our paper shows the customer viewpoint in establishing the safety framework and then setting his requirements for the project, to enable him to build the Safety Case.

For a large supplier of systems, we show how a safety culture was developed and integrated with the rest of the consortium.

For a relatively small supplier of voice switching equipment (the most critical element of the whole system), we show the considerable culture change needed to enhance the development processes for safety, and the benefits this brings.

Part 1: Customer Safety Requirements for the NERC Air Traffic Control System

Shoky Visram, NATS
United Kingdom

1 The New En-Route Centre (NERC)

NERC is the replacement for the Civil Airspace Operations Room (CASOR) at the London Area and Terminal Control Centre (LATCC), the controlling authority for the London Flight Information Region (FIR) currently situated at West Drayton.

LATCC is one of the busiest units in the world. Over 1.2 million movements are recorded each year. Due to the rapid growth in the 1980s, studies conducted by NATS showed that if the growth continued at the same rate and no improvements were made then by the turn of the century the demand could be so great that the safety could be threatened, efficiency reduced and airline operating costs due to ATC restrictions could become intolerable. The NERC Project is a response to that demand. Its justification is increased air traffic capacity while maintaining or improving safety - in brief, providing a better service to its customers.

The NERC Operational Systems Project started with the ITT in 1991. After a nine month Project Definition Phase (PD), the main contract for the Implementation Phase was awarded to the IBM International Federal Systems Division (now the Lockheed Martin Air Traffic Management Division) as the Prime Contractor at the end of 1992. The Prime Contract led to several subcontracts involving:

IBM (U.K.)

Siemens Plessey Systems (U.K.)

EDS (U.K.)

Logica (U.K.)

Frequentis (Austria)

2 The General Concept of ATC

The object of Air Traffic Control is to plan provide and operate a **'safe, efficient and expeditious flow of air traffic'**

SAFE means that aircraft separation is not infringed. In many flying conditions, it is not possible to see much from the flight deck and so the risk of separation infringement is severe. Modern flight decks have collision avoidance and ground proximity warning systems but these are still the minority and at very high speed (closing speed for a couple of opposite direction 747s is around 1200 mph) a free for all' is a poor way to manage the situation. Commercial interest makes it dangerous. Everybody wants to be first in the queue.

EFFICIENT AND EXPEDITIOUS is a reference to the operation of the aircraft. The 'Operator' might be an individual pilot, a military service or an airline and is the primary customer of air traffic control services.

Whether commercial, military or private each flight within controlled area operates to a 'flight plan' which defines its flight profile, more or less precisely, by time, route and level. The interest of the operator is in following the flight plan profile without delay or deviation.

Given the volume of traffic not everyone can do what they want, when they want without conflict. **SAFE** and **EFFICIENT** have to be balanced. The aim of air traffic control is to exercise minimum interference with the operation of flights to ensure safety.

3 The Operational Concept

The Operational Concept is the means by which improved technology is converted into increased air traffic capacity.

NERC sectors form a three dimensional jigsaw of 32 high and low level pieces.

The airspace is divided into sectors to manage controller workload. Controller workload is a function of the number of aircraft under control at any instant and the complexity of their interaction. If you divide the airspace sensibly then you also divide the workload. It is a complex relationship but a NERC rule of thumb is that a controller will probably have no more than 20 aircraft to deal with at a time.

4 The Elements of ATC

There are three common elements in the provision of an ATC service based upon the concept of 'warning'.

(a) **Flight Planning.** The earliest warning is the flight plan. This is filed by the operator and indicates his intentions hours, days, weeks or even months ahead. This information is communicated to all ATC units which will, or at least may be affected.

(b) **Co-ordination.** Co-ordination is the process by which controllers prepare the way ahead for a particular flight. When the aircraft is ready to depart, the operator advises the first ATC unit and seeks 'permission to proceed'. The controller considers the flight in the context of obstructions, restrictions and traffic already within his 'sector' and forms a plan which will guarantee safe, efficient passage. Permission is then granted, conditionally or unconditionally, in the form of an 'ATC Clearance'. Subsequently each controller in the chain advises the next, by passing an estimate and seeking permission in the same way to provide safe end to end passage of the flight from departure to destination. Co-ordination typically precedes the passage of the aircraft by 15 to 20 minutes.

(c) **Handover.** Two or three minutes before sector penetration, following both of the previous steps, the aircraft will be transferred to the executive control of each sector in turn, normally by contacting the controller on the appropriate R/T frequency. Each sector controller puts the previously agreed plan into effect by providing more detailed ATC clearances over the R/T. They manoeuvre the flight if necessary to maintain separation while following the flight profile and ultimately to comply with the acceptance conditions set by the next sector before the aircraft is transferred onward to their executive control.

5 Safety Requirements

The NERC system as a whole provides an ATC service. Control of aircraft is entirely the responsibility of the controllers, who base their decisions and actions on the information presented to them by the NERC ATC computer systems. The purpose of the NERC ATC computer systems is essentially advisory, although the air-ground-air voice communications facility also provides the only link between the controllers and the aircraft under their control.

The air-ground-air voice communications system is, by design, segregated from the other ATC computer systems within NERC, including the other voice communications systems. It is isolated from all other ATC computer systems such that it is not affected by failures of these systems. If any or all of the other systems fail, then the controllers will still be able to communicate with the aircraft.

Hence, many, if not most, of the functions which NERC equipment provides are safety related to some degree. This is unlike systems which ensure safety by employing protective device. Moreover, the continuance of safe air traffic control is also dependent on the correct processing of data by the NERC system. If the computer systems present data which has been processed incorrectly or the controllers misinterpret what is being presented, then incorrect instructions may be given to aircraft. Again, this could erode separation. Consequently, it is a fact that the majority of NERC requirements, as expressed in the System/Segment Specification, must be met in order for the system to be safe.

The System/Segment Specification also specified Process Requirements which expressed the customer expectations in provision of the Safety Case by the Prime Contractor.

6 Safety Processes

The process the Prime Contractor were asked to follow was as follows:

(a) To build on the preliminary hazard analysis conducted by the customer using HAZOP.

(b) Identify the major functional hazards attributed to ATC Operational Systems.

(c) Classify them as per the NATS Severity Classification.

(d) Analyse the hazards by the use of Fault Tree Analysis (Top-down) and Failure Mode Effect Critical Analysis (Bottom-up) and provide a quantitative assurance that the hazards were within the appropriate probability targets for the classifications.

7 NATS Severity Classifications

The Severity Classifications used for classifying hazards are as follows:

Category 1

Continued air traffic control prevented for longer than an acceptable period of time. Fallback, reversionary mode or procedure unable to compensate.

Category 2

Continued air traffic control severely compromised until fallback, reversionary mode or procedure able to compensate. System capacity severely restricted and/or controller workload increased beyond acceptable limits.

Category 3

Continued air traffic control impaired. System capacity affected and/or controller workload increased beyond normal.

Category 4

Continued air traffic control unaffected. System capacity unaffected and controller workload not increased.

In addition, implicit in this severity classification scheme is the concept of "no effect", which falls below Category 4. This represents the situation where alternative modes or procedures are not required, and there are no hazard implications.

The scheme grades the effects of hazards within an ATC system by considering the exposure time, availability of fallback systems, and the effects on system capacity and operator workloads.

Having assigned a severity classification to each hazard, each Configuration Item is assigned a safety category according to the severity of the hazards resulting from its functional failure modes.

8 BEYOND THE SAFETY CASE

Having identified the hazards and carried out the hazard analysis to ensure that they fall within the target error probabilities appropriate to the safety classification, the question is, is that the end of safety analysis?

The stringent development method alone does not totally remove all errors. A number of latent errors remain in the software. For a large project like NERC, where there are two million lines of code, of which one million are new, the time and cost pressures are such that to attempt removal of all latent errors would be impractical. With the stringent development used on NERC, it is predicted that at operational date, there will be an error per ksloc. This meets the safety requirements but the fact that 1,000 latent errors are predicted to remain in the system when it goes operational cannot be ignored. It would be extremely difficult to predict the severity of the remaining latent errors and hence a Fallback Strategy has been developed by NATS.

NERC is a fully redundant system and so a single failure would not impact the user. Fallback Modes are defined as those modes of operation where the user would have to provide service with reduced functionality and on NERC this would only happen due to multiple failures.

NATS in conjunction with Lockheed Martin have identified the Fallback Modes and documented the procedures and the associated MMI to continue providing Air Traffic Service. These have been walked through by the user to ensure that the remaining system can support them. Additional design has been implemented where necessary to support these Fallback Modes. The Fallback Modes will be tested thoroughly before the system goes operational and the users will be trained to work under the various Fallback Modes.

PART 2: BRINGING THE SAFETY DIMENSION INTO THE SYSTEM AND THE SOFTWARE DEVELOPMENT

Philip Marsden, EDS
United Kingdom

1 WHO ARE EDS?

EDS is a large IT company, with some 80,000 employees world-wide, which has recently become independent of General Motors. We can track the progression of the UK arm of the company through acquisitions and mergers to the today's leading role in both delivery of systems and major outsourcing agreements. Most relevant to this conference, we have a specialist high integrity systems group, and a healthy and very active Air Transport division from which this presentation comes.

2 EDS' ROLE IN NERC

The NERC system is provided by the prime contractor, Lockheed Martin Air Traffic Management (formerly Loral Federal Systems, and formerly part of IBM). EDS is one of several major subcontractors to Lockheed, providing four Computer Software Configuration Items (CSCIs) of which three are safety related. EDS is a member of the Project Safety Team, under the overall safety leadership of Lockheed Martin.

3 SAFETY FEATURES OF THE NERC SYSTEM DESIGN

The NERC system architecture is based on the following high level concepts:

Distributed Processing - The functionality of the system is distributed across multiple computers and uses a mix of centralised and distributed techniques to balance performance, availability, and data integrity. For example, the system dedicates extremely powerful workstation processing resources to each position to ensure system responsiveness. In addition, this permits critical functions to be allocated to the workstation, reducing the operational impact in the event of failures. In contrast, it uses centralised computing facilities for functions that are dependent on the coherence of critical data such as the Flight Data Subsystem.

Comprehensive Approach to Fault Tolerance - The system incorporates fault tolerance and availability management functions at the architectural level. These advanced functions manage the system to maintain the availability of the operational services rather than just the hardware. The availability management functions form a set of layered services that provide a well defined environment for higher level application functions. This approach simplifies the application software by encapsulating data and low level functions.

Segregation of Voice Communications Functions - The safety critical Voice Communications Subsystem is functionally independent of all other parts of the system.

LAN Based Interconnection - The basic connectivity between internal building blocks that compose the system is provided by a local area network (LAN). All LAN resources are at least duplicated for redundancy; some critical resources have as many as four levels of backup

Use of Open System Standards - The system architecture makes extensive use of open system standards to provide a safe and solid foundation for the NERC system, and for future growth and expansion. Use of standards and compliant products - such as communications services that conform to the ISO/OSI Model, a POSIX compliant operating system, the X-Windows standard, and the Ada and C programming languages - provides assurance in the system implementation, and contributes to the growth and expansion potential of this foundation. These standards are well defined and represent proven technology, thus reducing risk and providing assurance. They are also supported by a variety of manufacturers and products.

Integrated Control and Monitor Capability - The system provides an integrated facility for managing all the NERC systems resources.

4 Achieving Design Assurance

The design assurance process comprised a number of different activities, based primarily on the formal reviews and inspections described in MIL-STD 1521B.

System Requirements Review (SRR) - The SRR was conducted during the project definition phase to demonstrate the contractor's understanding of the requirements, to discuss the feasibility of the task within acceptable constraints and to determine initial direction and progress of the Systems Engineering effort.

Requirements Database - All project requirements are held on a relational database, which provides a comprehensive trace from customer requirements through A and B level requirements to test verification. The database also provides assurance of completeness and correctness of the requirements, with automated production of much of the documentation.

System Design Review (SDR) - The SDR was conducted when the project definition phase had proceeded to the point where system characteristics and the configuration items were identified. The SDR evaluated the optimisation, traceability, completeness and risks of the total system, including test requirements.

Software Specification Review (SSR) - The SSR was the final formal review held during the project definition phase. The SSR was a review of CSCI's functional, performance, interface and delivery requirements. It ensured that the requirements specifications formed a satisfactory basis for proceeding to the preliminary design.

Preliminary Design Review (PDR) - The PDR was conducted when the top level of the software design was available.

Critical Design Review (CDR) - As a result of the incremental build plan, there were a number of CDRs, each of which reviewed the detailed design of the associated build.

5 ACHIEVING SYSTEM ASSURANCE

5.1 FAULT TREE ANALYSIS AND FMECA

To demonstrate that the system will achieved the required level of residual risk, a series of high level hazards was agreed, each worded in terms of degradation of a system function supplied to an Air Traffic Controller. Each of these hazards was decomposed using Fault Tree Analysis (FTA). Base events were at the level of a hardware Line Replaceable Unit (LRU) or a whole CSCI (typically around 50,000 source lines of software code). Thus you can see that we have made some assumptions about the probability of software failure, both in terms of cessation of processing, and also in terms of undetected corruption. The residual risk is declared to the customer against the Severity Classifications described in Part 1 of this paper. The acceptability of each hazard is measured against limits set by the customer for each severity. The highest probability requirement in the customer's scheme is 10^{-7} events per hour.

Although the FTA was carried out by one company in the consortium, based on information supplied to the analysis team, responsibility for meeting the hazard probabilities lies with the Prime Contractor. The fault trees are reviewed by system engineers and development subcontractors.

5.2 ARM SUPPORT, AND ITS LINKAGE WITH SAFETY

Base event probabilities were provided by the ARM programme. For software, we assume that each CSCI has 1 latent error per 1000 source lines of code (ksloc) at operational date. This assumption is translated into Mean Time Between Failures (MTBF), again making assumptions about the rate of fixing faults and the number of instances of this item in the system. This method provides us with means of assessing the effect of the software on the top level hazards, as well as contributing to the reliability assessment.

During the development process, we collect software metrics, and the numbers of major errors found at each phase of the lifecycle (normalised to 1000 source lines of code) are compared with Rayleigh curves to predict the achieved number of latent faults in service. This process predicts latent fault rates of between around 0.5 and 2 errors per ksloc at the in service date.

The FTA is supported by a Failure Modes, Effects and Criticality Analysis (FMECA), with software FMEA conducted generally at the level immediately below the fault tree base events, providing, a qualitative analysis of CSCI behaviour. This has given the customer visibility of the software failure behaviour, and assurance that the system has the ability to detect, contain and recover from a wide range of faults and error conditions.

6 ACHIEVING SOFTWARE ASSURANCE

6.1 DEVELOPMENT PROCESSES

Given that the only way we know how to achieve the level of reliability and integrity appropriate for an application, the NERC software development was directed by a set of software development requirements. The top level requirements were originally set by NATS in 1992, owing much of their pedigree to the UK nuclear industry. Requirements are varied depending on the criticality of the software, and the resulting variation in development methods is similar to that in IEC1508 Part 3. Each CSCI is accorded a safety category, using the same set described in Part 1 of this paper, the category being defined by the most serious hazard which can be attributed to that CSCI. Thus, for example, the more critical (Category 1 and 2) CSCIs are required in Unit Test to demonstrate 100% coverage of all possible conditions (ie every condition in a decision takes both true and false states at least once), where the lower category (Category 3 and 4) units are only required to demonstrate 100% of the possible decision outcomes.

Within EDS, low level design was expressed with Real Time Structured Design, using Data Flow Diagrams and Structure Charts, as well as some State Transition Diagrams, supported by a Process Definition Language (PDL). All software units were subject to the full series of low level design inspections, code inspections (including static analysis) and test reviews, and many of these were supported by locally developed check-lists.

A variety of tools were used to automate some of these processes. EDS uses Software through Pictures (StP) for the design, flowing to Framemaker for document production, 'lint' and QA C for static analysis, Cantata for C code test coverage and Insight for dynamic analysis.

Local integration and test were carried out to both informal (first pass) and formal acceptance scripts, combining software units into Configuration Items (CSCIs).

System integration (leading to system acceptance) is conducted on a separate site and with a separate team to the software development, crafting their own independent test scripts. This independence has been taken as compliance with the requirement for Independent Verification and Validation. Parts of this phase of the testing is dedicated to specific safety testing, including tests of the safety features.

The customer is fully involved in many of the above processes, but, in order to assure himself formally that the software development and integration processes are working properly, he conducts a series of reviews at key project milestones.

A further "layer" of assurance is given by an Independent Safety Auditor (ISA), contracted to a company outside the main consortium. Independent audits are conducted according to part of UK Defence Standard 00-56.

6.2 COTS AND RE-USED SOFTWARE

A significant amount of the safety-related software in NERC is Commercial-off-the-Shelf (COTS) software and software re-used from existing applications. The advantage of using such software is that it is generally more mature than software specially developed for NERC: it will have been tested in operational use in a variety of applications, its performance and behaviour will be well documented and understood, and there will be already developed support. The disadvantage is that we cannot necessarily have access to development records and thereby assure ourselves of its integrity. For NERC, there a range of examples. For some, we claim the track record and large user base of the product (eg the operating system AIX); for others, we can influence and possibly enhance the development process (eg the availability management software).

In all cases, an analysis was conducted to identify which ones can directly affect the safety of the Air Traffic Control service, and data gathered to confirm and support its suitability and pedigree.

7 SAFETY IN THE EDS SOFTWARE DEVELOPMENT

NERC was the first major project in the EDS Transport division involving safety related software. Therefore, there was a need to establish clear procedures both at the local level, and to work with the Prime Contractor and the customer.

Our approach differed from other companies in the team, in that we decided that the most effective way of working was to vest the primary responsibility for safety at the detailed level with our Software Development Managers. Later we extended this to our Integration Managers. These managers carried out all design and code reviews for software units, and our approach has been to both educate them and to provide them and their teams with the necessary tools. By taking this approach, not only did we achieve excellent safety awareness in the development team, but we ensured that safety requirements were 'owned' by them, and not an external person.

Our Safety Consultant (the writer) established the procedures, encapsulating them in a local System Safety Plan and provided the necessary education, using support from our specialist High Integrity Systems Group. He also carried out his own spot checks and audits.

Fundamental to our approach, and our delivery of software safety assurance, was a very strong Quality Assurance role, working in close co-operation with the Safety Consultant. This paid enormous dividends, in that, for example, the customer could always find the safety assurance information he wanted during his reviews.

The Safety Consultant therefore set the policy and procedures at each stage, and acted as the single point of safety contact for the Prime Contractor and the rest of the consortium. He worked in the same rooms as the developers.

7.1 MMI Safety and Local Hazard Logging

Initially, during Project Definition, task analysis activities were carried out, resulting in an overall structure of the required Man Machine Interface (MMI) components (windows).

During implementation, MMI components were mocked up using the tool TeleUse. This mock-up, with its related behaviour, was reviewed extensively by the customer's ATC user staff and a Human Factors consultant, before coding was allowed to start. Some MMI components were designated as being safety related, and there was also a mechanism for recording MMI safety issues or hazards. These hazards then required formal closure either by a design change, or more usually by a recommended operator procedure or training.

As well as the consortium-wide hazard log, EDS also maintained its own local log. The mitigations for these hazards became testable safety features of the system.

8 Thoughts on the Future of Software Safety - A Personal View

Whilst we have for NERC made some contentions about the reliability of the software, the basis for the quantitative assessments is subject to some doubt, and the assumptions we have made are open to question. Nonetheless, the claims we make are fully visible, and they do at least provide some means of factoring software into a probabilistic assessment of top level hazards. Given that we have not attempted to do this at any granularity finer than a whole CSCI, and that, in general, there are several software base events in any one fault tree, the contentions we make are both reasonable and are presentable to a regulator. Our methods are fully visible to and accepted by our customer and his regulator.

The current internationally accepted way of achieving software safety integrity is by the application of "industry best practice" development methods, and these are articulated in such standards as IEC1508 and RTCA DO-178B. However, there is no conclusive proof that this assertion is correct. Indeed, there is some evidence showing that even Formal Methods may not deliver all that we might expect.

Given the doubts over the achievement and assessment of software reliability that we have in the world-wide software industry, who should have the final say in whether a system is safe enough? Each country puts in place a regulation framework within which service providers and suppliers are required by law to work. But, at the end of the day, will what we have been doing on the shop floor of software development prove to be enough to show that we have discharged ourselves sufficiently not to have to pay the damages?

PART 3: CHALLENGE AND EFFECT OF NERC ON AN AUSTRIAN COMPANY

Werner Artner, Frequentis GmbH
Vienna

1 THE SUBSYSTEM VCS

In March 1993 Frequentis was subcontracted to develop, deliver and integrate systems to provide NERC with operational voice communication facilities. The company has nearly 50 years experience in development and integration of systems for Ground/Ground and Air/Ground/Air Communication and other related equipment for the ATC environment.

The NERC Voice Communication System is based on our VCS3020 series, which was developed in 1989. The VCS3020 is a fully distributed micro-processor system with end-to-end digital voice switching. The NERC operational system consists of ~4000 printed circuit boards and ~4700 micro-processors and digital signal processors; approximately 90 % of the software is reused. The subsystem for Air/Ground/Air communication is defined as being in the most critical class according the NATS Safety Classification Schema described in part 1. This paper shows the cultural and formal changes in Frequentis caused by the involvement of the NERC project.

2 FREQUENTIS BEFORE NERC

Starting in 1991 the company implemented a quality management system in accordance to ISO9001 and was certified 1993. This process was mostly characterised by documenting already existing procedures. Because we already had a comprehensive QA System in place, we did not expect serious problems in getting certified according ISO9001. The audit raised no major observation and the company was certified without repetition of the audit.

With contract award for NERC we had to generate a mapping between the ISO9001 processes to the requirements of the several US Military Standards the project had to follow. On one hand, it was not difficult to establish a relationship between the existing ISO9001 procedures and that required for NERC. But on the other hand it turned out, that there were several gaps in Frequentis' Quality Management, specially for the phases and procedures of a Safety Programme (MIL-STD 882B).

3 FREQUENTIS DURING NERC

Apart from several functional system requirements identified as being safety related, NERC includes a set of explicit safety requirements on the development process to be followed. NATS' Safety Management decided to verify these requirements by means of conducting "Safety Reviews" on the development process of all subcontractors.

The first safety review in this series was held some months after starting the software development. It provided evidence that Frequentis has followed so far the required software development life cycle, however there was no explicit awareness of a safety programme (hazard identification, explicit description of mitigating features, sensible account to safety criticality). These concerns were embedded non-systematically in the traditional development phases. Due to this lack of consciousness of Safety also the assessment culture was missing to contribute to the necessary safety case. The customer (NATS) was seriously worried about this situation and expressed his concern, even suggesting that Frequentis might never meet the safety requirements for NERC.

The management philosophy of the company is based on small project teams with a very high motivation. So, the NERC development team took this challenge on board and tried to implement the necessary procedures in parallel with the daily work.

At this time Frequentis came to the conclusion, that ISO 9001 is mandatory to built up a comprehensive safety culture, but is insufficient to implement the procedures of a safety programme.

4 THE CONSOLIDATION AND IMPROVEMENT PHASE

4.1 STEP ONE: RESOLVING THE NERC SAFETY CONCERNS

Initially the contractual situation of the project helped to sort out the serious concern of the customer. The role of main contractor contributed to form explicit actions and to continue with the programme. This supported the Frequentis development team to overcome the first big gap of knowledge. The required activities to resolve the issues were well-balanced between feasibility and high challenge. Later, the close relationship to the project safety team including the safety analysis team, other experienced subcontractors and the customer supported the start of an improvement process within the Frequentis NERC development team.

For some time this process was supported by an external safety consultant. Also this measure contributed the cultural and formal change in the development team to some degree. However, the involvement of a consultant so late in the project (the development phase was near the end) was combined with the lack of knowledge about the NERC system and the VCS3020. It turned out, that the implementation of a comprehensive safety programme must be founded on a complete understanding of the system and specially the software. If such a consultant is involved from the very beginning of a project, this problem could be eliminated, but considering the amount of reuse and the complexity of the underlying system the effort to eliminate this lack of information might be bigger than training an existing system engineer in managing a safety programme.

So, as the project approached the end of the development phase, the software development manager was made more and more available to take over the responsibility of the safety programme. As this function in the project was mostly concerned with the safety issues anyway, the person was already trained in this area to some degree. Combined with the comprehensive system knowledge, it was the best approach to finally reverse the trend in the project, and meet the safety requirements.

4.2 Step Two: Management Commitment to Safety

At the end, the Voice Communication Subsystem was delivered in time and this project phase concluded within cost. After this successful delivery of the product the team approached the third safety review with the customer. With this milestone it got obvious, that this turn-over not only caused one of the best systems ever delivered, but also proved that Frequentis had successfully implemented the necessary procedures and produced the requested safety evidence. At the beginning of the project, Frequentis was considered the last one to meet the safety requirements; with the third safety review the customer expressed his surprise that Frequentis seems to be the first Subcontractor getting these requirements signed off.

It was obvious to the people involved in the project, that the implementation of the safety programme was a major fact for the outstanding success of this project. There were some other facts like improvements in the project management, but as they are outside the scope of this paper, they are not further described here.

Already before the delivery of the system it became obvious also to the rest of the company, specially all the other development teams, that the improvements the NERC team had made were very effective. This conviction was not driven by anybody, but the outcome led to a fundamental change in the whole company and to an immediate commitment of the top management to implement the improvement experienced with NERC in the whole company.

4.3 Step Three: The Improvements

It is not the central objective of the paper, which and how the single improvement actions were implemented. However a summarised list shows, how many of such single improvements were introduced in the company in a very short time:

(a) harmonisation of the design parts of the ISO9000 QMV's with MIL-STD 2167A, MIL-STD-1521, IEEE standards and V-Model (for system, hardware and software design)

(b) total revision of the software development manuals, including:

   (i) total new release of the handbook for design guidance
   (ii) creation of a Frequentis C-Coding Standard
   (iii) creation of a SW-Development Environment Handbook
   (iv) creation of a comprehensive SW Configuration Management Handbook
   (v) creation of a SW-Reuse Handbook

(c) Creation of a Document Configuration Management Handbook

(d) Development of a company-wide configuration item database

(e) Development of a standard for databases for tracking requirements

(f) Creation of Handbook for Independent Verification & Validation

(g) Implementation of company-wide database for tracking program trouble reports

(h) Creation of a Safety Management Handbook

(i) Implementation of a safety management department

(j) Implementation of a central configuration management department

(k) Implementation of an independent department for verification and validation

Later on several programmes have been started:

(a) Training for and implementation of safety management and safety analysis methods.

(b) Development of a SW-metrics standard.

(c) Training for and standardisation of ARM methods, specifically for SW.

(d) Development of standardised SW cost estimation.

It is not clear, which of these contributed most to the success described in the following chapter. However, one fact is beyond question: the whole improvement process was carried out first in the NERC development team including many lessons learned and revised processes. This project was like testing and optimising the whole improvement first in a classroom like environment.

This situation had two advantages:

(a) the improvements introduced into the whole company were already adjusted to the particular needs of the company;

(b) engineers outside the NERC project were convinced of value of the new procedures, because they could see them successfully applied to NERC.

5 IMPROVEMENT QUANTIFICATION

5.1 QUANTIFICATION BY QUALITY ASSESSMENT

Before this whole improvement process was put into action a BECO assessment was carried out on the whole company early in 1995. BECO is a method developed by the Austrian Company APAC (Austrian Product Assurance Company) and it is a combination of the Capability Maturity Model with ISO 9001. The results have shown, that the project team NERC was already 1 level higher than other project teams at this time.

After the introduction of the improvements into the whole company this assessment was repeated under the same conditions early 1996. And with this audit it was possible to quantify the success of the improvement, although the benefits are now only starting to be felt. The whole company improved its quality score from 1.7 to 2.5 on the 5 points scale. The management is convinced that the measures taken will bring the whole company a score of 3.0 by the end of this year.

5.2 QUANTIFICATION BY PROFIT INCREASE

Frequentis has experienced rapid growth over the last 5 years (turnover growth of 15 to 30 % per year). It is a well known fact that such a rapid growth is always accompanied by several problems. One such side effect is that it is very hard to meet the expectations on profit. This happened to Frequentis in the first years of this rapid growth; the achieved profit was marginal in these years.

Since the improvements were introduced in the NERC project and then in the whole company, profit will have nearly tripled each year. In this period also some other improvements were made such as a company wide resource planning/control system and professional training in project management. However, it is beyond question that introduction of this safety culture and the related improvements made in our quality procedures played a major role in our success.

Session 8

The Safety Case

Legal Aspects

Integrity Levels and their Application to Road Transport Systems

Peter H Jesty and Keith M Hobley
School of Computer Studies, University of Leeds
Leeds, LS2 9JT, UK

Abstract

This paper describes some of the modifications that are necessary to draft IEC 1508 in order to make it applicable to road transport systems. The assessment of the controllability of the safety of the situation after a failure provides a technique to identify the Safety Integrity Level in cases where the environment is continually changing. By defining a structured approach to gain confidence with increasing Safety Integrity Level, it has been possible to state the properties expected for both software and EMC in a consistent manner. This approach may also permit traffic engineers to "plug and play" with traffic control equipment.

1. Introduction

Although there is now an acceptance of the concept of safety integrity levels (SILs) as defined in draft IEC 1508 [1], there are still a number of outstanding questions as to how they might be used. The basic philosophy behind the use of SILs is that some failures are perceived as being more hazardous than others and thus, to some degree, as being less acceptable. Draft IEC 1508 has suggested some figures for the failure rates that might be considered to be acceptable for each SIL (see Table 1), though they have not yet been tested in the full light of a public debate. These figures should normally be taken to be the desirable minimum failure rates, and some industry sectors or companies may wish to aim for a better figure, especially when working to the lower SILs. For example if a probability of failure of 10^{-3} for SIL 2 is given to a system created in volumes of 10^{+6} per year, this implies that there are designed to be 1000 failures per year, which would almost certainly be considered commercially unacceptable. This situation can certainly occur in the automotive industry where millions of units are produced each year.

SIL	Demand Mode of Operation (Probability of failure to perform its design function on demand)	Continuous/High Demand Mode of Operation (Probability of dangerous failure/year)
4	$\geq 10^{-5}$ to $<10^{-4}$	$\geq 10^{-5}$ to $<10^{-4}$
3	$\geq 10^{-4}$ to $<10^{-3}$	$\geq 10^{-4}$ to $<10^{-3}$
2	$\geq 10^{-3}$ to $<10^{-2}$	$\geq 10^{-3}$ to $<10^{-2}$
1	$\geq 10^{-2}$ to $<10^{-1}$	$\geq 10^{-2}$ to $<10^{-1}$

Table 1 - Safety Integrity Levels and Target Failure Measures

Unfortunately a problem arises when Table 1 is applied to systems which might be subject to systematic faults, for example those associated with software and electromagnetic compatibility (EMC). Draft IEC 1508 recognises this problem but, as we will show below, its recommendations do not produce results of a nature such that they can be combined into logical arguments. This paper proposes an approach to the use of SILs based on a defined philosophy, which is an essential pre-requisite for their general use.

There are a four main issues associated with the use of SILs:

- Identification of the SIL required.
- The life-cycle implications of each SIL.
- The demonstration that a given SIL has been achieved.
- The integration of separate sub-systems to form a single system.

This paper will discuss these issues and show that a consistent philosophy can be defined, and make proposals for its use. Although the ideas expressed within this paper have emerged as a result of our experience with road transport and automotive systems, most of them will be applicable to all other industry sectors as well.

2. Definition of Safety Integrity Level

The basic definition of a SIL is in terms of the degree of risk reduction required. Draft IEC 1508 uses Figure 1 to describe this concept.

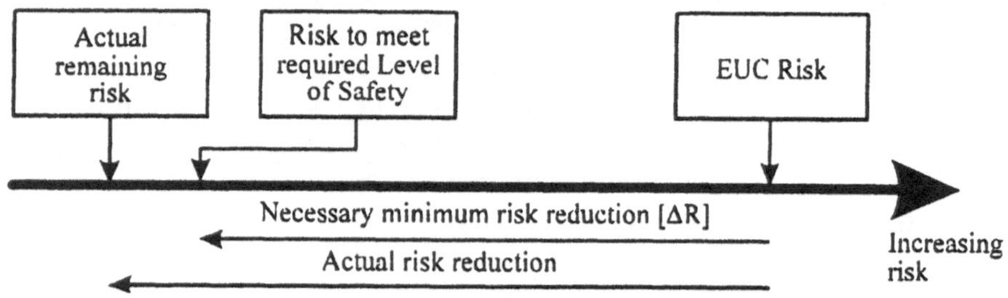

Figure 1 - Risk Reduction

Risk is usually defined by the relationship:

$$risk = probability\ of\ failure \times effect\ of\ failure$$

This definition shows that in order to reduce a risk either the *effect of failure* should be reduced by a suitable design, or the *probability of failure* should be reduced by suitable means. Since ΔR can usually only be used in a qualitative manner, an assessment has to be made of all the relevant factors, before the risk reduction required can be placed into one of four bands, or SILs. However, since it is possible to get different answers depending upon the basis used to categorise the level of risk, draft IEC 1508 recommends that each industry sector should produce its own interpretation.

3. Safety Integrity Levels for Road Transport Systems

Draft IEC 1508 suggests a number of different techniques for the identification of SILs. However in order to understand them it is almost essential to maintain a mental image of a protection system in, say, a chemical plant. We can thus identify the characteristics of the types of system for which these techniques are valid:

- protection system
- static, i.e. a known environment
- predictable failure → effect path

There are few road transport systems to which these characteristics apply. Most current and planned systems are either direct, or indirect, control systems. In addition vehicle based systems are obviously not static, and whilst road-side systems do not move, the road traffic environment adjacent to them varies continuously. The consequence of this is that when analysing a failure for its effect, one usually ends up saying "it depends on the situation at the time". Taking the worst case scenario is not much help either, because it is usually possible to dream up a situation where even what would normally be considered to be a 'mild' failure could produce the worst possible conceivable event. There are few statistics available on likely road traffic conditions and, anyway, the road user is often able to influence the final effect of the incident for both good or ill.

We therefore need a technique which is independent of the traffic conditions, and that can take into account the possible reactions of the road user to the failure. It also needs to be independent of the number of units deployed so that, say, a high volume vehicle manufacturer would use the same SIL for the same system in the same application, as a low volume manufacturer.

3.1. Controllability

A technique that satisfies all these conditions is known as *controllability* [2, 3]. This provides a qualitative assessment of the:

controllability of the safety of the situation after a failure

Note that no attempt is made to identify the final effect. The technique identifies the fact that in between a failure and a final event, there is a loss of control. The degree of loss of control is assessed by considering:

- The degree of control that the sub-system has on the safety of the system when it is working normally, and that therefore might be lost. This includes a factor for the case that other systems may be dependant on data being provided by the sub-system under consideration. (Some of the proposed road transport systems require communications between vehicles and/or road-side equipment.)

- The number, and type, of other sub-systems available to mitigate the loss of control caused by the failure.

- The speed with which it is necessary for a user to react with the back-up sub-system(s) in order to mitigate the loss of control.

Figure 2 shows how these factors can be combined to assess the controllability of any particular hazard [3]. (This figure is a generic road transport version of the one produced for road vehicles in [2].) Note that we have introduced a category for systems that are not safety-related in order to provide a complete continuum of loss of control.

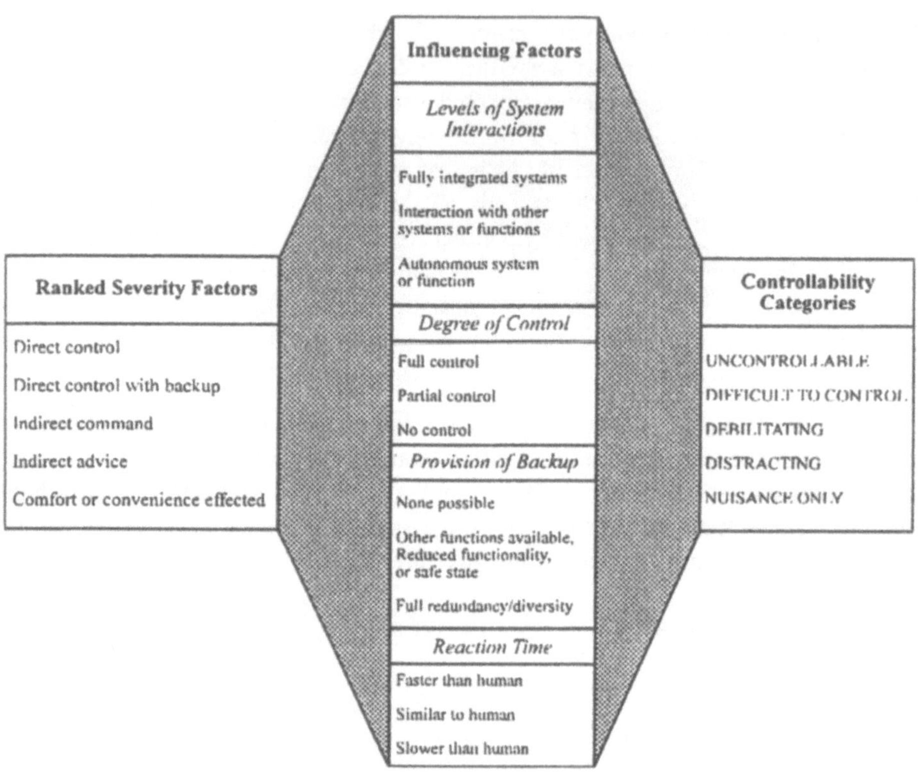

Figure 2 - Guide to assigning SILs

We are still learning how to use Figure 2 to produce repeatable results. The best technique identified so far is to allocate an alphabetic grade (A-E) to each of the factors described above. (An alphabetic grade is used to reduce the temptation simply to average the 'score' of each of the factors.) Using the "Ranked Severity Factors" as an initial estimate, the grades are then considered in conjunction with the full definition of the Controllability Levels (see Table 2), with particular attention being paid to any anomalous grade, especially a high peak. At some point during this process the grades are converted into a single SIL. The basic principle is to choose the lowest SIL necessary, rather than the highest SIL possible. This exercise should always be done in a meeting with a variety of experts present,

Controllability Categories	Definition	SIL
Uncontrollable	This relates to failures whose effects are not controllable by the road user(s), and which are most likely to lead to extremely severe outcomes. The outcome cannot be influenced by a human response.	4
Difficult to Control	This relates to failures whose effects are not normally controllable by the road user(s) but could, under favourable circumstances, be influenced by a mature human response. They are likely to lead to very severe outcomes.	3
Debilitating	This relates to failures whose effects are usually controllable by a sensible human response and, whilst there is a reduction in the safety margin, can usually be expected to lead to outcomes which are at worst severe.	2
Distracting	This relates to failures which produce operational limitations, but a normal human response will limit the outcome to no worse than minor.	1
Nuisance Only	This relates to failures where safety is not normally considered to be affected, and where customer satisfaction is the main consideration.	0

Table 2 - Definition of Controllability Categories

including those with a knowledge of human factors, and practical experience shows that a consensus can eventually be reached. It is usual to combine this exercise with that of hazard analysis [3].

4. Conformance Criteria

Once a SIL has been allocated to a system, or a sub-system, it is then necessary to choose a life-cycle such that the probability of failure of the final product is in accordance with Table 1. It is, however, generally accepted that it is not possible to demonstrate such low figures when the failure may be due to systematic faults. Whilst there has already been some considerable work done in this area for software faults [4], we can show by analogy that a similar argument should hold for EMC.

Although electromagnetic interference (EMI) will occur at random, the same EMI will produce identical effects if the system is in the same state each time it occurs; the same situation that holds for software when it responds to a given stimulus. In addition the normal complexity of electronic circuits, and of software, when combined with the very large number of possible stimuli, means that it will never be possible to fully test the EMC or the software of the system. Testing can therefore only be used to identify faults, not to prove correctness.

The various drafts of IEC 1508 have made suggestions as to how confidence can be gained in a safety-related or safety-critical programmable electronic system

(PES). Whilst the authors have no direct evidence of how the recommendations in draft IEC 1508 were obtained, we can infer the following process:

- List all the possible techniques that might be used in the development of a safety-related or safety-critical PES.
- Consider each technique separately, and decide whether it is to be Highly Recommended, Recommended, Disregarded or Not Recommended for use in development at each of the four SILs.
- Group the techniques into categories and decide which technique is essential and which are alternatives.

This approach is designed to give both the developer and independent assessor a warmer and warmer feeling about the product of the development process as the SIL increases. However there is nothing definite that one can say about a product that has been developed to a particular SIL, or of the differences between SILs. It is therefore not possible to begin to argue logically about combinations of products developed in accordance with draft IEC 1508. Problems will also arise in the future when new techniques emerge which are not specified in the Standard itself.

4.1. Levels of Confidence

The concept of a Level of Confidence is based on the fact that, for example, people travel in vehicles or fly in aeroplanes because that are confident that the equipment will behave in a safe manner during their journey. We therefore ask the question "what information does an Intelligent Transport System (ITS) developer or assessor need in order to gain sufficient confidence that it may be used safely by the general public?". When we combine the answer with the need to avoid unnecessary expense, the result is a set of qualitative requirements whose rigour increases with each SIL.

Confidence is built up by the developer of a product with two basic techniques:

- Quality - by working in accordance with a relevant Quality Plan a developer is assured that all phases of the life-cycle will be performed, and any data that needs to be passed back to an earlier phase (e.g. test results) will be acted upon.

- Knowledge of the System - most faults are introduced because of an inability to manage the actual, as opposed to the perceived, complexity of the system requirements. (Some of the proposed ITSs are likely to be the most complex systems yet considered with many and varied sub-systems being integrated together.) When it is not possible to mathematically prove the correctness of a system design (the usual condition) one should go as far down this path as is necessary; the higher the SIL the closer to a proof of correctness under all conditions one has to get. This requires greater and greater understanding of the system itself (see Figure 3).

SIL 0 Commercial considerations only
| Quality assurance procedures
| Structured approach
| Increasing justification
SIL 4 "Proof"

Figure 3 - Gaining Confidence

The practical effect of this approach is that, as the SIL increases, not only must additional processes be performed, but there must be increasing justification that the actions being undertaken within those processes are the correct ones for that particular system. The main features of this philosophy are as follows:

- Safety should not be added to a system as an afterthought, it must be designed in from the beginning.
- Testing can signal the presence of errors but not their absence, i.e. it cannot guarantee the safety of a system. Additional forms of verification and validation, performed under a Quality Assurance regime, e.g. ISO 9000, are therefore necessary to gain the confidence that the safety of the system has been developed correctly.
- Quality Assurance techniques are necessary to reduce the possibility of manufacturing faults compromising the safety of the system.

4.2. Software and EMC

The approach shown in Figure 3 has been applied to both software development and to the demonstration of EMC. The UK SafeIT project MISRA has made recommendations for the properties that should be demonstrated during the development of software for vehicle-based systems. An example is given in Table 3. Some of the recommendations for SIL 4 are ideal properties that may not be possible to achieve with current techniques. The knowledge of the system is gained through the creation of a formal specification and the confidence in the final product is achieved by a combination of rigorous testing and mathematical argument. The end result is not dissimilar from that proposed by draft IEC 1508, but the MISRA Guidelines leaves the choice of techniques to the developer.

A similar approach was taken for EMC in the CEC DG XIII Transport Telematics project EMCATT, and the results are summarised in Table 4. Since it is not possible to make an accurate prediction of the electromagnetic environment, as the SIL increases the developer should identify the EMI that the system will actually experience and them demonstrate that it will maintain EMC under all conditions. This is very different indeed from the way that EMC is currently assessed by a series of tests at various field strengths through a range of frequencies, as laid down in Standards [6]. Such tests are difficult to repeat exactly, being highly dependant on the layout of the test rig: the EMCATT project also demonstrated major differences in results between test facilities [7]. The need for EMC in accordance with IEC 1508 has been recognised by the formation of IEC

	Development Process	
SIL	*Specification and design*	*Verification and validation*
0	ISO 9001	
1	Structured method.	Show tests: are suitable; are acceptable; exercise safety features. Traceable correction.
2	Structured method supported by CASE tool.	Structured program review. Show no new faults after corrections.
3	Formal specification for those functions at this level.	Automated static analysis. Proof (argument) of safety properties. Justify test coverage.
4	Formal specification of complete system. Automated code generation (when available).	Proof (argument) of code against specification. Show object code reflects source code.

Table 3 - Extracts from MISRA Guidelines [5]

	Development Process	
SIL	*Specification and design*	*Validation*
0	"EMC Directives".	Commercial considerations only.
1	Design for: EMC; maintenance.	Show tests: are suitable; are acceptable. Traceable corrections.
2	As for 1.	As for 1.
3	Design for testing.	Justify: design; test conditions. Prove calibration of test equipment.
4	Design for actual susceptibility.	Proof (argument) of susceptibility. Justify test plan.

Table 4 - Extracts from EMCATT Proposals [7]

TC77 WG14. We therefore propose that the standard that emerges should be worded in terms of the properties given in Table 4 rather than in the traditional manner.

A study of Tables 3 and 4 will show that the requirements for SILs 1 and 2 can be obtained using 'traditional' engineering techniques. For SILs 3 and 4, however, a more formal approach should be taken combined with a justification of each process performed. This 'quantum leap' in the degree of effort required is sometimes recognised by calling systems at SILs 1 and 2 'safety-related' and those at SILs 3 and 4 'safety-critical'.

5. System Architecture Issues

There have been research programmes in Europe, USA and Japan on the use of telematic systems in road transport for about 10 years. It has therefore been possible to discover the types of ITS that are being planned to facilitate the flow and the management of traffic. There are two specific attributes that are common to many of the proposed systems.

1. There is a large amount of information that it is possible to collect about traffic and its environment, and there are many possible uses for this information (the Americans have currently identified 29 different services). Developers wish to take advantage of current communications technology to distribute the information between these services, and there is thus the potential for a medium/high degree of coupling between them.

2. Urban and inter-urban traffic management systems will be highly dynamic and variable throughout their lifetime. It is most unlikely that any such system will have a life-cycle that resembles well bounded products such as aircraft or vehicles. Instead the most likely scenario will be that a core system will be developed, often incorporating existing equipment, and then over the years it will be added to and modified both functionally and geographically.

Traffic engineers are therefore looking for a system architecture which will permit "plug and play" with traffic management equipment, and the normal approach to the modification of a safety-related system, i.e. that it should be treated as a new system, will not be practical. The costs will be seen to be too prohibitive for ITS systems, most of which are likely to have SILs of 1 or 2.

We have therefore been trying to identify the necessary attributes that a safety-related system should poses in order to permit "plug and play". It is clearly essential that the system architecture of an ITS should be structured to maintain as much independence as possible between the various building blocks, and we are working on this aspect in the current EU project CONVERGE-SA. One promising approach is to minimise the reliance on communications by safety-related functions. This already happens in the case of traffic light controllers which always maintain the basic safety of their own junction, even when receiving commands to change from a central traffic management system.

The approach to SILs described above will provide a certain degree of consistency between, say, two road-side controllers manufactured by different companies, but to the same SIL. This will be particularly important when there is a need to extend the ITS but the original controllers are no longer on the market.

The properties required by Tables 3 and 4 are those needed to avoid faults in the development process. The control of faults using diversity and redundancy is an architectural issue and it may be possible to agree an algebra of SILs such that, for example, two redundant SIL 1 items from different manufacturers could be used for a SIL 2 application. A similar argument could be made for SILs 3 and 4. However, as noted above, there is a major difference in the properties held by SIL 2 and SIL 3 systems, and thus it may not be sensible to combine SIL 1 and 2 systems in any manner to form SIL 3 and 4 systems.

Further work in this direction is necessary before an acceptable solution can be found. However we are being encouraged in this direction by the traffic engineers who want the ability to "plug and play". The ease with which it is possible to modify the function of an ITS is also proving to be extremely attractive, and we will have to find a solution to how to maintain a SIL in this situation as well, without making the cost prohibitive.

6. Conclusion

We have shown that it has been necessary to modify some of the details of draft IEC 1508 in order to make it applicable to road transport systems. The assessment of the controllability of the safety of the situation after a failure provides a technique to identify the SIL in cases where the environment is continually changing. By defining a structured approach to gain confidence with increasing SIL, it has been possible to state the properties expected for both software and EMC in a consistent manner. This approach may also permit traffic engineers to "plug an play" with traffic control equipment.

Acknowledgements

The work described in this paper was undertaken in the UK SafeIT project MISRA, the EU DRIVE I project DRIVE Safely, the EU DRIVE II projects PASSPORT and EMCATT, and the Framework IV Transport Telematics project CONVERGE-SA. The contents of this paper do not necessarily represent the views of the Commission.

References

1. Draft IEC 1508, Functional Safety - Safety-Related Systems, 1995.
2. MISRA, Development Guidelines for Vehicle Based Software, MIRA, Nuneaton, CV10 0TU, 1994, ISBN 0 9524156 0 7.
3. PASSPORT, Framework for Prospective System Safety Analysis, DRIVE II Project PASSPORT (V2058), 1995.
4. Brocklehurst S and Littlewood B, New Ways to Get Accurate Reliability Measures, IEEE Software, Vol. 9, N° 4, 1992.
5. MISRA, Integrity and Safety Engineering Practices, MISRA Report N° 2, MIRA, Nuneaton, CV10 0TU, 1994.
6. Hick G, Following Standards: The Importance of Keeping Up, IEE Review, Vol. 40, N°4, 1995.
7. EMCATT, Final Reports, DRIVE II Project EMCATT (V2064), 1995.

LEGAL SUFFICIENCY of TESTING PROCESSES

Clark Savage Turner, Esq.
Debra J. Richardson
John L. King
Department of Information and Computer Science, University of California, Irvine, CA., 92717 USA

Abstract

Software processes are executed for a purpose: to satisfy a set of process requirements and to meet process constraints [1]. This paper shows that there is a critical set of process constraints, often considered only implicitly or even ignored, that are derived externally from social expectations. This paper suggests an approach to determining this set of process constraints and a basic method for their consideration during safety-critical testing process design.

1 Introduction

Software is increasingly used to control systems that pose risks to the public. As dramatically demonstrated in the Therac-25 accidents [2], these risks are serious. When an innocent member of the public is injured by a new technology based on software, who bears the loss and how should software professionals deal with the issue?

Certainly, the behavior of the developer is under scrutiny. Did the development process include a reasonable set of precautions designed to avoid the predictable risks? If not, the developer might be expected to pay for the damage. If so, the developer is not at "fault" for the unfortunate accident, and the injured party may be left to suffer the damages uncompensated. This is the underlying philosophy of negligence law.

Ultimately, this discussion raises questions about externally imposed constraints on the technical requirements for safety-critical testing processes. This paper suggests that there is a set of testing process considerations that is important *beyond its ability to affect the product*, and is relevant to the long-term economic health of an organization. This set of constraints on the *process* of testing safety-critical software is imposed by the law of negligence. This paper also suggests an approach to determining a set of applicable constraints and integrating them into our safety-critical testing processes.

1.1 Why is this an important problem?

All major innovation requires investment. Investors are risk-sensitive; they avoid projects with a high risk of failure. Software engineering changes what's technically feasible while investors support the economically profitable. (See generally [3] for a thorough discussion of these issues.) Successful innovation therefore occurs at the intersection of the two worlds:

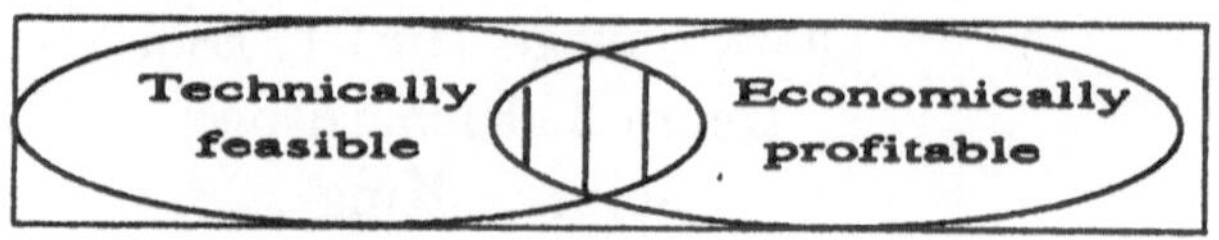

Social expectations affect the economically profitable through the law. In particular, negligence law defines a portion of the economic risk for innovative projects. The testing process may be designed to lower the risk of technical failure, but must also address the risk of economic failure in any organization with limited resources.

Detailed negligence constraints for some engineering processes are well known since they have had years to develop. Much general engineering experience and knowledge about accidents is built into these constraints. But software is new to negligence law. It not only changes the relative "size" of our projects, but their technological "reach". New social expectations and thus the applicable negligence constraints will develop in response to these changes. Our processes must be designed to meet these constraints, and to do that, we must have an explicit set to work with!

1.2 Technical constraints on the testing process

Software testing, like negligence law, deals with the inevitable risks of new software technologies. Since no one wants a software system to arbitrarily inflict harm on innocent persons, it typically goes through a stringent testing process designed to reduce or eliminate that damage potential. It is known that in realistic situations, risks cannot be entirely eliminated by testing [4]. Thus, testing processes must attempt to minimize risk within practical limits.

Work has shown that certain processes are necessary (though not sufficient) to an effective risk reduction effort. Improvement of testing processes is an important step in the maturation of software engineering. However, evaluation and analysis of any "improvements" can only be done by explicit reference to a complete and correct set of process requirements and constraints. Therefore, it is appropriate to ask, "what are the relevant process requirements and constraints?" This paper will show that there is an important set of process constraints derived from the law of negligence.

2 The law of negligence

The tort of negligence is based upon conduct that is socially unreasonable. It involves the endeavor to "... strike some reasonable balance between the plaintiff's claim to protection against damage and the defendant's claim to freedom of action for his own ends" [5, page 6]. Negligence imposes a set of expectations on behavior through the concept of *duty*. We all have a duty to act with "reasonable prudence"[1] whenever our conduct foreseeably creates a threat of injury to others.

Negligence generally defines a portion of the economic exposure of an industry for any foreseeable damage done by its products or services whenever it has not taken reasonable precautions to avoid such damages. Since the duty in negligence focuses on behavior, it yields constraints on development processes and not the products themselves.[2] Behavior that fails to meet these process constraints may be a basis for negligence liability. Behavior that satisfies these constraints will not be the basis for liability, even if the product was defective and caused harm to an innocent party.[3]

The law of negligence may be different, or may be applied differently, in common law jurisdictions (like the USA and England) and civil law jurisdictions (like other countries in the European Union). Indeed, it may be applied differently within those jurisdictions, but the basic tenets remain the same, society imposes constraints on our technical processes.

2.1 Principles of negligence under common and civil law

The basic feature of the civil law that distinguishes it from the common law is that its primary source is the written law as enacted by legislators. The common law is seen to develop from principles developed through written court decisions. However, with the increasing codification of the common law through legislation in the USA and England, and the practical necessity for civil law judges to interpret cases in order to make sound decisions, the differences do not appear to be great. See generally [7]. Regardless of the source for negligence law, these constraints are derived from the same social concerns. Use of a common law model for constraint development therefore does not result in consideration of substantially different issues for the development of process constraints.[4] The rest of this paper will assume a common law model for the development of the relevant constraints.

2.2 A common law duty to the public

In common law jurisdictions, the duty of "reasonable prudence" is determined by the court under the circumstances of the case at hand. A duty to act (or to refrain from acting) is found by a "balancing" test involving the risk[5] to the plaintiff and the burden[6] of preventing the harm. Evidence is adduced to give relative weights to the relevant factors, and if the risk of harm "outweighs" the

[1] "Negligence is the omission to do something which a reasonable man, guided upon those considerations which ordinarily regulate the conduct of human affairs, would do, or doing something which a prudent and reasonable man would not do" [6].

[2] Distinguish negligence from product based liability such as strict products liability.

[3] Though there may be other legal theories whereby liability is imposed

[4] Further, in an increasingly international marketplace, organizations must take note of the rules of the countries where they do business. The common law is observed in much of the international market. Any multinational organization should be aware of the common law of negligence for that reason.

[5] Risk is said to be the probability of the harm "multiplied by" the gravity of the harm.

[6] This is the burden on the defendant *and* the relative loss to society if the product has limited utility after it is made safer.

cost of prevention, then a duty is imposed. If the cost of prevention outweighs the risk of harm, then a duty is not imposed.

A prominent legal scholar explains in [8] that there is no question that software engineers owe a general duty of care in negligence to their customers and the general public: *to use such care as a reasonably prudent software developer would use under similar circumstances.* What might this mean to the individual testing organization?

One cannot yet research the common law and find a list of specific software testing process constraints known to meet this general duty in negligence. We thus develop a basic model to illustrate the process of developing constraints, giving us insight into constructing a complete and accurate set.

3 Legal sufficiency of testing process designs

Two things are noteworthy about the software engineer's general duty of care due under negligence law: (1) it is defined by reference to software engineering itself, and (2) perfection is not required.

3.1 The common law model

The development of specific legal rules by application of general principles to concrete situations can be modeled with a pair of interacting process control systems, where control is applied in the attempt to keep each process within certain accepted limits. See Figure 1.

Figure 1: Common Law Process Model

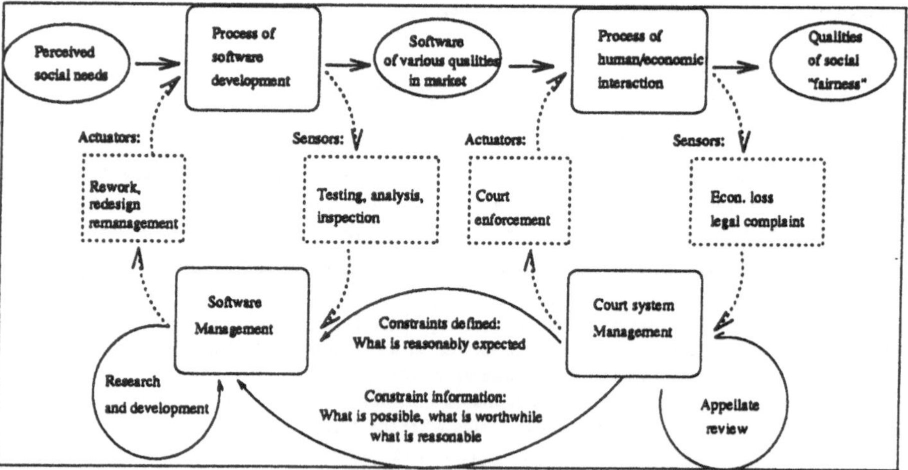

Constraints on the testing process are developed through the model: the legal rule is applied in the context of technical facts and circumstances proved in court. These technical facts about research results and accepted practices are derived from the testimony of experts and from software engineering literature. A given constraint will develop through repeated application of the general rule to similar situations. The new constraint is established firmly by appellate review and published. It must then be faithfully applied in subsequent cases by trial courts. It provides a public statement of the legal expectation - the constraint on the software testing process.

Use of this model to derive relevant constraints offers two main benefits to the testing process designer: (1) awareness of previously implicit (or unknown) constraints; and, (2) availability of the rich history of safety experience included in the constraints.

With this awareness, testing process designers must explicitly consider these constraints or knowingly suffer an additional risk of liability and economic loss. The general constraints may be refined given the particular circumstances existing at the time the process is to be executed. Consideration may then be given to their satisfaction and reasoned justification recorded for future use.

4 Consideration of legal process constraints

How can we explicitly include these process design constraints in our testing processes in a practical way? We suggest the use of a fault tree structure in order to trace the ways in which our testing process designs may fail to satisfy key constraints.

Fault tree analysis is a familiar tool to many involved in safety-critical systems. It is a deductive, effect - cause approach, traditionally used in hazard analysis. The fault tree is the logical model of a process with regard to some undesired event. It is represented graphically with a tree structure. See generally [9] for an industrial process approach, [10] for a software approach to fault tree construction.

Fault trees may be used to analyze our safety critical testing process designs. "Civil Liability" (money damage award) is the "undesired top event." The boxes indicate events, and the labels name those events. Events higher in the tree are the effects of the events lower in the tree combined with the logical operators such as AND and OR. When a box is connected to another by a single line with no logical operator, it indicates another name for the event that is critical to understanding the boundaries between the legal and software processes. Leaf nodes that are circles will indicate further development of the tree as a separate subtree with a root having the same label. These subtrees are broken off separately in order to focus on their individual properties.

The fault tree, once constructed, can be decomposed into its "minimal cutsets" (a combination of components whose simultaneous failure is just sufficient to cause the undesired event). Note that there may be several cutsets for a given undesired top event. Each fault tree is generally not unique. The events in the trees we construct are not generally independent of each other.

4.1 The basic software process fault tree

We present a process fault tree with the undesired top event "civil liability" in Figure 2.

Figure 2: Basic Process Fault Tree

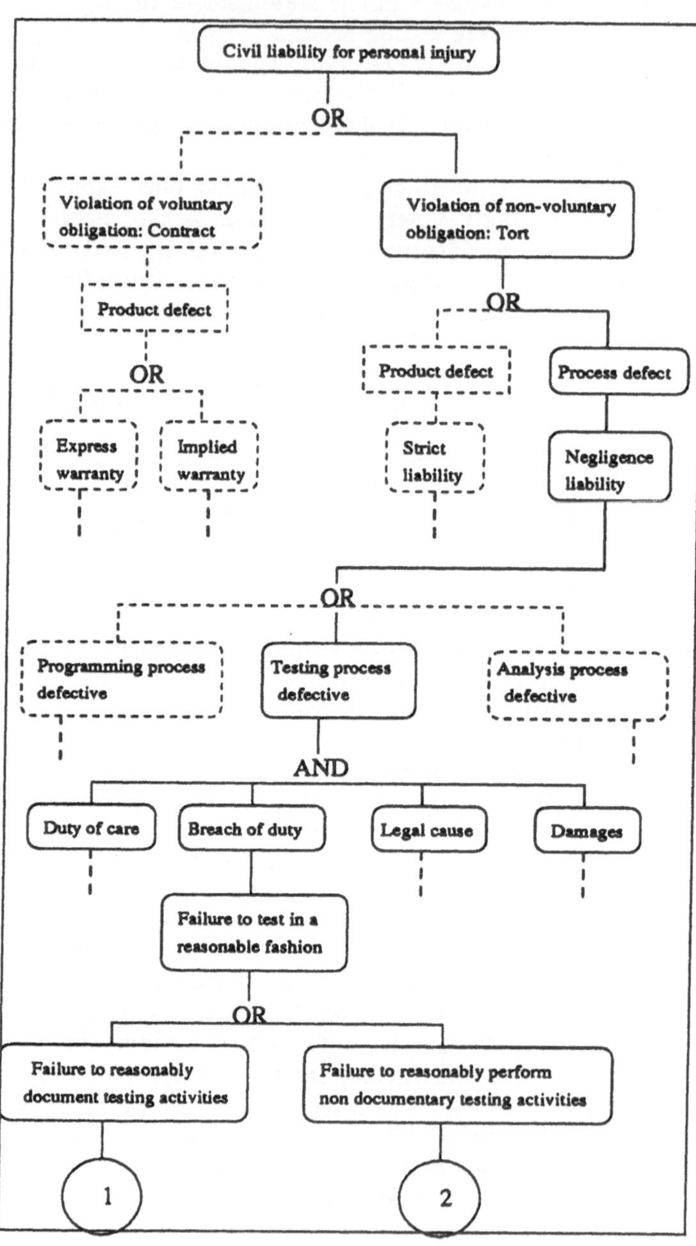

Figure 3: Documentary subtree

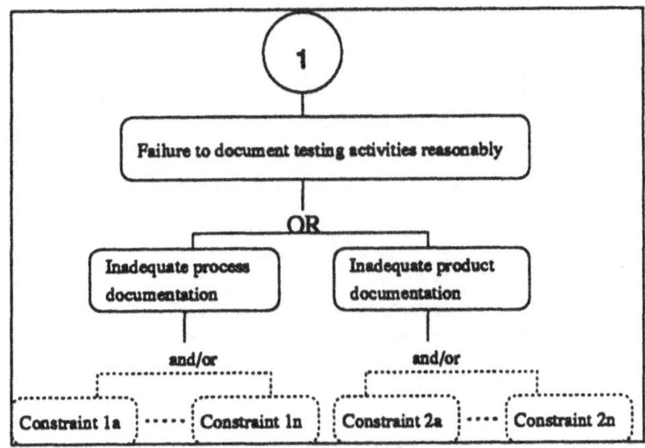

It gives a general picture of ways civil liability may be imposed. This paper is only concerned with paths leading through a negligence node and concerned with testing, but note that in order to add overall context, other nodes and possible paths are indicated by dotted lines and boxes.

Just below the "testing process defective" node, there is an AND connector and four nodes, which represent the elements that must be proven in any negligence case. Arguably, the most critical node here is the "breach of duty" node. It is this node we choose to expand to illustrate the concepts. It is this node where the organization's activities in constraint satisfaction will be the controlling factor. The other nodes are generally legal questions that have little to do with the particular testing process design used.

It is seen that civil liability may ultimately result from either failure to reasonably document testing activities, or a failure in other testing activities. This is one way to define the "failure to test in a reasonable fashion", which is a "breach of duty" in negligence law.

4.2 Development of the testing process fault tree

The tree is developed down to its leaves that indicate individual constraints on the process. Node number 1, the reasonability of the test documentation process is diagrammed in figure 3, while node number 2, the reasonableness of other process activities is diagrammed in figure 4.

4.3 Determine the constraints

What is a "reasonably prudent tester," where do we start? We may begin by looking at the literature of software testing, checking standards and industry customs to gather information about what is considered reasonably prudent testing by experts in the field. We might also check relevant legal literature to see what has been written on the topic, if anything. See, for example, [11].

Figure 4: Nondocumentary subtree

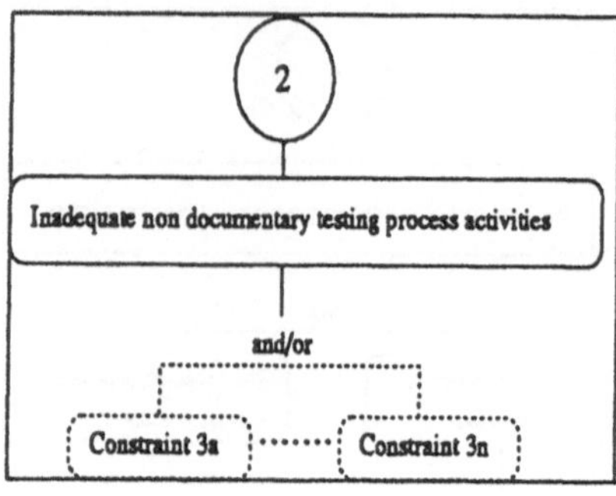

We do not propose a list of potential constraints as this is beyond the scope of the present work. We choose an example to illustrate our approach to determining and documenting reasonable efforts to consider them. In checking the testing literature, we find that one important issue is "independence." In fact, upon examination, the literature seems to be in agreement that the main testing effort for a safety-critical project should be independent of the rest of development. One very early cite to this conclusion is [12] and a recent one that speaks specifically of safety-critical projects is [13]. We further note that ISO 9000-3, though not specifically a standard for safety-critical processes, requires independence in testing and verification activities to some extent [14]. Thus, "independence" should appear as a constraint under node number 2. In any actual effort, information on all sides of the issue should be gathered and abstracted so that a well reasoned decision may be made to an approach to satisfaction or non satisfaction of the proposed constraint.

4.4 Constraint attributes

In addition to determining the minimal cutsets to find potential problem areas of our testing processes, we suggest attribution of the leaf nodes with pointers to critical constraint information. This may be crucial to a reasoned negligence analysis during process design. It is also a foundation for defense to a possible negligence suit (where we may need to prove reasonable satisfaction of the constraints, or that non-satisfaction was ultimately reasonable).

The attribution to the constraints is shown in Figure 5. *External* constraint information includes all sources used in deriving the given constraint such as caselaw, statutes, regulations and other literature. It may also include indirect sources such as software engineering research, expert opinions and standards which are used in courts to derive the constraints legally.

Internal constraint information is the record of the organization's response to the given constraint. How the known risks were actually considered and resolved is shown here. Process documentation and other information relevant to cost and risk analyses used by the organization can be located from here.

With our proposed independence constraint, how might this play out? First, we must gather the information that leads us to believe that the independence

Figure 5: Constraint Attribution

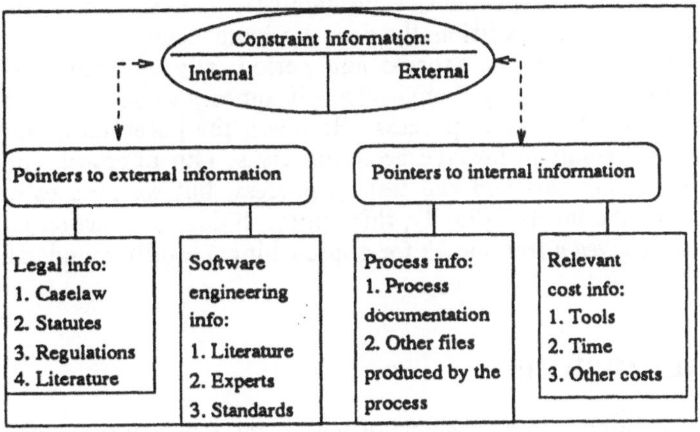

Constraint Information:
Internal
External
Pointers to external information
Pointers to internal information
Legal info:
1. Caselaw
2. Statutes
3. Regulations
4. Literature
Software engineering info:
1. Literature
2. Experts
3. Standards
Process info:
1. Process documentation
2. Other files produced by the process
Relevant cost info:
1. Tools
2. Time
3. Other costs

constraint is important. Above, we listed citations to two references from the field of software engineering and one international standard. All the main references in addition to these should be abstracted and referenced so that they can be part of the analysis. Further, legal references that may bear on this constraint must be listed as direct or indirect authority on the constraints in question.

Second, the testing organization needs to document its own response to the constraints with pointers to any relevant internal constraint information. To that end, the process must be documented and each part of the process that addressed the constraint in question must be noted. The particular test tools used, the extent and depth of the procedures, and the costs must be recorded for possible use in the future. In case of legal questions, the organization could then justify its own response with reasoned analyses, as a court would do. If it can be shown that the organization's behavior was reasonable in light of the foreseeable circumstances, negligence liability will not follow.

5 Advantages and limitations of the approach

There are benefits to this sort of explicit approach to testing process constraints. We may increase the actual safety of the product due to a higher quality process. A safer product exposes the organization to lower risks of liability. The approach certainly heightens awareness of the constraints on our processes and may also lower the risk of lawsuits due to increased attention to them. In any case, there is a higher probability of successful negligence defense (or settlement) whenever legal action is taken, due to solid awareness of the legal issues and how these issues were handled during design. There will also be lower legal costs since the process explicitly includes relevant constraints and justifications, hence part of the lawyer's investigative work is already done. In some cases, this sort of approach may prevent an award of punitive damages by recording the organization's good faith efforts. Much of this analysis may be reusable whenever a project is shown to involve similar constraints.

This model and the approach also have limitations. Since the nodes in the fault trees are not independent, a probabilistic analysis is not contemplated. Fault tree analysis is traditionally done by hand using trained personnel. It must be done early in the process and periodically updated, since the constraints and references may change. Thus it appears to be a costly approach in the beginning of the testing process. However, the potential is that the early investment will result in lowered economic risks. Our approach currently covers only negligence law and the testing process, but we plan to extend it to other areas in the future. Finally, this approach does not answer specific legal questions, but gives a framework for approaching a legally sound safety-critical testing process.

6 Conclusions

We have presented a model showing how legal constraints on safety-critical testing processes develop under negligence law. Even though there are no decisions published at this point for personal injury due to a software flaw, it is agreed that certain constraints may soon be found applicable to our testing processes [11], and those constraints may be postulated now and used in analysis and evaluation of our current testing processes.

We further presented an approach to analyze our safety-critical testing processes. The intent is to assure that our process designs do indeed consider a basically complete and correct set of process requirements and constraints. With the process fault trees attributed with constraint information, we can explicitly show how we addressed the relevant constraints. This allows us to later justify our particular approach to their satisfaction with the explicit consideration of key risk factors.

References

[1] Osterweil, Leon. *Software Processes are Software Too*, 9th Int'l Conf. on Soft. Eng., 1987.

[2] Leveson, Turner, *An Investigation of the Therac-25 Accidents*, IEEE Computer, Vol 26, No. 7, July, 1993.

[3] Nelson and Winter, An Evolutionary Theory of Economic Change, Belknap Press of Harvard University Press, 1982.

[4] Hamlet, *Are We Testing for True Reliability?*, IEEE Software, July 1992

[5] Prosser, Handbook of the Law of Torts, 4th Ed., West Publ., St Paul, Minn., 1971

[6] Blyth v. Birmingham Waterworks Co., 1856, 11 Ex. 781, 784, 156 Eng. Rep. 1047.

[7] Glendon, Gordon, Osakwe, Comparative Legal Traditions, West Publ. Co., St. Paul, MN., 1994

[8] Gemignani, Law and the Computer, CBI Publ., Boston, MA., 1981

[9] Lapp, Powers, *Computer-aided Synthesis of Fault-Trees*, IEEE Trans. on Reliability, April 1977, page 2.

[10] Leveson, Harvey, *Analyzing Software Safety*, IEEE Trans. on Soft. Eng., Vol. SE-9, No. 5, September, 1983, page 569.

[11] Fossett, *The Development of Negligence in Computer Law*, No. Ky. L. R. 14 No. 2, pps 289-310 (1987)

[12] Weyuker, *On Testing Non-Testable Programs*, The Computer Journal, Vol 25, No. 4, 1982

[13] Parnas, *Evaluation of Safety-Critical Software*, CACM Vol. 33, No. 6, June 1990

[14] ISO 9000-3 : 1991 (E), Quality management and quality assurance standards - Part 3: Guidelines for the application of ISO 9001 to the development, supply and maintenance of software.

Session 9
Security

Application of Formal Methods in the Scope of IT-Security

Frank Koob, Markus Ullmann, Stefan Wittmann
Bundesamt fuer Sicherheit in der Informationstechnik
Postfach 20 03 63, D - 53133 Bonn
Germany
Tel.: x49-228-9582-142
Fax.: x49-228-9582-455
Email: vse@bsi.de

Abstract

The *being trusted* of software systems is a growing concern not only for economical reasons but also a "must" for those systems whose demand for quality control means life and death. With each technical advance, the security and reliability of complex software systems is extremely important. As an example, the failure or misuse of computerised traffic systems can result in commuter gridlock or the collision of passenger trains. In the fields of communication, data administration and accounting, a miscalculation whether by a "bug" or a "hacker" can cause a company to lose billions. But these problems are not the only cost to a company, expensive correction of the software is also required. Cost overruns in bringing a technically advanced project to fruition is every companies nightmare. Ask officials at the Denver International Airport.

A method already recognised in the research area to help solve the problem to achieve better control of software systems is the formal program development as a precondition of the development of verifiable correct systems. Statements about correctness are possible because of strictly mathematical modelling. This method has been applied already for a long time in the classical fields of engineering. for software development it means the proof of security/safety properties as well as the verification of the program according to the specification. The German *Bundesamt fuer Sicherheit in der Informationstechnik* (BSI) recognised early the need of mathematically correct security sensitive IT systems and their components. To assist in applying the criteria BSI has constructed a development and evaluation tool called *Verification Support Environment* (VSE). VSE supports especially the use of formal methods for specification and verification of software systems according to the state of the art in current research.

1 ITSEC Requirements

For the assessment of the trustworthiness of technical systems in the area of Information Technology (IT) security evaluation criteria have been developed world-wide during the last years. In Europe currently the Information Technology Security Evaluation Criteria (ITSEC) [1] are valid and applied as a basis to certify technical security products in France, Germany, Great Britain and the Netherlands. According to the ITSEC the integration of security considerations into the system development means, besides specification of functional requirements, to define security properties and to verify that these properties are guaranteed by implementation. Therefore the ITSEC define generic terms, support the assessment of the effectiveness of security functions and provide a scale for trustworthiness and correctness by assessment of single aspects like

- development process
- development environment
- operational documentation
- operational environment.

This scale is divided into 7 levels (evaluation levels E0 - E6) where requirements increase with the level number. The demand on the development process is broken down into demands on

- requirements
- architectural design
- detailed design
- implementation.

Rating an IT system in the correctness levels E1 - E3 only informal or semiformal development methods/descriptions are required. Level E4 demands a formal security model of the implemented security functionality, i.e. a description of the security requirements of an IT system using a mathematical notation. An assessment according to level E6 requires in addition to the formal security model and formal architectural design (formal specification of the security functionality) the evidence of the consistency between security model and architectural design.
As a result the use of formal methods in specification, design and verification plays a central role within the development of software systems following the high correctness/quality levels.

2 VSE Methodology

VSE as a development tool for security-critical software systems provides components that enable developers to fulfil the higher ITSEC levels, in some respects even exceed the requirements. The following part describes the components. Details can be found in [2].
VSE is founded on the specification language VSE-SL based on First Order Predicate Logic. It includes parts for specification and implementation. Besides that

VSE-SL provides means that allow to verify consistency between two specifications (SATISFIES-relation) and the absence of contradictions within one specification (IMPLEMENTS-relation).

This means in actual applications that security properties are defined in a security model to be fulfilled by the functional description of the system, the top level specification. The system is specified using two different views. The result is two partly redundant specifications where the security model pretty much under specified. In addition specifications can be refined stepwise, should be in complex projects. An abstract specification is implemented using expressions of more concrete specifications. With this means the final absence of contradictions within the abstract specification is provided. The satisfaction of the security properties by the lowest implementation level, that means source code level is given by transitivity of the SATISFIES-relation (Fig. 1).

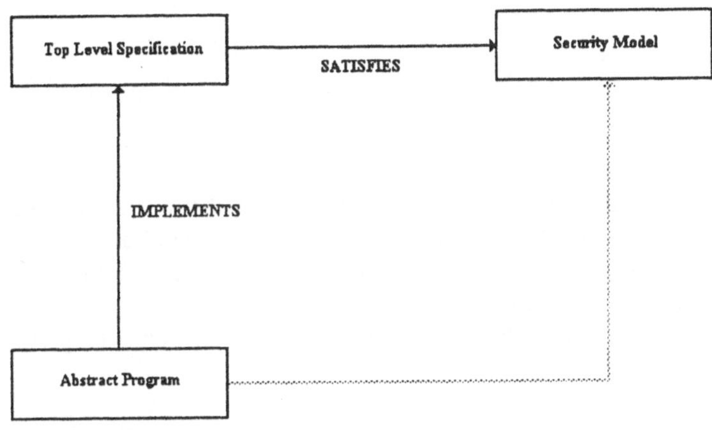

Fig. 1

The above mentioned evidence of consistency and absence of contradictions are provided using the VSE-system by verification of proof obligations automatically generated out of the specification. The verification itself is done semi-automatically by two integrated deduction systems, KIV (Karlsruhe Interactive Verifier) and INKA (INduction prover KArlsruhe).

Verification of large complex systems many times is not practicable or feasible. Therefore VSE-SL supports structuring concepts, horizontal (modularization) and vertical (refinement), that results in the compositionality of proofs. For the correctness of the entire system it is sufficient to prove the correctness of the single modules. There is only a linear growth of verification effort in relation to system complexity (Fig. 2).

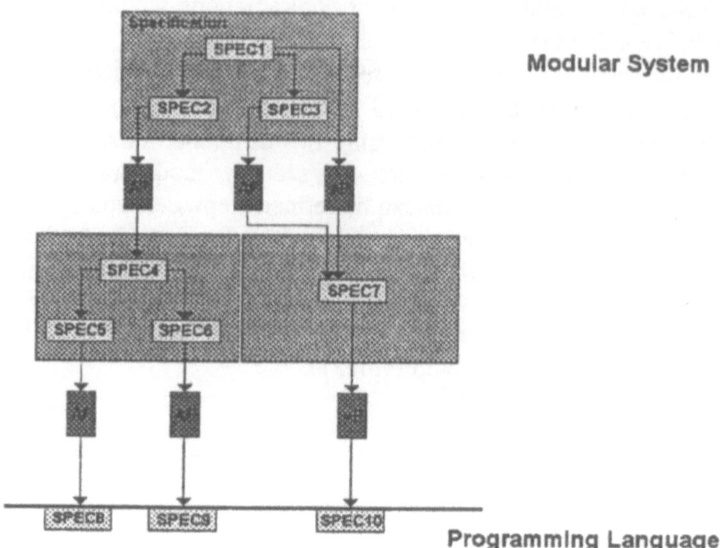

Fig. 2

With a practical and user oriented surface and the overall support of formal methods from formal specification through automatically generated proof obligations to semi-automatic, interactive verification the VSE-tool offers all means to make the fulfilment of the higher ITSEC-levels feasible.
The applicability especially in the development of large and complex systems was shown in two case studies [3].

3 Security Application: Information Filter

Since the official release of the VSE-system in March 1995 several pilot applications have been carried out by independent companies in order to evaluate the usability of the tool in different industrial areas (space flight systems, hospital administration systems, command control information systems, traffic control systems, smartcard systems, etc.).
The pilot project presented here was worked out with WTD 81, the evaluation center for IT-security of the German Federal Armed Forces in Greding, Germany. Subject was a module out of a communication system that has to control and monitor the exchange of information between ADP-systems with different security requirements. It had to be ensured that there is no uncontrolled data exchange (filter function) and that a trusted transfer of information is possible, even with different security classifications.

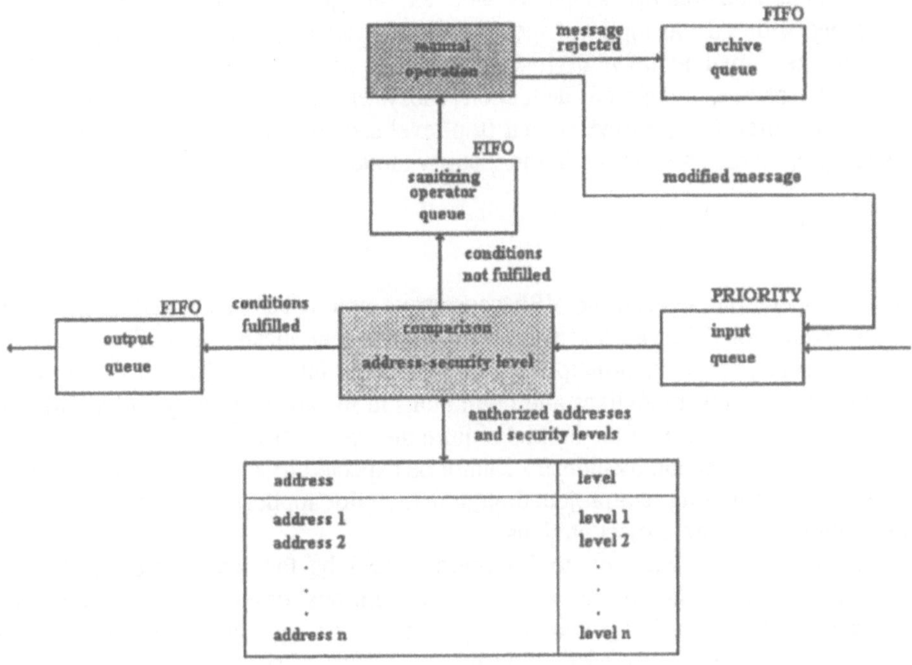

Fig. 3

Because of time constraints only a simplified model of the entire system could be realised. Message transfers from a high classified information system to a low classified system were the main tasks to be worked on formally. The following part explains that part model (see Fig. 3).

Within the entire system data of various degrees of classification are controlled. Input information is sorted into an input queue dependent on priority and processed following an algorithm that has to fulfil two main security properties:

Messages may only be passed on if

- the receiving address is an element of the list of authorised addresses.

- the degree of confidentiality of the sending authority must be greater
  or equal than the degree of the message.

Additionally in case of transfer rejection a special person (sanitising operator) may change the degree of confidentiality e.g. by modification or deletion of sensitive parts and put it back into the input queue.

The rough functionality scope as well as the global security properties were specified with VSE as top level specification and security model following the VSE development methodology. Both were modelled as state transition systems. With the help of the integrated deduction subsystem the mathematical proof was performed that the specified system (top level specification) guarantees the global security properties formalised in the security model.

4 Conclusion

The co-operation between the VSE-consortium and WTD 81 as well as the other pilot applications showed that formal methods are applicable within the software development process in principle if tool support is intensive. Now it is up to the providers of norms, guidelines and regulations in the areas of safety and security to adapt those to the state of the art and require the use of formal methods.
A voluntary conversion by industry cannot be expected because of the much higher costs at the beginning of a project though many pilot applications made advantages throughout the entire project obvious.
Furthermore future work on tool support should be focused on clearly defined interfaces of the applied formal method to conventional development environments in order to provide an integrated development method covering conventionally and formally developed system parts. Improvements in proof support and automation is essential, too. The VSE pilot applications showed that verification activities take over a big part in the use of formal methods.
With these future tasks VSE will be on a way to increase the industrial acceptance of formal methods in software development.

Literature

[1] Kriterien fuer die Bewertung der Sicherheit von Systemen der Informationstechnik, EGKS-EWG-EAG, Bruessel * Luxemburg, 1991

[2] F. Koob, M. Ullmann, S. Wittmann, Industrial Usage of Formal Development Methods - the VSE-Tool Applied in Pilot Projects, COMPASS '96, Computer Assurance Conference, Gaithersburg, MD, USA, 1996

[3] Begleitende Fallstudien VSE-1: SuD und PERSEUS, 1994, not published, availiable through the authors.

Reliability and Security in Communication Software: PBX Systems and CSTA Applications

Dipl.-Ing. Herwig Stöckl
Siemens AG Österreich
Vienna, Austria

Abstract

Communication software means real-time processing at a very high level of reliability. This paper shows some of the most important aspects of reliability, availability and also security in PBX systems and connected CSTA applications. Furthermore examples will be given to show how these problems could be solved.

1 PBX Systems and CSTA Applications

The use of computerized PBX systems (mostly based on ISDN) is state of the art at least since the early eighties [1]. Our society with its emerging technologies in communication and multimedia forces the development departments to implement more and more features in PBX systems and to make the system itself customizable. So most of the PBX manufacturers implemented interfaces to allow applications running on partner systems to influence the switching functions of the PBX itself and to generate improved statistics [2], [3]. There is already a wide supply of such (mostly PBX specific) applications, mainly called CSTA (due to an ECMA standard) or CTI applications [4], [5], [6]. Today these applications often enhance the functionality of PBX systems in a way that the user does not recognize that they are not implemented in the PBX itself [7], [8].

The most important influences on the design of hardware components for PBX systems are miniaturization, cost, safety, reliability, reusability and performance. The hardware for connected CSTA (or CTI) applications is mostly based on well known, good selling computer systems, increasingly on Intel PCs. The major influences for the software design of both PBX systems and CSTA applications are customer features, performance, reliability and security.

Reliability and security have essential influence on the software design of PBX systems and related CSTA applications. This paper shall point out in which way aspects of reliability and security have to be considered in real-time communication software for PBX systems, CSTA applications themselves and CSTA applications in connection with the PBX system (CSTA solutions) based on experiences with Siemens Hicom 300 and CSTA applications for Hicom 300. The importance of a good design of PBX systems and connected applications is especially evident in call centers where a sufficient higher revenue can be earned if there is no lack of availability [9].

2 Reliability in Communication Software

If we speak about reliability in communication software we have to distinguish between three different levels of reliability. At the first level there is the reliability of the PBX system and its software, whereas the second level is the reliability of the CSTA application and its platform. But the third level is the reliability which the customer really gets: the system reliability of the PBX system working together with the specific CSTA applications. Therefore I want to point out the different aspects and possible solutions step by step culminating in high sophisticated - and of course very reliable - CSTA solutions.

2.1 Reliability of PBX Systems

A PBX system is a real-time system, therefore there are very hard requirements to reliability not only of used hardware but also of the software. But according to the experience of every computer user there is no software without errors. Additionally software development cycles have to be very short because of the emerging market requirements and the hard competition. This trade-off is solved by high quality effort in software development and by the design of the product, which means both by hardware and software design.

The user of a PBX system cannot accept a situation in which any of the phones or trunk lines is not working because all people are used to a 24 hour availability of communication services. If there is such a situation the highest goal has to be that this problem is restricted and all other parts do not recognize it. The recovery has to be automatic and must happen within the shortest time which is possible. If the whole system is in trouble there is the following hierarchy for reliability:

- already existing connections must be kept
- it must be possible to collect correct call charge records for all calls if wanted
- it must be possible to establish new connections to the wanted party
- special features like call forwarding, conference call, ACD calls, etc. must be available
- historical call detail reporting must be available
- real-time supervision (e.g. of ACD agents) must be available
- supervisory interaction in configuration must be available

2.1.1 Modularization by Means of Hardware

Big companies or governmental authorities usually need an internal telephone network where all users shall get the impression that there is only one PBX to which all of them are connected. Indeed there is a network of PBXs which are interconnected. On the one hand this network is often necessary because of different locations on the other hand it is the result of cascading a number of single PBX systems to one large. Also within one single system there are cascades of single trunk line driver devices to a several number of trunk lines per 19 inch board, of such trunk line boards to shelves and of the shelves together with some other boards to PBX systems. For the administration of system resources like harddisk, floppy drive or interfaces there are some special boards in use. Each of these boards usually has its own CPU, modern phones also include microprocessors, there is one powerful CPU for the switching software and some CPUs for the mentioned specific tasks.

Of course based on this hardware modularization there is a related modularization of software necessary. Each board and each phone has its driver software download into its own RAM. In this way hardware defects or severe software errors do hardly influence the whole system. This hardware modularization needs some additional administration overhead in the software but it is also necessary to meet the hard performance requirements and it is the possibility which allows to produce small parts at a quantity with the economies of scale.

2.1.2 Modularization by Means of Software

The best way to guarantee a most reliable software is to design software in many tiny modules. Using this design principle it is possible to develop with a lot of engineers and to test the single modules simultaneously. The modules also do not interfere with each other in such an degree as monolithic software does. Last but not least it is possible to implement mechanisms for stabilization and runtime monitoring of the most important parts of software. An error which occurs in a less important software module does not influence most of the other features of the system. Almost all PBX systems' software is based on actual real-time operating systems. Therefore it is possible to use separate software processes for different tasks. Commonly only these parts of software are running which are currently necessary for the specific customer system.

The second level of modularization in software is that most PBX configurations allow to group distributed resources (e.g. single trunk lines on different boards are grouped to one logical trunk group with a single access code). Using such logical groups it is easy for the configuration personnel to configure the PBX system very reliable for the case of hardware defects. They only have to group trunk lines which reside on different trunk boards to such a logical trunk group.

2.1.3 Enhanced Techniques in PBX Systems

With the principles of modularization and based on different backup capabilities most PBX systems enhance system reliability by using at least one, mostly more of the following mechanisms. The number of mechanisms and the quality of the implementation of these principles vary from vendor to vendor. This depends on the quality requirements a specific vendor tries to fullfill and it also depends on the market segment the system is designed for. It is obvious that there are differences in small low-cost solutions which start at small companies hardly bigger than SOHOs and in powerful solutions for large emergency call centers like the US rescue 911 service. Usually the number of the applied mechanisms is also increased by the complexity of the specific system.

2.1.3.1 Distributed Systems

If one trunk line or board in a PBX or one PBX system in a network fails all other still work. This is an improvement to a breakdown or recovery of the whole system or the whole network. Mostly there are different levels of distribution which starts with - maybe redundant - microprocessors and ends up with several PBX systems in a network. This mechanism is hardware based and most effective if the software is modular.

Applying distributed systems the problem of partial breakdown resp. partial errors can be restricted to a small impact on the whole system. Most of the non affected systems do not even recognize such a partial problem.

2.1.3.2 Network Topology

There are rules for the configuration of both the corporate network of an company and the stand alone PBX system which lead to much higher reliability. Only two examples: The corporate network at least should have backup lines if the usual tie lines are not available. A published phone number should not be plugged to the PSTN only by one PBX and not at all only by lines which are on one single hardware board in a single PBX system. Using features like least cost and alternative routing such a network may create profit also in normal use by reducing cost depending on current charges of PSTN providers.

Finding a powerful network topology helps to master situations of regional disasters (e.g. breakdown of local PSTN, lack of personnel in branch department, etc.).

2.1.3.3 Restart Levels

Usually there are different levels of restart at least for the switching part. Depending on the severity of the problem a single functionality is cancelled, the RAM is reinitialized, the whole database is reinitialized or even downloaded software is reloaded. If a problem cannot be solved by the appropiate restart level this is usually repeated one or two times. In case of failure after repetition or in case of expiration of a timer the restart is escalated to the next level. Using this principle on the one hand most errors are solved with a minimum of influence to the other parts and on the other hand the system is reliable.

In other words restart levels help to optimize the availability of all parts in a system. Besides the impact to not affected parts is reduced to a minimum.

2.1.3.4 Modular Restarts

Only the affected module (mostly software process) restarts. If the problem cannot be solved within this error treatment related modules are also restarted, in worst case the whole system restarts. This is a consistent application of the same principle which is the basis for the restart levels. Modular restarts are a kind of distributed software systems.

Similar to the distributed systems philosophy modular restarts help to restrict the problems to the smallest scope which is possible.

2.1.3.5 Watchdog

The most important real-time processes have to notify a watchdog in intervals. If one notification is not in time the watchdog timer expiration restarts this process. This watchdog can be a software watchdog, i.e. usually a special task only waiting for all timers to be reset. If one of these timers is not reset the watchdog task causes a software restart of that task which did not notify in time.

Even more effective but also more effort in realisation is a hardware watchdog. Here the watchdog software is running on a separate CPU usually on a PC card and is also waiting for regular timer resets of the supervised software task(s). If one of these resets is too late the watchdog card usually switches off and on the power supply of the system (specific part) and causes a full reboot. The advantage of this solution is that it cannot be influenced by any software error.

The benefit of a watchdog is to assure that a system restarts in all cases of software errors, a dead-lock situation because of a software error is nearly impossible.

2.1.3.6 „Hot" Standby Server

In systems with very high reliability requirements the real-time servers are doubled in hardware (Of course the hardware of trunk lines etc. cannot be doubled .There is only one hardware line to a phone!) and software. The second server is only monitoring all actions of the first one and gets automatically control of the system if the first one fails and recovers. The mechanisms how to get control of system usually uses a kind of watchdog functionality. Sometimes there are even two systems working on the same problem, the are designed to be competitive and guarantee a real redundancy. In this case there is no server in a „hot" standby any longer. This is even more sophisticated.

„Hot" standby servers are a means of creating a high degree of redundancy to master all problem situations with nearly no lack of functionality. Of course this solutions are very expansive in design and implementation. But there is a certain price you have to pay for the highest quality.

2.1.3.7 Backup Media

For all kind of records automatic backups are made in PBX systems. Especially call charge records (but also ACD historical reports) are data which have to be kept with the same reliability as the switching functionality. The reason for this is that a PBX user earns a lot of money which is based on valid call charge records or ACD statistics. Additionally the call charge records can be compared with the bill of the PSTN provider which is usually not so detailled. But of course both have to be consistent. To guarantee the highest level of reliability almost all of these data are immediately stored to hard disc and are not only kept in runtime memory. In specified intervalls, often even sometimes a day, the data are additionally stored on a backup media. This backup media often is geographically seperated from the originating PBX system. Nowadays very often it is another hard disc at a LAN-server on another place. Formerly (and still well known) as backup media are a second hard disc in the PBX system itself, a floppy disc drive or other magnetic storage media.

To guarantee the availability of important statistical data these data are often kept at a backup media as well as at the origin itself.

2.2 Reliability of CSTA Applications

In general the reliability problems of CSTA applications do not differ from the problems within the PBX system. The additional aspect is that the CSTA application is running on a host system (often a PC) which is not included in the reliability concept of the PBX system and is often based on an operating system which is not really intended for real-time solutions. Further the chance that something fails is increased because there is a hardware link (LAN, V.24, S0, etc.) between PBX and host and last but not least especially PCs are often not very reliable. Depending on the degree of the real-time aspect and on the importance of the features of the CSTA application the principles of PBX systems can be used on the host system and its CSTA application.

If the CSTA application is an essential switching feature (e.g. an ACD routing) link or application failures may not lead to situations in which the switching function of the PBX does not really work. In case of ACD this often happens because therefore call receiving ports or queues are used which are only operated by the application and not by the PBX itself. Therefore an acceptable emergency concept is necessary.

Using more than one hardware link is a possible way to reduce the risk of emergency. But this method usually extends the complexity of the application because different hardware links may work unsynchronized. Further ideas are two applications working on the same system or the real „hot“ standby server.

2.3 Reliability of adapted CSTA Solutions

The best way to get a very reliable CSTA solution is to implement a PBX specific adapted CSTA application which really works together with the PBX. On a low level this means that the application is notified if there are any troubles in the PBX (e.g. module restarts, defective devices, the cause for a failed command, etc.). On a high level the PBX system and the CSTA application really work together.

An example for a high level adaptation is an automated emergency service for CSTA applications. As mentioned before it is essential to preconfigure an emergency service in the PBX system for the case of link or application failures. The automated emergency service allows the CSTA application to reconfigure this emergency service at runtime. In this way the emergency configuration in the PBX system can be updated at any time. Therefore e.g. an ACD application on a CSTA host can switch the PBX emergency service to current logged in agents' phones or to the current night service target. In case of a link or application failure the PBX at least routes to a currently attended target device and the loss of switching function is reduced to a minimum.

The PBX system itself detects that there is a link or application failure and starts to route following the current emergency rules. When the application succesfully reestablishes the connection it resynchronizes and stops the emergency service of the PBX. In this way the control goes back to the CSTA application. If it is necessary the CSTA application updates the current emergency configuration in the PBX.

3. Security

The security aspects for both PBX systems and CSTA applications are mostly the same. At first there must exist an absolute reliable password hierarchy which guarantees that only authorized service personnel is able to change the configuration. Especially this is true for the most common way to configure PBX systems and CSTA applications, for teleservice via modem. Password checks are often enhanced by callback mechanisms which shall reduce the danger that hackers get into the system.

The password hierarchy is necessary for different types of users. The highest level is for service personnel, the second one for administrators which have more or less restricted access rights to configuration. There is a third level for different types of supervisors which have restricted access rights to read configuration data but are not allowed to change them. The fourth level is for works council which is usually allowed to see all configuration data and sometimes additionally can lock the configuration for all other users except the service personnel. A fifth level exists for the permission to read and delete call charge records and other call statistics (e.g. call traffic measurement or ACD reporting). And of course there are some specific features which may be used only if certain passwords are entered (e.g. silent monitoring of calls, PIN identification for voice mail services, etc.).

In some countries the telecom authorities define a special procedure how the call statistics can be accessed. In Germany it is common that three independent passwords have to be entered. The idea is that one is entered by the operator, the second by management and the third one by works council. It has to be considered that in opposite to this trend in central and western continental Europe in UK., US and especially in Japan and southeast Asia security of (private) data is not required. In those countries only configuration must be locked so that only authorized personnel has access.

The CSTA link is another aspect of security. The link must be reliable so that the data which are transmitted are correct. Since ECMA-CSTA is a published standard it is easy to monitor this link. Sometimes the hardware for this link is an existing LAN. Therefore it must be ensured that this link is no security problem. For some applications the data transfer has to be encoded.

A CSTA application specific problem is to ensure that there are not unregistered copies of the application software in use. This is not so easy because the host system usually is a simple PC. Therefore most CSTA applications use a hardlock (dongle) to ensure this. Not only the reduced earnings forced the vendors to use hardlocks but also the problem of diagnosis in error situations if there is unregistered software connected to the PBX and the PBX customer does not tell this fact. And different hardlocks can be used to ensure the authorized availability of special features of a CSTA application.

4. Future Trends in Communication Software

The main future trend in communication software is an integration of different hardware and software applications because of the emerging multimedia market, the rapidly growing importance of call center solutions and the fact that there will be many specialists for specific problems or features in future. These specialists will grow up because the market demands for high sophisticated features and short release cycles.

On the other hand this situation has negative impact on both reliability and security. A future multimedia call center will be an arrangement of hardware parts of different vendors and at least as many software providers. The only chance to keep the current level of reliability and security in PBX systems and CSTA applications is that standardized interfaces between all parts are used. The trend to standardization (e.g. ECMA-CSTA, TAPI, etc.) has already been started but is mainly concentrated on functionality so far; security aspects are mostly considered to a very small extend. It will be a great challenge for the next years to enlarge these interfaces especially by reliability aspects. If this goal can be reached „plug and play“ with the best specific products of different vendors will work. Of course it is not the intent of big PBX providers to push this trend but it will soon be a requirement for the market.

Abbreviations

ACD	Automatic Call Distribution
CPU	Central Processing Unit
CSTA	Computer Supported Telecommunications Applications, standardized protocol for CTI applications by ECMA
CTI	Computer and Telephony Integration
ECMA	European Computer Manufacturer Association
HW	HardWare
PBX	Private automatic Branch eXchange
PC	Personal Computer
PIN	Personal Identification Number
PSTN	Public Switching Telephone Network
RAM	Random Access Memory
SOHO	Small Office Home Office
SW	SoftWare

References

1. Siegmund G. Grundlagen der Vermittlungstechnik, lecture paper. Berufsakademie Stuttgart, 1990
2. Todesca M. Integration von Computersystemen und Telefonanlagen, diploma paper. Technical University Vienna, 1990
3. Walters R. Computer Telephone Integration. Artech House, Boston-London, 1993
4. ECMA. Computer-Supported Telecommunications Applications, ECMA TR/52. Geneve, 1990
5. ECMA. Protocol for Computer-Supported Telecommunications Applications, ECMA/TC32-TG11/95/10. Geneve, 1995
6. ECMA. Services for Computer-Supported Telecommunications Applications Phase II, ECMA/TC32-TG11/95/11. Geneve, 1995
7. Stöckl H. CSTA - Computer Supported Telecommunications Applications, Die funktionelle Integration von Telefon und Computer auf dem Weg von herstellerspezifischen Insellösungen zu einer standardisierten Anwendungsvielfalt, diploma paper. Technical University Vienna, 1992
8. Kohlhammer P, Baum G, Conrad P et al. CIT' 90, congress paper for telematica. Messe Stuttgart, 1990
9. Gable R A. Inbound Call Centers: Design, Implementation and Management. Artech House, Boston-London, 1993

Byzantine Agreement with Limited Authentication

Malte Borcherding

Institute of Computer Design and Fault Tolerance
University of Karlsruhe
76128 Karlsruhe, Germany
malte.borcherding@informatik.uni-karlsruhe.de

Abstract. Reaching agreement in the presence of faults is a fundamental problem in distributed systems. One of the strongest kinds of agreement is *Byzantine agreement*. It requires that a set of nodes (processors) agree on a message sent by one of them, despite the presence of arbitrarily faulty nodes.

It has been shown that message authentication is useful in order to reach Byzantine agreement with any number of arbitrarily faulty nodes. Unfortunately, this approach imposes the additional problem of key distribution. Key distribution can be regarded as a pre-agreement on the public keys of the participants.

In the past, agreement protocols were either designed for no authentication or for complete agreement on the public keys. Because this pre-agreement has to rely on techniques outside the system (e.g., trusted servers which never fail), it is useful to consider weaker levels of key distribution which need as few additional assumptions as possible.

In this paper, we investigate the achievable fault tolerance for Byzantine agreement under the assumption of different kinds of key distribution.

Keywords: Byzantine agreement, authentication, distributed systems, fault tolerance

1 Introduction

The problem of Byzantine agreement arises when a set of nodes in a distributed system needs to have a consistent view of a message sent by one of them, despite the presence of arbitrarily faulty participants. This message could be a transaction in a replicated database, or the rules of a game (where all players have to agree on the same rules).

Tolerating arbitrary (*Byzantine*) faults instead of simpler faults can have two motivations: The first is that some participants may have an interest to disturb the protocol (e.g., by distributing different rules in order to gain an advantage). The second motivation is that, even if the participants are benign, one can be sure to cover *all* possible fault classes.

The exact definition of Byzantine agreement is as follows:

(B1) All correct nodes decide for the same value.
(B2) If the sender is correct, all correct nodes decide for the value of the sender.
(B3) Each correct node eventually decides for a value.

Protocols for Byzantine agreement are generally divided into two classes: authenticated protocols and non-authenticated protocols. In authenticated protocols, all messages are signed digitally in a way that the signatures cannot be forged and a signed message can be unambiguously assigned to its signer. This mechanism allows a node to prove to others that it has received a certain message from a certain node. Authenticated protocols can tolerate an arbitrary number of faulty nodes. In non-authenticated protocols, no messages are signed. These protocols require more than two thirds of the participating nodes to be correct ([LSP82]).

While authenticated protocols offer the best fault tolerance, it is not at all trivial to distribute the public keys of all participants of an agreement protocol consistently amongst each other. Typically, key distribution requires one trusted entity, or a group of entities which is completely reliable as a whole ([Gon93]). Since these assumptions restrict the usual assumptions about the participant's behaviour, they should be kept as weak as possible. Hence, it is useful to have a look at possible scenarios with different kinds of key distribution and different common knowledge about these distributions.

2 Model of Computation

In this section we describe the model of computation used throughout this paper. Our world consists of a fully interconnected network with n nodes (processors), t of which may be faulty. The nodes operate at a known minimal speed, and messages are transmitted reliably in bounded time. Furthermore, the receiver of a message can identify its immediate sender.

The nodes communicate in successive *rounds*. In each round, a node may send messages to other nodes and receives all messages sent to it in the current round. The actions a node takes in the next round depend solely on the messages it has received so far. We make no assumptions about the *type of failures* that occur. If a node is faulty, it may behave in an arbitrary manner. This type of behaviour is usually referred to as *Byzantine fault*.

In addition, we assume the existence of an unforgeable signature scheme. Examples for signature schemes which are unforgeable with a sufficiently high probability (given today's state of the art) are DSA and RSA [Nat92, RSA78]. In these schemes, a prospective signer has a pair of keys, namely a private key and a public key. The private key is used for signing, while the public key can be used to verify a signature made with the private key.

The assumption of a signature scheme alone is not very strong (see section 3.1). It is often necessary to make assumptions about the distribution of the public keys. The strongest assumption is that of *complete authentication*. It comprises the following four properties:

(A1) If a correct node assigns a signed message to a correct node P, then P has signed the message.
(A2) A message signed by a correct node P is assigned to P by all correct nodes.
(A3) If a message is assigned to a (possibly faulty) node P by a correct node, then all correct nodes assign it to P.
(A4) All correct nodes can sign messages.

In terms of private/public keys, properties (A1) and (A2) state that there is agreement on the public keys of the correct nodes, and the correct nodes keep their secret keys secret. Property (A3) extends the agreement to the public keys of the faulty nodes. Due to the Byzantine failures, it cannot be assumed that the faulty nodes keep their secret keys secret. Fault models which assume that faulty nodes do not give their secrets away (or sign messages on behalf of others) are used in [GLR95, EM96].

3 Limited Authentication

Traditionally, there have been only two different kinds of authentication in Byzantine agreement: no authentication and complete authentication. These levels mark the two extreme points in a spectrum of possible scenarios. In this section, we will identify several different kinds of authentication, give situations in which they arise, and present their properties with regard to the achievable fault tolerance.

3.1 Local Authentication

This type of agreement can be reached when no means of agreement on the keys is provided and each node distributes its public key by itself, using a signature scheme. With a challenge-response key distribution protocol, properties (A1) and (A2) can be enforced if (A4) holds and a signature scheme exists ([Bor95]). That is, a faulty node can distribute different public keys to different nodes, but it can not claim a public key of a correct node for itself.

Although no agreement on the public keys of the *faulty* nodes can be guaranteed, this level of authentication has been shown to be useful for Failure Discovery, a sub-problem of Byzantine agreement ([HH93, Bor95]). Using local authentication, Byzantine agreement with few messages in the failure-free runs is possible. Unfortunately, in this setting the impossibility of Byzantine agreement for $n \leq 3t$ cannot be overcome, as stated in the following theorem.

Theorem 1. *It is not possible to reach Byzantine agreement if one third of the nodes is faulty and only local authentication is assumed.*

Proof. (Sketch) This proof is a variation of the proof in [LSP82]. Consider three nodes A, B, and C, and three scenarios σ_1, σ_2, and σ_3. In σ_1, B is faulty, and A sends 0. In σ_2, C is faulty, and A sends 1. In σ_3, A is faulty, and A sends 0 to C and 1 to B.

Hence, in σ_1, C has to decide for 0, and in σ_2, B has to decide for 1. The faulty nodes will behave in a way that in σ_3, B receives the same messages as in σ_2, while C receives the same messages as in σ_1. So, in σ_3, B and C have to decide for different values, which contradicts (B1).

For the sake of brevity, we will only show that Byzantine agreement is impossible in two rounds; this result can be extended to arbitrarily many rounds. In the first round, the sender sends its value to all nodes (including itself). In the second round, all nodes repeat to the other nodes what they have received.

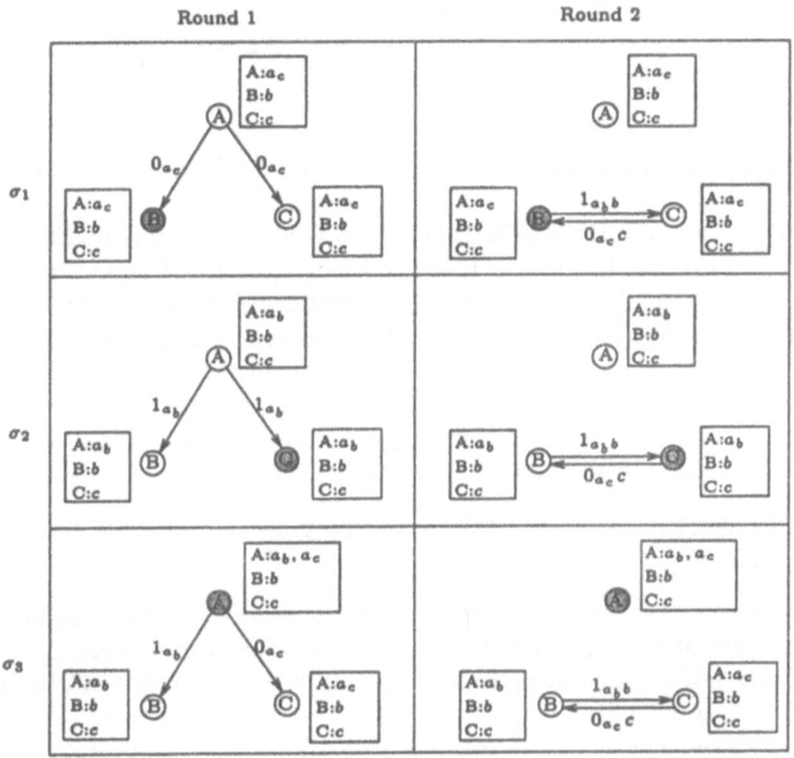

Fig. 1. Impossibility of Byzantine agreement with local authentication

Figure 1 shows this protocol (we have omitted messages to and from A in the second round). The boxes at each node show the respective views of the public keys, and faulty nodes are shadowed. Signatures are denoted with subscripts.

In σ_1, B behaves towards C as if it had accepted a_b as the public key of A, although A's key is a_c (B may have created a key pair a_b by itself). Furthermore, it pretends to have received 1_{a_b} in the first round. In σ_2, C displays the inverse faulty behaviour towards B.

In σ_3, A has actually distributed different public keys and sent different values in the first round. From C's point of view, σ_1 is indistinguishable from σ_3. Hence, it has to decide for 0 in both cases. On the other hand, B cannot distinguish

σ_3 from σ_2 where C is cheating and A wants to transmit 1. Hence, it has to decide for 1 in σ_3, which leads to an inconsistency. For $n \leq 3t$, this reasoning can be extended to three sets of nodes, each comprising at most t nodes. Hence, Byzantine agreement is impossible for $n \leq 3t$. □

3.2 Sound, but Incomplete Authentication

For this kind of authentication we assume that no two correct nodes have different public keys of a third node. However, a node may not know the public keys of all other nodes. In this setting, two cases can be distinguished. In the first case, there is no agreement on who knows which keys, while in the second case, this agreement is assumed.

No Agreement on Distribution of Public Keys (Crusader Authentication)

We assume that only the public keys of the faulty nodes are distributed incompletely. This can happen with the following (common) approach to key-distribution: A globally trusted server signs certificates containing (at least) the name and public key of a node. The server gives such a certificate only to the holder of the certified key. It is then the responsibility of each individual node to send its certificate to the others. If a node is faulty, it may give its certificate only to a part of the other nodes, but it cannot produce false certificates. Instead of (A3), (A3)′ holds in this context:

(A3)′ All correct nodes which assign a certain message to a node assign it to the same node.

We will refer to this level of authentication as *crusader authentication*, since the public keys of the nodes are agreed upon like values sent by *crusader agreement* ([Dol82]): If a correct node A has a key of B, it is the correct one. Otherwise, A knows that B is faulty.

Byzantine agreement can be reached with $n \geq 2t + 1$. The protocol in Figure 2 has this fault tolerance. It is a variation of the Exponential Information Gathering *(EIG)* protocol which was introduced by Bar-Noy *et al.* [BDDS87], based on the protocol in [LSP82]. In this protocol, the sender starts by sending its value to all other nodes. In the following t rounds, each node signs and forwards the messages received in the previous round to the other nodes.

During protocol execution, each node maintains an *EIG* tree which contains the received information in a structured manner (see Fig 3). Such a tree has $t+1$ levels, one level per communication round. The root has $n-1$ children, and in each of the following levels, the vertices have one child less than those of the previous level. Hence, on level r, each vertex has $n-r$ children (we consider the root level as level 1). A part of such a tree for $n = 7$ is shown in Fig. 3.

The vertices have labels which are assigned in the following manner: The root is labeled with the sender's name. In the following levels, the children of a vertex are labeled with the names of the nodes not yet on the path from the root. We identify a vertex in the tree by the the sequence of labels from the root to the vertex in question. Note that in no such sequence a node's name appears twice. A vertex labeled with the name of a correct (faulty) node will be called a correct (faulty) vertex. From the construction, a node on level r has at most t faulty children and at least $n - r - t$ correct children.

In the first round of the protocol, each node stores the value received from the sender in the root of its *EIG* tree. In the following rounds, each correct node broadcasts the contents of the level of its tree most recently filled in, and fills the next level with messages it receives. If a node X receives a message from Y claiming that it has stored v in vertex $ABCDE$, X stores v in vertex $ABCDEY$ of its *EIG* tree. Hence, a value v in vertex $ABCDEY$ is interpreted as "Y said E said ... B said A said v". If a node failed to send a value, a default value is stored.

Due to the structure of the tree, not all received messages are stored. Those messages in which a node reports about a message which was once sent by itself are ignored. In Fig. 3, labels and stored values are separated by a colon. The faulty vertices are shadowed.

When a node has completed its tree (after round $t + 1$), it uses the collected data to decide for the outcome of the protocol. This is done by *resolving* each vertex to a certain value, depending on the resolved values of its children. The exact rules are given in steps 2 and 3 of the protocol. A vertex which is resolved to the same value by all correct nodes will be called *common*. The following example will demonstrate the rules for resolving.

1. The nodes fill their EIG trees for $t+1$ rounds. They only consider messages which carry the recognizable signature of the immediate sender.
2. A leaf is resolved to its value with the last signature removed.
3. For a non-leaf on level $r < t+1$ with label X, two cases are distinguished:
   - The signature of X is known: Consider the set of resolved children which are signed correctly. If it has at least $t - r + 1$ members, take the (relative) majority value. The vertex in question is then resolved to that value without the last signature. If the set has fewer members, the vertex is resolved to a default value.
   - Signature of X is unknown: Take a set of maximum size of resolved children which carry the same signature. If it contains at least $t - r + 1$ elements, take the relative majority. The vertex in question is then resolved to that value, with the last signature removed. If no such set exists, the vertex is resolved to a default value.
4. The result of the protocol is the resolved value of the root.

Fig. 2. Byzantine agreement for crusader authentication

Example 1. Let $n = 7$ and $t = 3$ (see Fig. 3). Let us further assume that the five children of a vertex at level $r = 2$ are resolved to the following values: default,

default, $1_{ab}, 2_{ab'}, 2_{ab'}$. A reducing node who knows the signature of b notices that only one of these values has been signed correctly. Since $1 < t - r + 1 = 2$, it will choose the default value.

A node which does not know the signature of b uses values $2_{ab'}$ and $2_{ab'}$ for its decision, since they constitute the largest set of values with the same signature, and the set has two elements. The majority value is $2_{ab'}$, and the vertex in question will be resolved to 2_a (this implies that the vertex is not common).

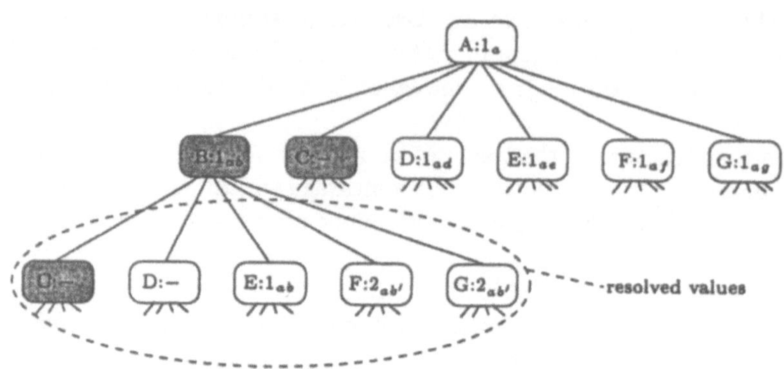

Fig. 3. Part of an EIG tree

For the proof of correctness of the protocol, we will first prove two Lemmas:

Lemma 2. *Assuming crusader authentication and $n \geq 2t+1$, the following holds: A correct vertex is resolved to its stored value, with the last signature removed.*

Proof. We omit the proof here for brevity. It is by induction on the levels of the tree, from the leaves to the root. □

Lemma 3. *Assuming crusader authentication and $n \geq 2t+1$, the following holds: A faulty vertex which has only faulty ascendants*[1] *is common.*

Proof. The proof is again by induction on the levels of the tree, from the leaves to the root. □

Theorem 4. *Assuming crusader authentication and $n \geq 2t+1$, the following holds: The algorithm in Figure 2 reaches Byzantine agreement.*

Proof. If the sender is correct, the root of the tree will be resolved to the value received in the first round by all correct nodes, due to Lemma 2. Hence, all correct nodes agree on the correct value.

If the sender is faulty, the root is common due to Lemma 3. Hence, all correct nodes agree on the same value. □

[1] An ascendant is a vertex on the direct path to the root.

Known Set of Signers (Partial Authentication)
In this setting it is common knowledge whose public keys are distributed completely. For simplicity, we will only regard the case in which a node's public key is either known to all nodes or to no-one. This situation arises when some of the nodes are not able to sign messages, or when the keys are distributed by a globally trusted server which does not know all keys. There is no relationship between the state of a node (faulty/correct) and the distribution of its public key. Here, (A3) holds, but (A4) becomes

(A4)″ Only a known set of s nodes can sign messages.

We will call this level of authentication *partial authentication*.

A node whose public key is known is called *signer*. We can distinguish the two cases of a signing sender and a non-signing sender.

Sender is no signer If $n > s \geq 2t+1$, the sender sends its value to the signers in the first round. Then the signers execute a completely authenticated protocol (cf. [DS83]) with only t rounds with all other non-signing nodes as bystanders. The result value is the majority of the results of the $2t+1$ sub-protocols.

A sketch of the proof is as follows: If the sender is faulty, there are at most $t-1$ faulty signers left and all processors will agree on the values of the s sub-protocols. Hence, all correct nodes will eventually decide for the same value.

If the sender is correct, all correct signers will send the same values in their respective sub-protocols. These protocols have the property that the nodes reach agreement on the values of the correct nodes, even if the round number is not greater than the number of faulty nodes. Hence, a majority of the results of the sub-protocols will be the correct value.

For $s \leq 2t$, no better fault tolerance than in the unsigned case can be achieved. This is shown in the following theorem:

Theorem 5. *If the sender of a protocol does not sign its messages and $s \leq 2t$ holds, then Byzantine agreement can not be reached with $n \leq 3t$, assuming partial authentication.*

Proof. For simplicity, we will again demonstrate the idea of the proof with three nodes, one of which may be faulty. The sending node, A, does not sign its messages. Consider the following three scenarios:

Scenario σ_1 (B faulty): A sends 0 in the first round. When B receives a message from A without any signatures (i.e., the message was only echoed by A itself), it forwards 1_b to all nodes. All other messages are correctly signed and forwarded.

Scenario σ_2 (C faulty): A sends 1 in the first round. When C receives a message from A without any signatures, it forwards 0_c to all nodes. All other messages are correctly signed and forwarded.

Scenario σ_3 (A faulty): In the first round, A sends 0 to C and 1 to B. In the rest of the protocol, A forwards unsigned messages unchanged to all nodes. Unsigned messages are passed to B as 1 and to C as 0.

1. The nodes fill their EIG trees for $t+1$ rounds.
2. A leaf is resolved to its value. If the last sender was a signer, the signature is removed.
3. For a vertex on level $r < t+1$ with label X, two cases are distinguished:
   - X is signer: Take the majority of the vertice's resolved children which carry X's signature.
   - X is no signer: Take the majority of the vertice's resolved children.
4. The result of the protocol is the resolved value of the root.

Fig. 4. Protocol for Byzantine agreement with partial authentication

In σ_1, C has to decide for 0, since the sender is correct. For the same reason, B has to decide for 1 in σ_2. Scenario σ_3 has been constructed in a way that it is indistinguishable from σ_1 by C and from σ_2 by B. Hence, B and C decide for different values in σ_3, contradicting (B1).

For $n \leq 3t$, this reasoning can be extended to three sets of nodes, each comprising at most t nodes. Furthermore, in the set of the sender, no node can sign messages. Hence, Byzantine agreement is impossible for $n \leq 3t$. □

Sender is signer With an *EIG*-Protocol, it is possible to reach agreement with $n > s \geq 2t$ (for $n = s$, any number of faults can be tolerated).

Theorem 6. *Assuming partial authentication, the following holds: If the sender signs and $n > s \geq 2t$ holds, the protocol in Fig. 4 reaches Byzantine agreement.*

Proof. We distinguish whether the sender is correct or faulty:

Sender correct: There is exactly one value signed by the sender. Since there are at least $t+1$ correct nodes, at least t of which are signers, at least one leaf is reduced to the sender's value signed by t correct nodes. This value will be considered recursively in the process of resolving. Hence, there is exactly one value for the result of the protocol, which is the value signed by the sender.

Sender faulty: Here, we show that all vertices in the second level are common. The vertices which correspond to correct signers are common with the same argumentation as in the case of a correct sender.

The vertices which correspond to correct non-signers are common, because they have at least t correct and signed children. Since these vertices are common (same argumentation as above), there is always a majority of correctly resolved children.

Finally, there are the faulty vertices (signed or not). Their correct children are common, as can be shown as above. For their faulty children to be common, it is (by recursion) necessary that all faulty vertices at level $t+1$ with only faulty ascendants be common. This is the case, since there are no such vertices.

Hence, all correct nodes take the same set of values as a basis for the final decision. This leads to a common value. □

4 Summary

In this paper, we have shown that there is a trade-off between the "quality of authentication" and the achievable fault tolerance for Byzantine agreement. We have identified several authentication scenarios which extend the usual all-or-nothing semantics.

Using a weak kind of authentication adds fault tolerance to the process of distributing public keys. This is desirable, because key distribution is a reliability bottleneck of authenticated agreement protocols.

References

[BDDS87] Amotz Bar-Noy, Danny Dolev, Cynthia Dwork, and H. Raymond Strong. Shifting gears: Changing algorithms on the fly to expedite Byzantine agreement. In *Proceedings of the 6th ACM Symposium on Principles of Distributed Computing (PODC)*, pages 42–51, Vancouver, Canada, 1987.

[Bor95] Malte Borcherding. Efficient failure discovery with limited authentication. In *Proceedings of the 15th International Conference on Distributed Computing Systems (ICDCS)*, pages 78–82, Vancouver, Canada, 1995. IEEE Computer Society Press.

[Dol82] Danny Dolev. The Byzantine generals strike again. *Journal of Algorithms*, 3(1):14–30, 1982.

[DS83] Danny Dolev and Raymond Strong. Authenticated algorithms for Byzantine agreement. *SIAM Journal of Computing*, 12(5):656–666, November 1983.

[EM96] Klaus Echtle and Asif Masum. A mutiple bus braodcast protocol resilient to non-cooperative Byzantine faults. In *Proceedings of the 26th International Symposium on Fault-Tolerant Computing (FTCS)*. IEEE Computer Society Press, 1996.

[GLR95] Li Gong, Patrick Lincoln, and John Rushby. Byzantine agreement with authentication: Observations and applications in tolerating hybrid and link faults. In *Proceedings of the Fifth Dependable Computing for Critical Applications (DCCA-5)*, 1995.

[Gon93] Li Gong. Lower bounds on messages and rounds for network authentication protocols. In *Proceedings of the 1st Conf. Computer and Comm. Security*, pages 26–36. ACM, 1993.

[HH93] Vassos Hadzilacos and Joseph Y. Halpern. The Failure Discovery problem. *Math. Systems Theory*, 26:103–129, 1993.

[LSP82] Leslie Lamport, Robert Shostak, and Marshall Pease. The Byzantine Generals problem. *ACM Transactions on Programming Languages and Systems*, 4(3):382–401, 1982.

[Nat92] National Institute of Standards and Technology. The Digital Signature Standard. *Communications of the ACM*, 35(7):36–40, July 1992.

[RSA78] R. Rivest, A. Shamir, and L. Adleman. A method for obtaining digital signatures and public-key cryptosystems. *Communications of the ACM*, 21(2):120–126, February 1978.

This article was processed using the LaTeX macro package with LLNCS style

Implementation of a Security Policy in Distributed Safety Related I&C Systems
A Case Study

Ferdinand J. Dafelmair
TÜV Bayern Sachsen
München, Germany
fdaf@tuev.spacenet.de

Abstract

Today's I&C Systems usually are distributed computer systems with interfaces to sensors and actuators for process control. Once someone has access to a any part of this distributed system he possibly is in control of part of the process. Networks of distributed I&C systems provide an ideal infrastructure for remote, quiet and inconspicuous hacking attacks against process computers in I&C systems. It is vital for safe process operation to avoid any subversion of the security of the I&C system jeopardizing it's integrity, availability and confidentiality. This paper starts from a rather generic model of a distributed computerized safety system, identifies important information-storage, -flow and -processing, gives an example of how to set up a security policy and finally discusses the implementation of this policy including service and maintenance issues.

1 Introduction

Whenever a technical process bears the risk of dangerous hazards, the operator of such a process has to take provisions to protect humans, the environment or investment from disasters. A major contribution to this goal is a well designed Instrumentation and Control (I&C) System to assist the human operators to keep the process in a safe state. A modern, computerized I&C system has two major tasks namely to run the process in a most efficient way and to keep important process parameters within safe limits. State of the art safety system designs honor this ambivalence and decompose the I&C system into two clearly separated subsystems-one for operational control and one for safety [Halang et al., 1993]. Security also relies on well designed system structures.

2 Modeling a generic I&C system

In many cases when security questions arise, the I&C system's principle structure is already given. A qualified system model showing system decomposition, functional allocation, data processing and communication - to mention only some parts of the required contents - is mandatory to start security planing. If such information may not be retrieved from system design documentation it has to be collected applying reverse engineering techniques. An important modeling vehicle is an overview schematic showing the main subsystems and network connections. Basically it should reflect both the local and logical allocation of subsystems in such a way, that entities with functional, operational or other attributes in common are grouped with the logical allocation being the primary goal.

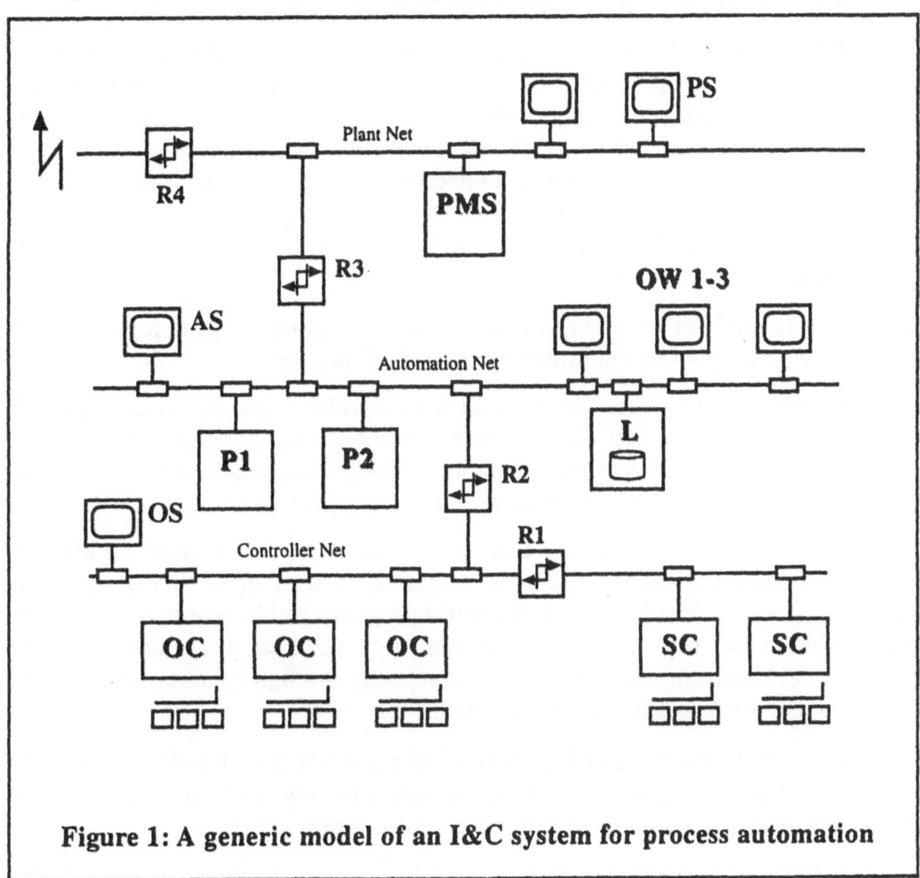

Figure 1: A generic model of an I&C system for process automation

Figure 1 shows a generic model of an I&C system for automation of a safety related process. At the bottom are several controllers with interfaces to process sensors or actuators either using field busses for intelligent process components or traditional multiwire cables. These controllers communicate across a network named *Controller Net* split by Router R1 into a safety critical part to the right and the operational part to the left. The controllers being part of the former are in charge of the safety functions, the ones in the latter carry out operational tasks only. Router R2 connects the controller net to the high speed *Automation Net* with process computers P1 and P2, operator workstations (OW1-3), a data logging system (L) etc. necessary for efficient process control. The automation net itself has a connection to the general purpose *Plant Net* through router R3. The plant management system (PMS) and other plant net computers are in charge of general purpose plant management functions like logistic-control, accounting, research etc. The plant net is considered outside the scope of the I&C system but the external links to it are security relevant. Wide Area Network (WAN) Router R4 finally is the gate all communication with the outside world passes through.

3 Information handling analysis

All functionality of I&C systems, as it is with every other computer - is based on information - or more precisely - data-handling. At first this data has to be secured against unauthorized modification and in many cases also against disclosure during storage, communication (data-flow) and processing. These three areas of data-handling should be carefully analyzed covering not only the process data itself but the data of all the process computer systems including programs, parameters and maintenance log data as access to any of this data may be more or less used to compromise computer systems.

To give an example of the results of such an analysis using our generic model from chapter 2 we may assume the following simplified situation:

The operational controllers (OC) monitor their related process entity, transmit acquired process data to the superordinate primary process computers (P1, P2) and perform local control tasks self-acting in closed-loop operation. To do that they also may need to exchange data among each other peer to peer.

The primary process computers use the process data from the controllers together with the manual commands from the operators, process them according to the control algorithms stored within their control programs and issue proper commands and set-points or new operation characteristics back to the controllers. The operators communicate with the process computers through operator workstations that also receive process data directly from the controllers.

The safety system controllers (SC) operate independently from the operational controllers (OC) or the primary process computers and only send safety messages or status information to the process computers or operator workstations.

A logging system be connected to the overall I&C system that logs any kind of data reaching from process data and operator commands to maintenance data (e.g. controller error and status messages).

Communication with the plant net be used to supply the plant management system (PMS) with relevant process data and receive quantitative or qualitative demands for process operation.

Separate service systems (OS, AS, PS) be attached to either the controller net, the automation net or the plant net. These service systems provide centralized network management for each network pertinent to them as well as centralized system management through remote access to the systems within their affiliated management zone, here identical to the particular subnet. System management includes all maintenance and supervision tasks from failure diagnosis to tuning and program update service.

Let's furthermore assume that a management demand exists to centralize maintenance (**CMR**) for the whole plant to reduce maintenance costs. This demand includes consideration on how to include third party maintenance support via remote maintenance. This requires various bi-directional dataflow from and to any computer system from a central plant service center or even remote site.

Based on these assumptions the outcome of an information handling analysis could be the following:

3.1 Data storage:

S1: Programs and parameters are stored in each computer system
S2: Process data, operator commands and maintenance data are stored in the logging system

3.2 Data flow:

F1: Process data transmission from controllers to

- **FC1**: peer controllers (only among non safety system controllers)
- **FC2**: process computers,
- **FC3**: operator workstations
- **FC4**: logging system

F2: Transmission of operator commands from the operator workstation to the process computers.
F3: Transmission of automatic commands and set-points from process computers to the controllers.
F4: Transmission of selected process data from process computers to the plant net.
F5: Transmission of global set-points from the plant net to the process computers.
F6: Transmission of maintenance log data from every system to the logging system.
F7: Interactive system management session data-flow between a service system and each computer system pertinent to it.
F8: Interactive system management session data-flow between remote maintenance sites and each computer system.

3.3 Data processing:

The analysis of where what data gets processed is especially necessary to evaluate the risk of deliberate and systematic data manipulation carrying out intelligent manipulation hacks. Within the limitations of this rather short conference contribution integrity of data processing may be assumed equally covered discussing integrity of data storage for programs (see **S1**).

4 Security planning

Knowing where security relevant data in the I&C system is stored, transmitted and processed, it is the next step to start security planning with the constitution of a security policy jointly agreed upon among the security task force and the plant management. A security policy is the set of decisions, that, collectively, determines an organization's posture toward security. More precisely, a security policy determines the limits of acceptable behavior and what response to violations should be [Cheswick and Bellovin, 1994]. The requirements of this policy need further refinement and interpretation through plant-wide security standards and guidelines. At the last level of security planning specific procedures need to be created that describe how to implement the policy, standards and guidelines in an operating environment. As this paper intends to focus more on implementation of a security policy rather than finding it, no further discussion concerning the framework of security planning is given. The interested reader will find extensive guidance in literature [Arthur E.Hutt, 1995].

Let's assume a security framework is given stating the following simplified security policy with it's four key requirements.

- **RQ1**: To keep the process in a safe state integrity is mandatory for both the safety relevant process data and the programs of the safety system controllers. **Criticality Level 1**
- **RQ2**: For efficient process operation integrity shall be assured for the programs and the related process data of all systems within the operational part of the I&C system. **Criticality level 2**
- **RQ3**: Process technology data and processing methods implicitly stored in the process computers internal algorithms and in the logging system are trade-secrets and shall be kept strictly under non-disclosure. **Criticality level 2**
- **RQ4**: The information processing assets of the plant shall be protected against unauthorized access from the outside world. **Criticality level 3**

4.1 Security Zones

With the given key requirements for security and their different criticality levels it is quite obvious to continue with security design allocating systems with common security criticality into separated security zones following the traditional principle of successive physical enclosure (Figure 2). Each of the three zones Z1-Z3 is delimited against the adjacent one through security gateways (e.g. Z1/Z2) also called firewalls systems. Presuming increasing penetration effort towards the center of such an onion skin model Z1 should be the most secure computing environment.

4.2 Defense in Depth concept

A well designed safety related I&C system consists of several staggered layers of defense against hazardous process events. Such a safety design concept is called defense in depth and may be directly applied to a security design concept where the security gateways mark the defense lines to provide the innermost zone with the highest level of protection against security threats. Ideally the defense lines from a safety related defense in depth concept should coincide with the borders of the corresponding security zones leading to clear interfaces between different areas of criticality within an I&C system. To defend the integrity properties of a system against deliberate attack it's mostly feasible to reach this coincidence. In response to additional requirements for confidentiality however, zones of critical security could need to be extended beyond safety defense lines.

In general, safety and security of an I&C system only benefit from a defense in depth concept, if the different layers of defense share no common mode or common cause failures. Translated into security terminology this means that e.g. the security gateways have no flaws in common, like identical error prone software components, or don't suffer from common system management negligence. Technical solutions that provide security protection should be diverse as far as feasible and in general they should be preferred to administrative regulations to free operating staff from non task focused workload and increase the reliability of the protection by eliminating human errors.

5 Implementing Security

5.1 The Safety-System Zone Z1

From a safety engineering viewpoint, the safety system shall be subjected to top ranked security and share as few common resources as possible with other parts of an I&C system to avoid any kind of safety function degradation by failure propagation. The same principle also counts for computer security. Another general principle that is applicable from safety design is to keep safety functions as simple as possible to reach a very high degree of test coverage and a higher probability to detect sabotage code.

Referring back to our model, the programs and parameters stored in the safety controllers as well as the safety-related process data (**S1**, **S2**) are critical according to **RQ1**. Access to this data has definitively to be blocked. For process control no dataflow is intended from Z2 into Z1 but the security gateway Z1/Z2 has to grant safety system message transmission to the operators to the logging system and the process computers (**FC2**, **FC3**, **FC4**).

A suitable solution for the Z1/Z2 firewall could be a firewall router FR1 and an application gateway GW1/2 inside Z1. The firewall router FR1 protects the entire zone Z1. It blocks any traffic from and to systems of zone Z1 besides a well defined connection service that's granted outbound of the gateway. This operation policy keeps configuration complexity very low at FR1. GW1/2 acts as a data dissemination agent implemented as a one way application-level gateway that collects and forwards any data Z1 systems intend to transmit to outside their zone. GW1/2 could even accept this data through stateless datagram communication services to avoid fault-related blocking of Z1 systems' communication resources and then forward it through a reliable virtual circuit to it's destination.

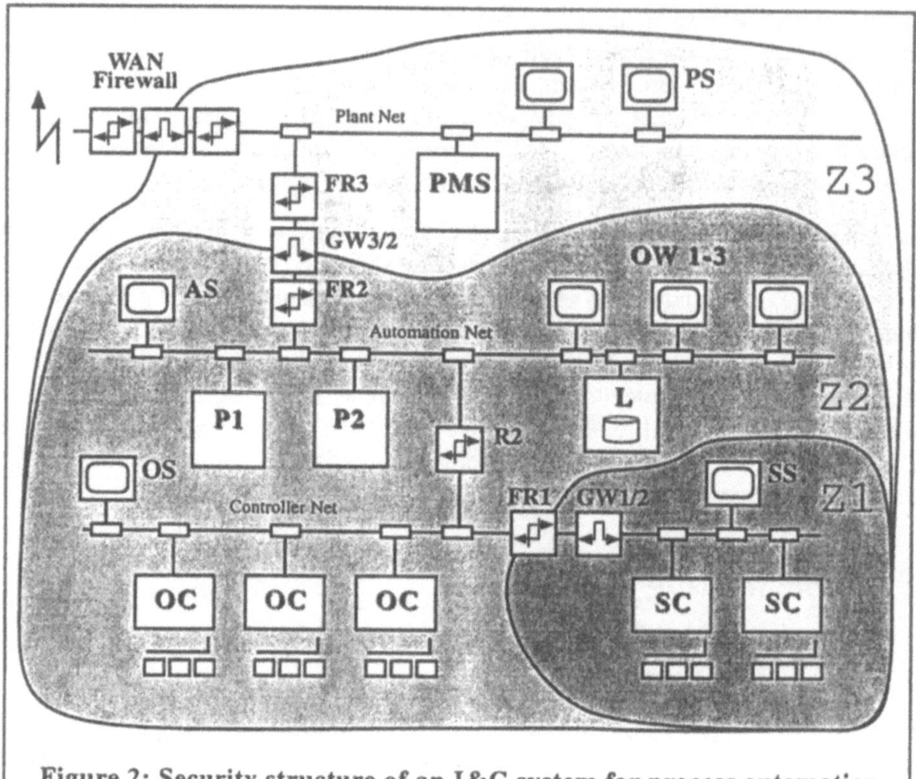

Figure 2: Security structure of an I&C system for process automation

5.2 The Conventional I&C Zone Z2

The major part of an I&C system is the operational one. Our model assumes that no partitioning of this net is required, which might indeed not be the case with huge installations. From requirements **RQ2** and **RQ3** we learn that the systems across the automation net share security requirements of the same criticality level. This suggests the definition of a second security zone Z2 covering the entire automation net. Strict access control between the systems of Z2 communicating among each other seems not to be mandatory at first glance but is required especially to fulfill **RQ3**. Confidentiality is the more stringent requirement because in general read access is much easier to obtain than write access and such attacks are very hard to detect as they rarely leave any marks. To avoid the fall of all systems if only one gets compromised through a successful attack systems trusting each other should be a well considered exception. Inter-system communication using authentication is more complex to realize and administrate but this effort pays off as it also strengthens the preventive layers of the defense in depth system design for safe process operation.

Due to the sometimes huge physical extension of a typical automation net equal physical security across the entire automation net may not always be assumed. This increases the vulnerability for physical attacks affecting data-flow **F1-F8** if not encrypted. MAC address level monitoring [Smythe, 1994] or dataflow encryption should be used to secure communication across physically insecure network segments inside Z2 e.g. communication links across outside plant areas. Encryption has certain limits when it comes to low level protocols of communication stacks. These protocols rely on open and standardized designs to enable communication in heterogeneous environments so these standard protocols have to feature common security protocol add-ons for each vendors implementation - yet not available industry wide. Encryption thus is limited to the upper layers of communication securing application data only. Again a security gateway Z2/Z3 protects Z2 baring low level access to Z2 systems. The Z2/Z3 security gateway should consist of 2 routers, an outer router FR3 in Z3 and an inner router FR2 in Z2 and an application-level gateway GW2/3 in the so-called *demilitarized zone* (DMZ) in between the two routers. The outer router FR3 protects Z2 against address-spoofing attacks from Z3, obstructs subversion of GW2/3 and provides additional defense against low level access to Z2 as now FR3 and FR2 have to be cracked [Cheswick and Bellovin, 1994]. For the communication tasks **F4**, **F5** and **F8** the gateway runs tailored handlers that could provide secure communication to outside target hosts using encrypted virtual circuit techniques regardless of the security of underlying transport services and provide extended logging and hide away all internals of Z2 by address translation.

Extra costs for such a solution may be justified referring to the increased risk for attacks out of Z3, the benefits from secure virtual circuits and the concentration of major security issues at just a single point - the security gateway solution rather than across a number of different systems.

5.3 The Plant Net Zone Z3

From a safety point of view, this zone is already outside the scope of the I&C system and as such not considered a trusted zone in any way. Data affecting I&C systems from **F5** communication are by no means safety critical but could be confidential. Requirement **RQ4** is fulfilled for Z2 already and concerning systems in Z3 **RQ4** is no safety related requirement but nevertheless the WAN Router R4 from Picture 1 should be substituted by a firewalls solution similar to that between Z2 and Z3 for commercial security reasons.

6 The Service and Maintenance Issues

The above security zone concept and the accompanying security measures comply with the key requirements for secure data storage, data-flow and data processing but seem to be in contradiction with the demand for centralized system maintenance (**CMR**). Protecting systems from remote hacking is contrary to envisage remote maintenance that most often claims unlimited access to systems and network components.

6.1 System- and Network-Management Techniques

When using centralized network management one of the widely used protocols is SNMP (simple network management protocol). SNMP in its generic form is usually based on unreliable datagram services, easy to implement and powerful, but with no protection against security threats [Case et al., 1990]. The architectural model of SNMP knows network management stations that run applications capable of controlling network elements like hosts, gateways, terminal servers etc. Each network element runs an agent communicating management information between itself and a network management station. As SNMP models all management agent functions as alterations or inspections of variables using simple datagram services, it's fairly easy to gain sensitive information by eavesdropping on the traffic [Garfinkel and Spafford, 1991] or even to take control over a wide range of important system functions by generating own SNMP messages.

For installations concerned about security it is mandatory to use SNMP security protocols to secure network management communication [Galvin et. al, 1992]. It's noteworthy to mention that especially in heterogeneous computing environments it's a challenge to select and find suitable protocol implementations for every system. Another problem are superuser logins across the network. At least challenge response authentication for such logins should be used if virtual circuit level encryption of sessions is not feasible. Despite all the benefit resulting from centralized network management the security drawbacks from insecure network management protocols shall not be neglected.

6.2 Servicing Zone Z1

First Zone Z1 obviously needs it's own service computer system for network management and maintenance separated from that of the general controller net as the security gateway GW1/2 bars any access to the Z1 systems from outside Z1. A centralized network management system SS for a safety system shall deploy security protocols even when operating in secure areas to authenticate origin of management data and thus to avoid accidental misconfiguration of systems or misinterpretation of status data. Maintenance includes program and parameter changes also intended to be carried out from a centralized service computer using interactive sessions to the particular target system. They have to be classified most critical for safety systems according to **RQ1**. From this viewpoint it is highly recommended to avoid any active maintenance from outside Z1 - a zone with already high physical security.

6.3 Servicing Zone Z2

For zone Z2 with it's reduced safety and security constraints the service question could be solved more easily. Due to it's rather big extension and complexity network management and centralized service increase maintainability and thus reliability of the entire operational part of the I&C system. On the other side the central service workstation AS is one of the most vulnerable points of Z2. It needs extra physical security and proper monitoring attention. Given this, and using a secure network management environment, centralized service and management is feasible as Z2 already has to be a secure zone as stated above. Should the centralized service and management of zone Z2 be carried out from a plant service center PS within Z3, this service center forms a security enclave inside Z3 with the same security requirements than Z2. Communication between both Z2 level security areas shall then be effectively secured by encryption already mentioned with the GW2/3 security gateway. Routing the entire network management traffic securely across the security gateway GW2/3 requires complex and costly gateway applications. It seems to be more effective to allocate the plant service center inside Z2 and implement a proxy server for the network management application and the interactive session protocol on GW2/3.

6.4 Remote Maintenance Support

Remote maintenance and support is necessary, if a plant has not sufficient qualified maintenance staff for it's own. External specialists usually access the systems from the outside world through dialup or network connections to provide fault diagnosis support as well as error correction and change support.

Diagnosis support is based on evaluating diagnosis data that could be provided securely, even for Z1 systems, as read only data using the mechanisms of data dissemination already mentioned for **FC2-4** communication from Z1 systems.

Using cryptographic methods data transfer and even interactive sessions for corrective support from remote sites could of course be secured properly, but this immediately rises the question: how secure is the entire remote site? The answer to this question is a matter of confidence in the security of the remote end which has to have the same level as the local end. For Z1 safety systems it is usually hardly possible to furnish proof of adequate security. Regulative restrictions as well as liability questions also prohibit such kind of access to safety systems.

For Z2 systems unrestricted remote maintenance access, even diagnosis access only, would not violate safety but confidentiality requirement **RQ3**. To circumvent this problem remote diagnosis data needs to be filtered by an application process on GW2/3 gateway forwarding non classified data only.

In reverse direction and in response to diagnosis new software versions or patches could be worked out externally, afterwards digitally signed to confirm correctness and then transferred to local Z2 service staff. In any case, digital signatures should be used to confirm correct code and secure transmission. Initially signed for correctness by manufacturer's QA responsibles it should be co-signed after independent verification by plant specialists or independent third parties if necessary providing the maintenance staff with sufficient proof of authenticity [Russel and Gangemi, 1991] for the software they install.

7 Conclusion

This case study showed that security for distributed safety related I&C systems requires substantial planning and implementation as well as maintenance efforts. It also demonstrated various ways on how safety related I&C systems could benefit from the achievements of modern communication technology without compromising security if only this new technologies were used carefully. This could mean, that at critical areas superficial economy promised from some new applications or outsourcing strategies doesn't pay off if evaluated in the whole context of safety and security.

References

[Case et al., 1990] J.Case, M.Fedor, M.Schoffstall, J.Davin. A simple network protocol (SNMP). RFC 1157, May 1990

[Galvin et. al, 1992] J.Galvin, K.McCloghrie, J.Davin. SNMP Security Protocols. RFC 1352, July 1992

[Cheswick and Bellovin, 1994] William R. Cheswick, Steven M. Bellovin. Firewalls and Internet Security : repelling the wily hacker. Addison Wesley, 1994

[Smythe, 1994] Colin Smythe, Internetworking: designing the right architectures. Addison Wesley, 1994

[Halang et al., 1993] Wolfgang A. Halang, Soon-Key Jung, Bernd J. Krämer, Johan J. Scheepstra. A safety licensable computing architecture, 1993

[Russel and Gangemi, 1991] Deborah Russel, G.T. Gangemi Sr.. Computer Security Basics. O'Reilly & Associates, 1991

[Garfinkel and Spafford, 1991] Simson Garfinkel, Gene Spafford. Practical Unix Security. O'Reilly & Associates, 1991

[Arthur E.Hutt, 1995] Arthur E. Hutt (editor). Computer Security Handbook. John Wiley & Sons, 1995

Author Index